Microstructure, Characterization and Mechanical Properties of Coal and Coal-Like Materials

Microstructure, Characterization and Mechanical Properties of Coal and Coal-Like Materials

Editors

Xuesheng Liu
Yunliang Tan
Yunhao Wu
Xuebin Li

MDPI • Basel • Beijing • Wuhan • Barcelona • Belgrade • Manchester • Tokyo • Cluj • Tianjin

Editors

Xuesheng Liu
College of Energy and Mining
Engineering
Shandong University of
Science and Technology
Qingdao
China

Yunliang Tan
College of Energy and Mining
Engineering
Shandong University of
Science and Technology
Qingdao
China

Yunhao Wu
College of Energy and Mining
Engineering
Shandong University of
Science and Technology
Qingdao
China

Xuebin Li
College of Energy and Mining
Engineering
Shandong University of
Science and Technology
Qingdao
China

Editorial Office
MDPI
St. Alban-Anlage 66
4052 Basel, Switzerland

This is a reprint of articles from the Special Issue published online in the open access journal *Materials* (ISSN 1996-1944) (available at: www.mdpi.com/journal/materials/special_issues/Microstruct_Charact_Mech_Prop_Coal_Coal_Like_Mater).

For citation purposes, cite each article independently as indicated on the article page online and as indicated below:

LastName, A.A.; LastName, B.B.; LastName, C.C. Article Title. *Journal Name* **Year**, *Volume Number*, Page Range.

ISBN 978-3-0365-7551-3 (Hbk)
ISBN 978-3-0365-7550-6 (PDF)

© 2023 by the authors. Articles in this book are Open Access and distributed under the Creative Commons Attribution (CC BY) license, which allows users to download, copy and build upon published articles, as long as the author and publisher are properly credited, which ensures maximum dissemination and a wider impact of our publications.

The book as a whole is distributed by MDPI under the terms and conditions of the Creative Commons license CC BY-NC-ND.

Contents

About the Editors vii

Xuesheng Liu, Yunliang Tan, Yunhao Wu and Xuebin Li
Microstructure, Characterization and Mechanical Properties of Coal and Coal-like Materials
Reprinted from: *Materials* **2023**, *16*, 1913, doi:10.3390/ma16051913 1

Yang Yang, Yao Zhang, Qingliang Zeng, Lirong Wan and Qiang Zhang
Simulation Research on Impact Contact Behavior between Coal Gangue Particle and the Hydraulic Support: Contact Response Differences Induced by the Difference in Impacted Location and Impact Material
Reprinted from: *Materials* **2022**, *15*, 3890, doi:10.3390/ma15113890 7

Shuo Wu, Guangpeng Qin and Jing Cao
Deformation, Failure, and Acoustic Emission Characteristics under Different Lithological Confining Pressures
Reprinted from: *Materials* **2022**, *15*, 4257, doi:10.3390/ma15124257 27

Yanchun Yin, Guangyan Liu, Tongbin Zhao, Qinwei Ma, Lu Wang and Yubao Zhang
Inversion Method of the Young's Modulus Field and Poisson's Ratio Field for Rock and Its Test Application
Reprinted from: *Materials* **2022**, *15*, 5463, doi:10.3390/ma15155463 43

Xicai Gao, Shuai Liu, Cheng Zhao, Jianhui Yin and Kai Fan
Damage Evolution Characteristics of Back-Filling Concrete in Gob-Side Entry Retaining Subjected to Cyclical Loading
Reprinted from: *Materials* **2022**, *15*, 5772, doi:10.3390/ma15165772 57

Wei Zheng, Linlin Gu, Zhen Wang, Junnan Ma, Hujun Li and Hang Zhou
Experimental Study of Energy Evolution at a Discontinuity in Rock under Cyclic Loading and Unloading
Reprinted from: *Materials* **2022**, *15*, 5784, doi:10.3390/ma15165784 71

Ke Ding, Lianguo Wang, Zhaolin Li, Jiaxing Guo, Bo Ren and Chongyang Jiang et al.
Comparative Study on the Seepage Characteristics of Gas-Containing Briquette and Raw Coal in Complete Stress–Strain Process
Reprinted from: *Materials* **2022**, *15*, 6205, doi:10.3390/ma15186205 87

Yunhao Wu, Xuesheng Liu, Yunliang Tan, Qing Ma, Deyuan Fan and Mingjie Yang et al.
Mechanical Properties and Failure Mechanism of Anchored Bedding Rock Material under Impact Loading
Reprinted from: *Materials* **2022**, *15*, 6560, doi:10.3390/ma15196560 99

Hengbin Chu, Guoqing Li, Zhijun Liu, Xuesheng Liu, Yunhao Wu and Shenglong Yang
Multi-Level Support Technology and Application of Deep Roadway Surrounding Rock in the Suncun Coal Mine, China
Reprinted from: *Materials* **2022**, *15*, 8665, doi:10.3390/ma15238665 117

Valentina Zubkova and Andrzej Strojwas
The Influence of Coal Tar Pitches on Thermal Behaviour of a High-Volatile Bituminous Polish Coal
Reprinted from: *Materials* **2022**, *15*, 9027, doi:10.3390/ma15249027 129

Junce Xu, Hai Pu and Ziheng Sha
Influence of Microstructure on Dynamic Mechanical Behavior and Damage Evolution of Frozen–Thawed Sandstone Using Computed Tomography
Reprinted from: *Materials* **2022**, *16*, 119, doi:10.3390/ma16010119 . . . . **149**

Ruojun Zhu, Xizhan Yue, Xuesheng Liu, Zhihan Shi and Xuebin Li
Study on Influencing Factors of Ground Pressure Behavior in Roadway-Concentrated Areas under Super-Thick Nappe
Reprinted from: *Materials* **2022**, *16*, 89, doi:10.3390/ma16010089 . . . . **173**

Qinghai Deng, Jiaqi Liu, Junchao Wang and Xianzhou Lyu
Mechanical and Microcrack Evolution Characteristics of Roof Rock of Coal Seam with Different Angle of Defects Based on Particle Flow Code
Reprinted from: *Materials* **2023**, *16*, 1401, doi:10.3390/ma16041401 . . . . **193**

Yong-Ki Lee, Chae-Soon Choi, Seungbeom Choi and Kyung-Woo Park
A New Digital Analysis Technique for the Mechanical Aperture and Contact Area of Rock Fractures
Reprinted from: *Materials* **2023**, *16*, 1538, doi:10.3390/ma16041538 . . . . **207**

About the Editors

Xuesheng Liu

Liu Xuesheng (1988–), male, professor, doctor, Shandong University of Science and Technology. Mainly engaged in teaching and research in rock mechanics and rock formation control in deep mines, published more than 60 academic papers in important journals at home and abroad, and authorized more than 40 invention patents at home and abroad.

Yunliang Tan

Tan Yunliang (1964–), male, professor, doctor, Shandong University of Science and Technology. More than 240 academic papers were published, 24 national and provincial scientific and technological progress awards were obtained, and 58 invention patents were authorized.

The main research directions include rock burst prevention, rock control, and roadway support.

Yunhao Wu

Wu Yunhao (1999–), male, doctor, Shandong University of Science and Technology. Mainly engaged in teaching and research in rock mechanics and rock formation control in deep mines, published over 10 academic papers in important journals at home and abroad.

Xuebin Li

Li Xuebin (1998–), male, doctor, Shandong University of Science and Technology. Mainly engaged in teaching and research in rock mechanics and rock formation control in deep mines, published over 10 academic papers in important journals at home and abroad.

Editorial

Microstructure, Characterization and Mechanical Properties of Coal and Coal-like Materials

Xuesheng Liu [1,2,*], Yunliang Tan [1,2], Yunhao Wu [1,2] and Xuebin Li [1,2]

[1] College of Energy and Mining Engineering, Shandong University of Science and Technology, Qingdao 266590, China
[2] State Key Laboratory of Mining Disaster Prevention and Control Co-Founded by Shandong Province and the Ministry of Science and Technology, Qingdao 266590, China
* Correspondence: xuesheng1134@163.com

Energy is the most basic driving force for world development and economic growth and the basis for human survival. Coal has long been a major component of global fuel supply, accounting for 27% of global energy consumption and 38% of the world's total power generation [1]. Although the share of coal in global energy consumption is gradually decreasing, for most developing countries, the dominant position of coal in their energy field cannot be changed in the short term. As the main source of energy in industrial development and people's daily lives, coal is widely used in various countries and regions [2,3].

With the gradual development of the coal industry, many scientists have found that different types of coal, such as coking coal, lignite and anthracite, have great differences in their mechanical properties and microstructure, and the mechanical response mechanisms and failure forms under different environments and conditions also have great differences [4,5]. Especially with the gradual increase of coal mining depth, coal is extremely vulnerable to damage and instability under the action of high temperature, high stress, high gas and mining disturbance, which endangers the safety and stability of underground production and the life safety of operators [6,7].

At present, the research related to coal structure mainly includes the macroscopic and microscopic characteristics of coal body and its classification, description method, pore structure, etc. [8]. The research of coal chemistry shows that the structure of coal determines the physical and chemical properties of coal, while the structure of coal rock mass controls the failure mechanism of coal rock mass; that is, the structural control of the catastrophic failure of coal rock in a general sense, and its failure form and strength are related to the shape of the structural plane and the physical properties of the materials.

Over the years, many scholars have conducted in-depth theoretical and experimental research on the structural behavior of coal rock failure. Based on the interaction between the shear strength of rock mass and the shear strength of the structural plane, Jaeger JC [9] explored the failure mode and strength index of a rock mass structure. Through the uniaxial compression test of the rock mass with penetrating joints, it was found that the rock mass with penetrating joints had slipped along the weak plane, causing vertical splitting and two forms of failure characteristics. The dip angle of the weak plane of the joint affected the failure form and overall strength of the structure. Nasseri [10] conducted a triaxial loading test on shale, and obtained the slip, shear and composite failure characteristics of a rock mass containing structural planes. Sun [11] believed that rock mass structure affects the mechanical properties of a rock mass and controls its deformation and failure. The control effect of a structure is much greater than that of rock material. This theory has been widely recognized by experts worldwide. Liu and Ma [12,13] found that the rock properties of the roof and floor slabs also have a great impact on the mechanical properties of coal, proposing a test method of coal mechanical properties considering the influence of roof lithology, and building a damage-constitutive model under different load conditions.

In addition, the high value-added utilization of tectonic coal resources can be effectively improved by studying the macroscopic and microscopic characteristics, classification and mechanical properties of coal and coal-like materials. In recent years, Paz-Ferreiro J [14] evaluated the multifractal characteristics of soil pores based on the mercury intrusion method and nitrogen adsorption method. Maria Mastal-erz and Sharon M Swanson [15,16] carried out a large number of pore determination experiments of coal samples. It was found that the pores in coal were mainly micropores and dominated the adsorption characteristics of coal samples. Larsen and Rahman [17,18] studied the adsorption process of organic compounds on the surface of a coal matrix and calculated the adsorption heat. Based on the mercury intrusion method and multifractal theory, Li analyzed the pore size distribution characteristics of tectonically deformed coals with different deformation degrees. The results showed that tectonic deformation leads to narrower pore size distribution, lower pore connectivity and more complex distribution of permeable pores in coal. These studies have explored the development direction of new coal sample materials, which is in line with the current trend of the global energy industry towards green, low-carbon and efficient utilization.

In order to further explore the structural characteristics, physical and chemical properties of coal and coal-like materials with the support of many experts and scholars worldwide, a Special Issue of *Materials* was launched, entitled 'Microstructure, Characterization and Mechanical Properties of Coal and Coal-Like Materials'. The aim is to show the latest scientific and technological achievements and cutting-edge testing techniques in the research of coal and coal-like materials, and to explore their microstructure characteristics, structural change rules and mechanical properties under various influencing factors. The Special Issue was launched on the 22 March, and 13 high-level papers have been included so far. Among them, five papers focus on the damage evolution and failure characteristics of coal and coal-like materials under different conditions. Four papers have studied the physical and chemical properties of new rock mass materials (coal gangue, coal tar pitch, freeze–thaw sandstone and anchored bedding rock mass). Three papers studied the crack propagation and seepage characteristics of coal rock mass, focusing on revealing its microstructure characteristics. In addition, some experts and scholars have used new experimental devices (CSS-1950 Biaxial Rheological Testing machine, the Split Hopkinson Pressure Bar system, Zeiss Xradia 510 Versa high-resolution CT system) to analyze the structural characteristics, physical and chemical properties of coal and coal-like materials, demonstrating great innovation. Promoting the development of cutting-edge science in the professional field, some experts and scholars have also studied the problems encountered in the field of coal mines and rock mass engineering based on the field engineering background, put forward a guiding field construction scheme and carried out industrial experiments, which have great engineering reference value.

The following are the main contents of this Special Issue:

1. Simulation Research on Impact Contact Behavior between Coal Gangue Particle and the Hydraulic Support: Contact Response Differences Induced by the Difference in Impacted Location and Impact Material (Yang Yang, Yao Zhang, Qingliang Zeng et al.) [19]

This paper studies the problem of coal gangue impacting hydraulic support in the process of top coal caving. The rigid–flexible coupling impact contact dynamic model between coal gangue and hydraulic support was established and the contact response difference in the support induced by the difference in impacted components and coal/gangue properties was compared and studied. The results show that the number of collisions, contact force, velocity and acceleration of impacted parts were different when the same single coal particle impacted different parts of the support. Various contact responses during gangue impact were more than 40% larger than that of coal, and the difference ratio could even reach 190%.

2. Deformation, Failure, and Acoustic Emission Characteristics under Different Lithological Confining Pressures (Shuo Wu, Guangpeng Qin and Jing Cao) [20]

In this paper, the influence of different surrounding rock pressure on the deformation and rock failure characteristics of a deep well drilling roadway is studied by uniaxial and triaxial compression testing of rocks. The results show that: the surrounding pressure has a significant effect on the damage deformation characteristics of the rock, and the change of the surrounding pressure directly affects the strength, damage form and elastic modulus of the rock; the strength limit of the rock increases with the surrounding pressure, and the damage form of the rock gradually changes to ductile damage with increase of the surrounding pressure; and the elastic modulus of the rock increases non-linearly with the increase of the surrounding pressure.

3. Inversion Method of the Young's Modulus Field and Poisson's Ratio Field for Rock and Its Test Application (Yanchun Yin, Guangyan Liu, Tongbin Zhao et al.) [21]

As a typical heterogeneous material, the heterogeneity of microscopic parameters of rock has an important influence on its macroscopic mechanical behavior. Based on digital image correlation (DIC) and the finite element method (FEM), a parameter field inversion method, namely the DF-PF inversion method, is proposed in this paper. The inversion accuracy was verified by numerical simulation and an indoor uniaxial compression test. The results show that compared with the traditional measurement method, the errors of macroscopic Young's modulus and Poisson's ratio calculated by the DF-PF inversion method were less than 2.8% and 9.07%, respectively. Based on the statistical analysis of Young's modulus field and Poisson's ratio field, the parameter uniformity and quantitative function relationship between microscopic parameters and principal strain can also be obtained in laboratory experiments. The DF-PF inversion method provides a new effective method for testing the Young's modulus field and Poisson's ratio field of rocks under complex stress conditions.

4. Damage Evolution Characteristics of Back-Filling Concrete in Gob-Side Entry Retaining Subjected to Cyclical Loading (Xicai Gao, Shuai Liu, Cheng Zhao et al.) [22]

Focusing on the influence of mining disturbances on the roadside support of gob-side entries being retained in deep coal mines, in this paper, uniaxial and cyclical tests of back-filling concrete samples were carried out under laboratory conditions to study damage evolution characteristics with respect to microscopic hydration. Understanding the characteristics of plastic strain, damage evolution and energy dissipation rate of filling samples has important reference value for real-time monitoring and failure warning of filling concrete in gob-side entry retaining.

5. Experimental Study of Energy Evolution at a Discontinuity in Rock under Cyclic Loading and Unloading (Wei Zheng, Linlin Gu, Zhen Wang et al.) [23]

To study the energy evolution of rock discontinuities under cyclic loading and unloading, cement mortar was used as a rock material and a CSS-1950 rock biaxial rheological testing machine was used to conduct graded cyclic loading and unloading tests on Barton's standard profile line discontinuities with different joint roughness coefficients (JRCs). According to the deformation characteristics of the rock discontinuity sample, the change of internal energy was calculated and analyzed.

6. Comparative Study on the Seepage Characteristics of Gas-Containing Briquette and Raw Coal in Complete Stress–Strain Process (Ke Ding, Lianguo Wang, Zhaolin Li) [24]

Through triaxial compression and seepage experiments, the different damage forms of the two coal samples and the effect of their deformation and damage on their permeability were analyzed from the perspective of fine-scale damage mechanics. This is of great significance to explore the real law of coalbed methane migration.

7. Mechanical Properties and Failure Mechanism of Anchored Bedding Rock Material under Impact Loading (Yunhao Wu, Xuesheng Liu, Yunliang Tan et al.) [25]

In view of the problem that anchored bedding rock material is prone to instability and failure under impact loading in the process of deep coal mining, the instability criterion of compression and shear failure of anchored bedding rock material was established, and the instability criterion of compression and shear failure of anchored bedding rock material was obtained.

8. Multi-Level Support Technology and Application of Deep Roadway Surrounding Rock in the Suncun Coal Mine, China (Hengbin Chu, Guoqing Li, Zhijun Liu et al.) [26]

To solve these problems of poor supporting effect, serious deformation and failure of surrounding rock of roadways under deep mining stress, the authors analyzed the deformation and failure law of roadways surrounding rock under multi-stage support by numerical simulation and obtained the key parameters of multi-stage support. At the same time, industrial test verification was carried out on site. The research can provide reference and technical support for surrounding rock control of deep high-stress roadways.

9. The Influence of Coal Tar Pitches on Thermal Behaviour of a High-Volatile Bituminous Polish Coal (Valentina Zubkova and Andrzej Strojwas) [27]

In this paper, the influence of three coal tar pitches (CTPs), with softening points at 86, 94 and 103 °C, on the thermal behaviour of a defrosted high-volatile coal during co-carbonization and co-pyrolysis was studied. It was determined that CTP additives change the structure of the coal plastic layer, the thickness of its zones and the ordering degree of the structure of semi-cokes to a different extent and independently from their softening points.

10. Study on Influencing Factors of Ground Pressure Behavior in Roadway-Concentrated Areas under Super-Thick Nappe (Ruojun Zhu, Xizhan Yue, Xuesheng Liu et al.) [28]

In this paper, during the mining activity under the super-thick nappe formed by thrust fault, the law of mine pressure behavior was complex, and it was difficult to control the deformation and failure of the surrounding rock. Combined with the actual engineering conditions, the influence of different roof lithology conditions, the thickness of nappe, the mining height, the size of the barrier coal pillar and the creep time on mine pressure behavior was studied by UDEC numerical simulation software.

11. Influence of Microstructure on Dynamic Mechanical Behavior and Damage Evolution of Frozen–Thawed Sandstone Using Computed Tomography (Junce Xu, Hai Pu, Ziheng Sha) [29]

Frost-induced microstructure degradation of rocks is one of the main reasons for the changes in their dynamic mechanical behavior in cold environments. Computed tomography (CT) was performed to quantify the changes in the microstructure of yellow sandstone after freeze–thaw (F–T) action. In addition, the influence of the microscopic parameters on the dynamic mechanical behavior was studied. The research results can be a reference for constructing and maintaining rock structures in cold regions.

12. Mechanical and Microcrack Evolution Characteristics of Roof Rock of Coal Seam with Different Angle of Defects Based on Particle Flow Code (Qinghai Deng, Jiaqi Liu, Junchao Wang et al.) [30]

The angle of the defects has a significant influence on the mechanical characteristics and crack evolution of coal seam roof rock. In this paper, multi-scale numerical simulation software PFC2D was adapted to realize the crack propagation and coalescence process in the roof rock of a coal seam with different angles of defects under uniaxial compression. The study provides a certain reference for the use of various analysis methods in practical engineering to evaluate the safety and stability of rock samples with pre-existing defects.

13. A New Digital Analysis Technique for the Mechanical Aperture and Contact Area of Rock Fractures (Yong-Ki Lee, Chae-Soon Choi, Seungbeom Choi et al.) [31]

In this study, a new digital technique for the analysis of the mechanical aperture and contact area of rock fractures under various normal stresses is proposed. The proposed technique has the advantage of being able to analyze changes in the mechanical aperture and contact area under various normal stresses without multiple experiments. In addition, the change in the contact area on the fracture surface according to the normal stress can be analyzed in detail.

Conflicts of Interest: The authors declare no conflict of interest.

References

1. International Energy Agency, *Coal Information 2019*; Organization for Economic Cooperation and Development: Paris, France, 2019.
2. Yuan, L. Theory and practice of integrated coal production and gas extraction. *Int. J. Coal Sci. Technol.* **2015**, *2*, 3–11. [CrossRef]
3. Xuebin, L.; Xuesheng, L.; Yunliang, T.; Qing, M.; Baoyang, W.; Honglei, W. Creep Constitutive Model and Numerical Realization of Coal-Rock Combination Deteriorated by Immersion. *Minerals* **2022**, *12*, 292.
4. Liu, X.S.; Ning, J.G.; Tan, Y.L.; Gu, Q.H. Damage constitutive model based on energy dissipation for intact rock subjected to cyclic loading. *Int. J. Rock Mech. Min. Sci.* **2016**, *85*, 27–32. [CrossRef]
5. Shilin, S.; Xuesheng, L.; Yunliang, T.; Deyuan, F.; Qing, M.; Honglei, W. Study on failure modes and energy evolution of coal–rock combination under cyclic loading. *Shock. Vib.* **2020**, *2020*, 5731721.
6. Xuesheng, L.; Qingheng, G.; Yunliang, T.; Jianguo, N.; Zhichuang, J. Mechanical characteristics and failure prediction of cement mortar with a sandwich structure. *Minerals* **2019**, *9*, 143.
7. Cheng, Q.; Guo, Y.; Dong, C.; Xu, J.; Lai, W.; Du, B. Mechanical Properties of Clay Based Cemented Paste Backfill for Coal Recovery from Deep Mines. *Energies* **2021**, *14*, 5764. [CrossRef]
8. Xu, Z.; Wu, J.; Zhao, M.; Bai, Z.; Wang, K.; Miao, J.; Tan, Z. Mechanical and microscopic properties of fiber-reinforced coal gangue-based geopolymer concrete. *Nanotechnol. Rev.* **2022**, *11*, 526–543. [CrossRef]
9. Jaeger, J.C. Shear failure of anisotropic rocks. *Geol. Mag.* **1960**, *97*, 65–72. [CrossRef]
10. Nasseri, M.H.; Rao, K.S.; Ramamurthy, T. Failure mechanism in schistose rocks. *Int. J. Rock Mech. Min. Sci.* **1997**, *34*, 460. [CrossRef]
11. Sun, G.Z. *Principles of Rock Mechanics*; Science Press: Beijing, China, 2011.
12. Liu, X.S.; Tan, Y.L.; Ning, J.G.; Lu, Y.W.; Gu, Q.H. Mechanical properties and damage constitutive model of coal in coal-rock combined body. *Int. J. Rock Mech. Min. Sci.* **2018**, *110*, 140–150. [CrossRef]
13. Qing, M.; Yunliang, T.; Xuesheng, L.; Qingheng, G.; Xuebin, L. Effect of coal thicknesses on energy evolution characteristics of roof rock-coal-floor rock sandwich composite structure and its damage constitutive model. *Compos. Part B* **2020**, *198*, 1–11.
14. Paz-Ferreiro, J.; Lu, H.; Fu, S.; Méndez, A.; Gascó, G. Use of phytoremediation and biochar to remediate heavy metal polluted soils: A review. *Solid Earth* **2014**, *5*, 65–75. [CrossRef]
15. Mastalerz, M.; Drobniak, A.; Strapoć, D.; Acosta, W.S.; Rupp, J. Variations in pore characteristics in high volatile bituminous coals: Implications for coal bed gas content. *Int. J. Coal Geol.* **2008**, *76*, 205–216. [CrossRef]
16. Swanson, S.M.; Mastalerz, M.D.; Engle, M.A.; Valentine, B.J.; Warwick, P.D.; Hackley, P.C.; Belkin, H.E. Pore characteristics of wilcox group coal, U.S. gulf coast region: Im- plications for the occurrence of coalbed gas. *Int. J. Coal Geol.* **2015**, *139*, 80–94. [CrossRef]
17. Larsen, J.W.; Kennard, L.; Kuemmerle, E.W. Thermodynamics of adsorption of organic compounds on the surface of Bruceton coal measured by gas chromatography. *Fuel* **2018**, *57*, 309–313. [CrossRef]
18. Rahman, K.A.; Chakraborty, A.; Saha, B.B.; Ng, K.C. On thermodynamics of methane+carbonaceous materials adsorption. *Int. J. Heat Mass Transf.* **2012**, *55*, 565–573. [CrossRef]
19. Yang, Y.; Zhang, Y.; Zeng, Q.; Wan, L.; Zhang, Q. Simulation Research on Impact Contact Behavior between Coal Gangue Particle and the Hydraulic Support: Contact Response Differences Induced by the Difference in Impacted Location and Impact Material. *Materials* **2022**, *15*, 3890. [CrossRef]
20. Wu, S.; Qin, G.; Cao, J. Deformation, Failure, and Acoustic Emission Characteristics under Different Lithological Confining Pressures. *Materials* **2022**, *15*, 4257. [CrossRef]
21. Yin, Y.; Liu, G.; Zhao, T.; Ma, Q.; Wang, L.; Zhang, Y. Inversion Method of the Young's Modulus Field and Poisson's Ratio Field for Rock and Its Test Application. *Materials* **2022**, *15*, 5463. [CrossRef]
22. Gao, X.; Liu, S.; Zhao, C.; Yin, J.; Fan, K. Damage Evolution Characteristics of Back-Filling Concrete in Gob-Side Entry Retaining Subjected to Cyclical Loading. *Materials* **2022**, *15*, 5772. [CrossRef]
23. Zheng, W.; Gu, L.; Wang, Z.; Ma, J.; Li, H.; Zhou, H. Experimental Study of Energy Evolution at a Discontinuity in Rock under Cyclic Loading and Unloading. *Materials* **2022**, *15*, 5784. [CrossRef] [PubMed]

24. Ding, K.; Wang, L.; Li, Z.; Guo, J.; Ren, B.; Jiang, C.; Wang, S. Comparative Study on the Seepage Characteristics of Gas-Containing Briquette and Raw Coal in Complete Stress–Strain Process. *Materials* **2022**, *15*, 6205. [CrossRef]
25. Wu, Y.; Liu, X.; Tan, Y.; Ma, Q.; Fan, D.; Yang, M.; Wang, X.; Li, G. Mechanical Properties and Failure Mechanism of Anchored Bedding Rock Material under Impact Loading. *Materials* **2022**, *15*, 6560. [CrossRef]
26. Chu, H.; Li, G.; Liu, Z.; Liu, X.; Wu, Y.; Yang, S. Multi-Level Support Technology and Application of Deep Roadway Surrounding Rock in the Suncun Coal Mine, China. *Materials* **2022**, *15*, 8665. [CrossRef] [PubMed]
27. Zubkova, V.; Strojwas, A. The Influence of Coal Tar Pitches on Thermal Behaviour of a High-Volatile Bituminous Polish Coal. *Materials* **2022**, *15*, 9027. [CrossRef]
28. Xu, J.; Pu, H.; Sha, Z. Influence of Microstructure on Dynamic Mechanical Behavior and Damage Evolution of Frozen–Thawed Sandstone Using Computed Tomography. *Materials* **2022**, *16*, 119. [CrossRef]
29. Zhu, R.; Yue, X.; Liu, X.; Shi, Z.; Li, X. Study on Influencing Factors of Ground Pressure Behavior in Roadway-Concentrated Areas under Super-Thick Nappe. *Materials* **2022**, *16*, 89. [CrossRef] [PubMed]
30. Deng, Q.; Liu, J.; Wang, J.; Lyu, X. Mechanical and Microcrack Evolution Characteristics of Roof Rock of Coal Seam with Different Angle of Defects Based on Particle Flow Code. *Materials* **2023**, *16*, 1401. [CrossRef]
31. Lee, Y.-K.; Choi, C.-S.; Choi, S.; Park, K.-W. A New Digital Analysis Technique for the Mechanical Aperture and Contact Area of Rock Fractures. *Materials* **2023**, *16*, 1538. [CrossRef]

Disclaimer/Publisher's Note: The statements, opinions and data contained in all publications are solely those of the individual author(s) and contributor(s) and not of MDPI and/or the editor(s). MDPI and/or the editor(s) disclaim responsibility for any injury to people or property resulting from any ideas, methods, instructions or products referred to in the content.

Article

Simulation Research on Impact Contact Behavior between Coal Gangue Particle and the Hydraulic Support: Contact Response Differences Induced by the Difference in Impacted Location and Impact Material

Yang Yang [1,2,*], Yao Zhang [1], Qingliang Zeng [1,2,3,*], Lirong Wan [1] and Qiang Zhang [1]

1. College of Mechanical and Electrical Engineering, Shandong University of Science and Technology, Qingdao 266590, China; sdkdzhangyao@126.com (Y.Z.); sdkdlirongwan@126.com (L.W.); sdkdzhangqiang@126.com (Q.Z.)
2. Shandong Provincial Key Laboratory of Mining Mechanical Engineering, Qingdao 266590, China
3. College of Information Science and Engineering, Shandong Normal University, Jinan 250358, China
* Correspondence: yang.yang@sdust.edu.cn or sdkdyangyang@126.com (Y.Y.); qlzeng@sdust.edu.cn (Q.Z.)

Abstract: In the process of top coal caving, coal gangue particles may impact on various parts of the hydraulic support. However, at present, the contact mechanism between coal gangue and hydraulic support is not entirely clear. Therefore, this paper first constructed the accurate mathematical model of the hydraulic cylinder equivalent spring stiffness forming by the equivalent series of different parts of emulsion and hydraulic cylinder, and then built the mesh model of the coal gangue particles and the support's force transmission components; on this basis, the rigid–flexible coupling impact contact dynamic model between coal gangue and hydraulic support was established. After deducing contact parameters and setting impact mode, contact simulations were carried out for coal particles impacting at the different parts of the support and coal/gangue particles impacting at the same component of the support, and the contact response difference in the support induced by the difference in impacted component and coal/gangue properties was compared and studied. The results show that the number of collisions, contact force, velocity and acceleration of impacted part are different when the same single coal particle impact different parts of the support. Various contact responses during gangue impact are more than 40% larger than that of coal, and the difference ratio can even reach 190%.

Keywords: coal and gangue; impact contact; response differences; equivalent stiffness; rigid–flexible coupling; different parts

1. Introduction

Top coal caving mining is the important mining method for thick and extra-thick coal seams: in the coal dropping stage of the top coal caving, the hydraulic support is parceled in the floor rock and coal gangue granule space body [1–7]. The large amount of particles and the continuous coal gangue dropping process will bring the impact contact of coal gangue and the hydraulic support. Due to the presence of flexible devices such as the hydraulic cylinder in the hydraulic support structure, the local impact contact behavior can cause the vibration of the whole hydraulic support. Grasp the impact contact characteristics and the contact response differences between coal gangue and the hydraulic support, is the foundation to study the interaction law between coal gangue or the surrounding rock and the hydraulic support and the precondition of coal gangue recognition in top coal caving. It is significant for the improvement of the hydraulic support application performance and advancement of top coal caving technology.

In the early stage, many studies have been carried out on the working characteristics of the hydraulic support and coal gangue recognition. Zhang et al. [8] studied the bearing characteristics of the hydraulic support by carrying out internal and external loading tests under different heel block contact conditions. Wang et al. [9] completed the stress and stability analysis of the prop by applying the load to the prop of the hydraulic support. Wang et al. [10] theoretically studied the coupling relationship of strength, stiffness and stability between the hydraulic support and surrounding rock. Ma et al. [11] studied the displacement and stress distribution of hydraulic support by stress testing and finite element analysis. Sun et al. [12] investigated load-bearing characters of hydraulic-powered roof support with dual telescopic legs. Xie et al. [13] studied the load-bearing characteristics of the top beam in a full range by constructing the spatial mechanical model of the top beam separation body of the four-column support. Wang et al. [14] established the plane mechanics model of four-column supporting shield support to study its adaptation. Ren et al. [15] designed the 1:2 reduced-scale hydraulic support to carry out the dynamic impact test and to study the response characteristics of hydraulic support under dynamic impact load. Gao et al. [16] added a step load to simulate the impact of the roof on the hydraulic support, so as to simulate and study the force transfer characteristics of the hydraulic support under the impact of roof pressure or coal lost gangue. Zhai et al. [17] used Amesim to establish the impact model of the column hydraulic system of the hydraulic support composed of the force principal model, safety valve model and weight model, then studied the impact response characteristics of the column hydraulic system. Luo et al. [18] studied the relationship between the impact force and the response speed of the hydraulic system by building the simulation model of the column hydraulic system of the hydraulic support. According to the existing research findings, one of the existing problems in the current study is that the interaction between coal gangue or the surrounding rock and the hydraulic support, or the working characteristics of the hydraulic support under dynamic load, are mainly studied by applying an equal load. The load is obtained through estimation, but the contact course between coal gangue or the surrounding rock and the support is ignored. The research scheme has a large error. The second problem existing in the current study is that there are few related researches on the interaction differences between coal gangue and the support. According to the Flores contact theory and the flexible energy absorption characteristics of the metal plate, Yang et al. [19,20] constructed the impact contact dynamics theoretical model between the single particle and the metal plate, the influence of impact velocity, recovery coefficient, material parameters, structure size, impact position and supporting spring stiffness on the dynamic contact response of the system was studied. However, the structural system is simple, and the study is not involved in the interaction between coal gangue and the multi-body combination hydraulic device such as the hydraulic support. Wan et al. [21,22] and Yang et al. [23] took the dynamic impact contact behavior of coal gangue and the tail beam as the research object, constructed the rigid–flexible coupling contact dynamic model of coal gangue impacting the tail beam structure and the dynamic contact finite element simulation model between the coal gangue particle and the tail beam under different constraints, respectively. The influence of impact angle, impact height, impact position, particle radius and particle material on dynamic contact response of the tail beam is studied. However, only the contact between the particle and the tail beam or equivalent structure of the tail beam is constructed, and the influence of the top beam, shield beam, front and rear connecting rod of the hydraulic support on the contact behaviors is not considered. Zeng et al. [24] established the numerical simulation model of the hydraulic support, considering the whole structure of the hydraulic support, but when studying the response of the hinge point under the condition of the single coal gangue particle impacting the hydraulic support, the contact behavior between the particles and the hydraulic support is imposed by the calculated contact force. At present, the contact response and response difference when the single coal gangue particle impact the different parts of the hydraulic support have not been studied in a reliable way. The third problem existing in current research is that some existing research carries out research on the hydraulic support by replacing the hydraulic cylinder with the equivalent spring damping module, but the spring stiffness model is not accurate. For example, Liang et al. [25,26] and

Wan et al. [27] establish the equivalent stiffness model of the hydraulic cylinder just with the consideration of the hydraulic oil compressibility, Liu et al. [28] and Yang et al. [29] only consider the elasticity of the hydraulic oil and the cylinder body to establish elastic equivalent stiffness model of the hydraulic cylinder. In fact, the elasticity of the piston rod and other structures will also affect the equivalent stiffness of the hydraulic cylinder. The fourth problem existing in current research is that although there are many related studies in the field of coal gangue recognition, the selection of effective coal gangue recognition parameters and the research of selection basis are rarely involved. Dou et al. [30] identified coal and gangue with four kinds of working space by image analysis and Relief-SVM. Lai et al. [31] applied the multispectral technology and two-dimensional autoencoder in their research on coal gangue recognition. Liu et al. [32] carried on the Hilbert spectrum analysis to the vibration signal of the tail beam so as to study the coal gangue interface identification technology. Song et al. [33] collected the vibration signal and the sound signal, and then proposed an effective minimum enclosing ball (MEB) algorithm plus the support vector machine (SVM) to coal gangue rapid detection in top coal caving. Pu et al. [34] conducted coal gangue image identification by the convolutional neural network and transfer learning. Hou et al. [35] established coal gangue classification system based on the difference between surface texture and gray scale characteristics, and identified coal gangue by image feature extraction and artificial neural network. Zhang et al. [36] analyzed the distribution characteristics of natural gamma rays in roof coal and rock, established the relationship between radiation intensity and coal gangue content, and identified mixed coal and gangue by radiation signal detection. Alfarzaeai et al. [37] studied coal gangue identification by convolutional neural networks and thermal images. Zhang et al. [38] used infrared imager with low emissivity to improve the coal gangue recognition accuracy based on liquid intervention. Yang et al. [39] used vibration, sound, pressure and other signals to classify and identify the mixing ratio of coal and gangue mixture. Yan et al. [40] used multi-spectral imaging technology and YOLOv5.1 target detection method to conduct the intelligent recognition and classification of coal and gangue. Yuan et al. [41] used six different classification methods to classify and identify coal and gangue by constructing the sample library of top coal caving sound signals. Wang et al. [42] constructed the lightweight accurate and fast recognition model of gangue rate, developed the intelligent image acquisition system and enhanced dust removal algorithm, and realized the image identification of coal gangue in the process of coal discharging. These studies directly applied the above methods in coal gangue recognition, but did not clarify the root causes or gist of recognition parameters or method selection, and did not analyze and discuss coal gangue identifiability at the theoretical level.

In top coal caving, the large number of coal gangue particles and the distribution characteristics of drawing space lead to the contact between coal gangue and various parts of the hydraulic support. Contact behavior between coal gangue particles and the hydraulic support involves the evolution of contact state and the real-time transmission of force. Overall, due to the lack of research method that can accurately describe the whole contact process between particles and hydraulic support, the precise contact characteristics and contact difference characteristics between coal gangue and the hydraulic support are not completely clear, in particular, the contact characteristics and contact difference between coal gangue and the different parts of the hydraulic support have not been studied. As a result, the selection of coal gangue recognition media lacks the theoretical basis in the research of coal gangue recognition technology. In view of the existing problems and deficiencies in the present studies and in order to further clarify the contact response characteristics and the contact response differences law between coal gangue and the hydraulic support, this paper proposed the idea of quantitative research, and the impact contact behavior between single particle coal gangue and the hydraulic support are taken as the research target. A rigid–flexible coupling impact contact simulation model between coal gangue and the hydraulic support is proposed to study the system dynamic response. For this purpose, an accurate model of the hydraulic cylinder liquid stiffness (equivalent spring stiffness) series by five different parts stiffness is firstly established. The rigid–flexible coupling impact contact simulation model between the coal gangue particle and the hydraulic support is established by combining the mesh model of the particle and the main force transmission components of the hydraulic support as well as the multi-body rigid dynamics simulation model of the hydraulic support. On the basis of determining the contact parameters and impact modes, the contact dynamics simulation analysis will be carried out when coal gangue impacts the different components of the hydraulic support with the same height and when the same size coal/gangue impacts the same position of the hydraulic support, respectively. Impact contact response characteristics when coal gangue impacts the different parts of the hydraulic support will be studied. Through comparative analysis, the difference rule of system impact contact responses caused by the difference in the impacted component and the particle material property will be determined, so as to explore the theoretical basis for the selection of coal gangue recognition media.

Our contributions in this paper are fourfold.

(1) Considering the compression elasticity of the emulsified liquid, piston rod, tail of the piston rod, bottom of the cylinder and the circumferential extension stiffness of the cylinder, a more accurate equivalent spring stiffness mathematical model of the hydraulic cylinder is established.

(2) The traditional scheme that studied the interaction between coal rock and the hydraulic support by replacing the contact process between coal gangue and the hydraulic support with force load is cancelled. The quantitative research method is put forward. Through the establishment of the rigid–flexible coupling impact contact dynamics simulation model between coal gangue and the hydraulic support, the contact response between particles and the hydraulic support is studied, which provides the direct and effective research method for the interaction between coal gangue or surrounding rock and the hydraulic support.

(3) Through the contact characteristics analysis between coal particles and the different parts of the hydraulic support, the variation rule of the system contact response caused by the change in the impact position is obtained.

(4) Through the contact response study of the direct contact parts, indirect related parts and the parts connection units after coal gangue impact, the contact difference characteristics between coal gangue and the hydraulic support are clarified, and the available parameters for coal gangue recognition are determined accordingly.

2. Establishment of the Hydraulic Cylinder Equivalent Stiffness Mathematical Model

In the hydraulic system of the hydraulic support, the props and the tail beam jack (collectively referred to as the hydraulic cylinder) are solid–liquid coupling-compressible devices of the steel structure and high pressure emulsion, and the hydraulic oil in the hydraulic cylinder cavity has the compressibility. According to previous studies, when dynamic software is used to analyze the dynamic characteristics of the hydraulic support, the equivalent spring damping module is usually used to replace the hydraulic cylinder.

In order to obtain accurate impact contact dynamic response between coal gangue and the hydraulic support, the equivalent stiffness of the hydraulic cylinder in the working process should be determined accurately first when using spring damping to analyze the dynamic characteristics of the hydraulic support. The prop and the tail beam jack in this paper are all single telescopic hydraulic cylinder. Taking the tail beam jack as the example, as shown in Figure 1, observation shows that when bearing the external load, not only the hydraulic oil in the hydraulic cylinder and part of the cylinder body contacted with oil are stressed, but also the piston rod, the tail of the piston rod and the bottom of the hydraulic cylinder are stressed. High pressure emulsion is filled between the piston rod and the prop cylinder body; the piston rod, the internal oil liquid and the cylinder body are regarded as the spring respectively; each part of springs are in series with each other. The equivalent spring stiffness of the hydraulic cylinder actually consists of the equivalent compression stiffness of the emulsified liquid, the equivalent compression stiffness of the piston rod, the equivalent compression stiffness of piston rod tail, the circumferential extension stiffness of the hydraulic cylinder and the axial compressive stiffness of the hydraulic cylinder bottom. The stiffness coefficient of each part is as follows:

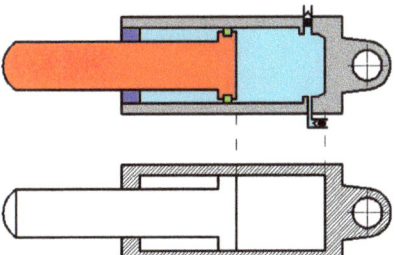

Figure 1. Stiffness calculation model of the hydraulic cylinder.

Equivalent compression stiffness of the emulsified liquid:

$$\begin{cases} K_Y = \frac{\Delta F_H}{\Delta l} = \frac{\Delta p \cdot S_H}{\Delta l} \\ E_r = \frac{\Delta p \cdot V_0}{\Delta V} \\ V_0 = l \cdot S_H \\ \Delta V = \Delta l \cdot S_H \\ S_H = \frac{\pi d^2}{4} \end{cases} \quad (1)$$

$$K_Y = \frac{E_r \cdot S_H}{l} = \frac{\pi d^2}{4} \cdot \frac{E_r}{l} \quad (2)$$

where ΔF_H is the variation of the hydraulic oil pressure, S_H is the liquid column cross-sectional area, Δp is the variation of the pressure intensity of the hydraulic oil, E_r is the volume elastic modulus of the emulsion liquid, l is the height of the emulsion liquid column, d is the inside diameter of the hydraulic cylinder and the diameter of the emulsion liquid column.

Circumferential extension stiffness of the cylinder body:

$$\begin{cases} \int_0^\pi \Delta p \cdot \frac{d}{2} \cdot l \cdot \sin\theta_Z d\theta_Z = 2\sigma_Z \cdot l \cdot \frac{(D-d)}{2} \\ \sigma_Z = E_{GT} \cdot \varepsilon_{GT} \\ \varepsilon_{GT} = \frac{\Delta d/2}{d/2} \\ K_{GZ} = \frac{\Delta F_Z}{\Delta d} = \frac{\Delta p \cdot \pi \cdot d \cdot l}{\Delta d} \end{cases} \quad (3)$$

$$K_{GZ} = \frac{\pi \cdot l \cdot (D_d - d) \cdot E_{GT}}{d} \quad (4)$$

where D_d is the outside diameter of the cylinder body, σ_Z is the circumferential stress of the cylinder body, ε_{GT} is the circumferential strain of the cylinder body, θ_Z is the circumferential pressure angle, E_{GT} is the elastic modulus of the cylinder body, Δd is the inner diameter variation of the hydraulic cylinder body, ΔF_Z is the variation of the circumferential pressure in the inner wall of the cylinder body.

Axial compressive stiffness of the cylinder bottom:

$$\begin{cases} K_{GD} = \frac{\Delta F_{GD}}{\Delta l_{GD}} \\ \sigma_{GD} = E_{GD} \cdot \varepsilon_{GDH} \\ \varepsilon_{GDH} = \frac{\Delta l_{GD}}{l_{GD}} \\ S_H \cdot \sigma_{GD} = \Delta F_{GD} \\ S_H = \frac{\pi d^2}{4} \end{cases} \quad (5)$$

$$K_{GD} = \frac{\pi d^2}{4} \cdot \frac{E_{GD}}{l_{GD}} \qquad (6)$$

where l_{GD} is the length of the hydraulic cylinder bottom, E_{GD} is the elastic modulus of the hydraulic cylinder bottom, ε_{GDH} is the axial strain of the hydraulic cylinder bottom, σ_{GD} is the axial stress of the hydraulic cylinder bottom, Δl_{GD} is the axial elongation of l_{GD}.

Equivalent compression stiffness of the piston rod:

$$\begin{cases} K_{GG} = \frac{\Delta F_{GG}}{\Delta l_{GG}} \\ \sigma_{GG} = E_{GG} \cdot \varepsilon_{GGH} \\ \varepsilon_{GGH} = \frac{\Delta l_{GG}}{l_{GG}} \\ S_{GG} \cdot \sigma_{GG} = \Delta F_{GG} \\ S_{GG} = \frac{\pi d_1^2}{4} \end{cases} \qquad (7)$$

$$K_{GG} = \frac{E \cdot S_{GG}}{l_{GG}} = \frac{\pi d_1^2}{4} \cdot \frac{E_{GG}}{l_{GG}} \qquad (8)$$

where l_{GG} is the length of the piston rod, E_{GG} is the elastic modulus of the piston rod, d_1 is the diameter of the piston rod, Δl_{GG} is the elongation of the piston rod, σ_{GG} is the axial stress of the piston rod, ε_{GGH} is the axial strain of the piston rod, S_{GG} is the cross-sectional area of the piston rod.

Equivalent compression stiffness of the piston rod tail:

$$\begin{cases} K_{GGW} = \frac{\Delta F_H}{\Delta l_{GGW}} \\ \sigma_{GGW} = E_{GGW} \cdot \varepsilon_{GGWH} \\ \varepsilon_{GGWH} = \frac{\Delta l_{GGW}}{l_{GGW}} \\ S_H \cdot \sigma_{GGW} = \Delta F_H \\ S_H = \frac{\pi d^2}{4} \end{cases} \qquad (9)$$

$$K_{GGW} = \frac{\pi d^2}{4} \cdot \frac{E_{GGW}}{l_{GGW}} \qquad (10)$$

where l_{GGW} is the length of the piston rod tail, E_{GGW} is the elastic modulus of the piston rod tail, ε_{GGWH} is the axial strain of the piston rod tail, σ_{GGW} is the axial stress of the piston rod tail, Δl_{GGW} is the axial elongation of l_{GGW}.

After each part is connected in series, the equivalent spring stiffness of the hydraulic cylinder can be obtained as follows:

$$K_Q = 1 / \left(\frac{1}{K_Y} + \frac{1}{K_{GZ}} + \frac{1}{K_{GD}} + \frac{1}{K_{GG}} + \frac{1}{K_{GGW}} \right)$$
$$= \frac{K_Y K_{GZ} K_{GD} K_{GG} K_{GGW}}{K_{GZ} K_{GD} K_{GG} K_{GGW} + K_Y K_{GD} K_{GG} K_{GGW} + K_Y K_{GZ} K_{GG} K_{GGW} + K_Y K_{GZ} K_{GD} K_{GGW} + K_Y K_{GZ} K_{GD} K_{GG}} \qquad (11)$$

3. Construction of the Rigid–Flexible Coupling Impact Contact Dynamic Model between Coal Gangue and the Hydraulic Support

The purpose of this paper is to determine the contact response differences between coal gangue and the hydraulic support through the impact contact behavior analysis of coal gangue and the hydraulic support, and to reveal the impact contact characteristics and response difference law when coal gangue impacts the different parts of the support. Due to the large number and complex shape of the underground coal gangue particles as well as the existence of anisotropy, direct theoretical or simulation modeling is difficult to achieve. Moreover, the contact position of irregular shaped particles will cause the change in the equivalent contact radius and the contact responses, so it is impossible to study the influence parameters and the changing law of the contact characteristics qualitatively by using irregular shape. In order to conduct a qualitative study and reveal coal gangue contact difference characteristics caused by their own attribute differences, this paper conducted regular treatment to coal and gangue, uniformly treating coal and gangue particles as spheres, ignoring the plastic deformation and brittle damage of particles, and ignoring the influence of rock micro-cracks [43] on contact response. For quantitative analysis, only the impact behavior of single coal gangue particles and the hydraulic support was studied.

In order to improve the accuracy of simulation results, the rigid–flexible coupling impact contact dynamic model between coal gangue and the hydraulic support is established to study the impact behavior. Particles and the main impacted part of the top coal caving hydraulic support such as top beam, the shield beam and the tail beam are mesh first, the structure after grid division is shown in Figure 2. Due to the movement and force transfer of the connection area of each component in the working process, so the grids of the connection areas such as the column nest surface of the top beam and the base and the pinhole inner surface of the other components are defined as rigid. In order to ensure the effect of force transmission, the front and rear connecting rods are defined as rigid bodies.

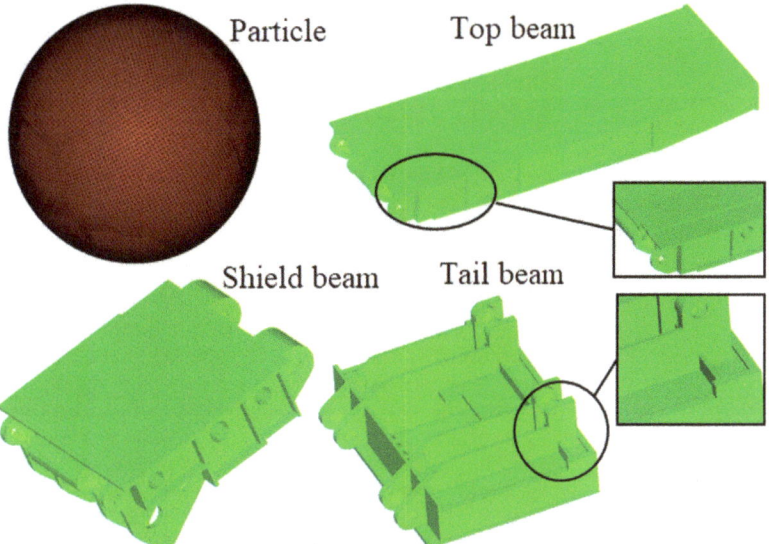

Figure 2. Mesh generation of sphere and the main parts of the hydraulic support.

After the 3D model of coal gangue particles and top coal caving hydraulic support is introduced into Adams, the meshing particles, top beam, shield beam and tail beam files are introduced, respectively, into Adams to replace the corresponding solid file. The rigid area of the pin hole in the top beam, the shield beam, the tail beam, the front and rear connecting rods and the base is connected by the revolute pair. The props and the tail beam jack are equivalent replaced by spring damping modules. The rigid insert plate and the insert plate jack are fixed on the tail beam, and the rigid base is fixed in the space coordinate system. The impact position of coal gangue particles can be adjusted according to requirements, so as to realize the impact of coal gangue on different parts or positions of the support. The completed rigid–flexible impact dynamic model of coal gangue and the hydraulic support is shown in Figure 3.

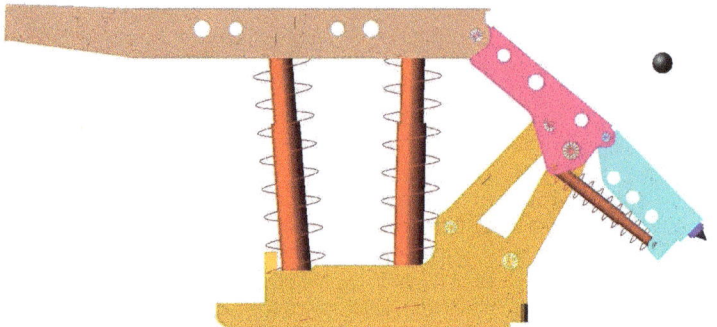

Figure 3. Dynamic model when coal gangue impacting the hydraulic support.

4. Simulation of the Impact Contact Characteristics between Coal Gangue and Different Parts of the Hydraulic Support

4.1. Setting of the Contact Parameters

The radius of coal gangue particles is 2.5×10^{-2} m. The contact stiffness between coal gangue and the hydraulic support can be calculated by the contact stiffness calculation formula [44–49]:

$$K = \frac{4\sqrt{R}}{3} \cdot E \quad (12)$$

Equivalent elastic modulus $\frac{1}{E} = \frac{1-\mu_1^2}{E_1} + \frac{1-\mu_2^2}{E_2}$, equivalent contact radius $\frac{1}{R} = \frac{1}{R_1} + \frac{1}{R_2}$, E_1, μ_1, E_2 and μ_2 are the elastic modulus and Poisson's ratio of the two contact body, respectively, R_1 and R_2 are the contact radius of the contact position on two contact bodies, respectively. For the contact between the spherical particle and the surface of the hydraulic support, the surface can be equivalent to a spherical particle with infinite radius, i.e., $R_2 \to \infty$, then the equivalent contact radius is $R = R_1$.

The props and the tail beam jacks are the single telescopic hydraulic cylinder, and the liquid column height in each cylinder is associated with the position of the hydraulic support. When the working height of the hydraulic support is 2.5 m and the coal dropping angle of the tail beam is 45°, the sizes of the props and the tail beam jack are shown in Table 1, respectively. The equivalent stiffness of the front row prop, back row prop and the tail beam jack are calculated as 9.5×10^7 N/m, 9.6×10^7 N/m and 1.1×10^8 N/m according to Equation (11).

Table 1. Dimensions of the prop and the tail beam jack.

Oil Cylinder	Front Row Prop (mm)	Back Row Prop (mm)	Tail Beam Jack (mm)
Length of the liquid column	836.2	829.7	257.1
Thickness of the cylinder bottom	32	32	50
Inside diameter of the hydraulic cylinder	230	230	140
Diameter of the liquid column	230	230	140
Outer diameter of the hydraulic cylinder	273	273	168
Diameter of the piston rod	210	210	105
Length of the piston rod	1214.5	1214.5	595
Length of the piston rod tail	131.5	131.5	110

4.2. Contact Model in the Simulation

The essence of the impact contact behavior between the coal gangue particle and the hydraulic support is the nonlinear contact between the particle and the metal plate plane, so the contact model based on nonlinear spring damping in Adams is applied to its definition, the normal contact force can be combined describing by the Hertz contact theory-based elastic contact force and the system damping dissipative force, as shown in Equation (13) [50–52]:

$$F_n = \begin{cases} K \cdot \delta^n + STEP(\delta, 0, 0, \lambda_{max}, c_{max}) \cdot \dot{\delta}, \delta > 0 \\ 0, \delta \leq 0 \end{cases} \quad (13)$$

where δ is the relative deformation (particle compression), λ_{max} is the a positive real value specifying the boundary penetration to apply the maximum damping coefficient c_{max}, $\dot{\delta}$ is the relative velocities of two bodies in contact, n is the nonlinear exponent of Hertz contact force.

The step function describing the system damping dissipative force in Adams can be further described as follows:

$$STEP(\delta, 0, 0, \lambda_{max}, c_{max}) = \begin{cases} 0, \delta \leq 0 \\ c_{max}(\frac{\delta}{\lambda_{max}})^2 (3 - \frac{2\delta}{\lambda_{max}}), 0 < \delta < \lambda_{max} \\ c_{max}, \delta \geq \lambda_{max} \end{cases} \quad (14)$$

4.3. Impact Mode of Particles and Different Parts of and Hydraulic Support

The impact position of coal gangue on the top beam was located at the axial center line of the top beam, as shown in Figure 4a. The center of the spherical particle was aligned with the axial midpoint of the front and rear props. In the process of coal gangue dropping, the particles collide frequently with the hydraulic support; however, restricted by the coal gangue dropping space, relative impact velocity between particles and the support is limited. The falling height of the particle free falling and impacting the top beam is defined as 0.8 m (the vertical distance between the center of the sphere and the top surface of the top beam); in this way, the impact contact velocity between particles and the support is limited. After the main structure is divided into grids and the jack is replaced by the spring damping module, the support under the action of gravity will generate structural deformation and vibration self-stability. When the particles fall from a height of 0.8 m and impact on the top beam, the support has not been self-stabilized when the particles contact with the top surface of the top beam, and the residual vibration cannot be ignored. We use "deceleration + free falling" compound movements to realize coal gangue low-velocity impact simulation with the hydraulic support, namely, to extend the fall time and weaken the influence degree of the system vibration on the simulation results by changing the initial relative position and applying an upward initial velocity. Here, we use the initial speed of 2.94 m/s, which is increased by 0.3 s for self-stable equilibrium time at the support.

When coal gangue impacts the shield beam, the center of the spherical particle and the midpoint of the axial centerline of the shield beam are located on the same vertical line, and the distance between the center of the spherical particle and the midpoint of the axial centerline of the shield beam is 0.8 m. The relative positions of the coal gangue and shield beam are shown in Figure 4b. The dip angle of the shield beam will lead to the possibility of collision contact between particles and the tail beam after the multiple collisions and separation between the particles and the shield beam, so contact is added between the particles and the shield beam as well as between the particles and the tail beam. When the coal gangue impacts the tail beam, the center of the spherical particle and the midpoint of the axial center line of the tail beam are located on the same vertical line, and the center of the spherical particle and the midpoint of the axial center line of the tail beam is 0.8 m. The relative positions of the coal gangue and tail beam are shown in Figure 4c.

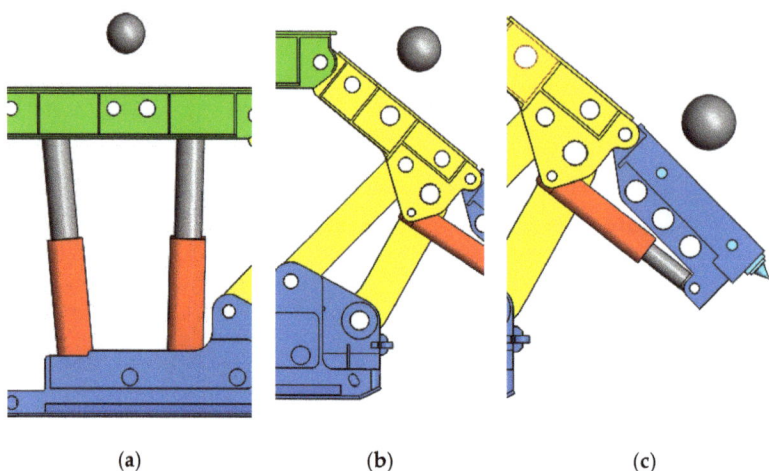

Figure 4. (**a**) Impact top beam; (**b**) Impact shield beam; (**c**) Impact tail beam; Impact modes of the particles and the different parts of the hydraulic support.

4.4. Impact Contact Response Analysis between Coal Particles and the Different Parts of the Hydraulic Support

The impact behavior of particles and the different parts of the hydraulic support will lead to the change in the contact state and the force transfer effect, and further lead to a change in the contact response law. When coal particles free fall from the height of 0.8 m and impact on the designated positions of the top beam, the shield beam and the tail beam, the impact contact response of the different parts of the hydraulic support can be obtained, as shown in Figure 5.

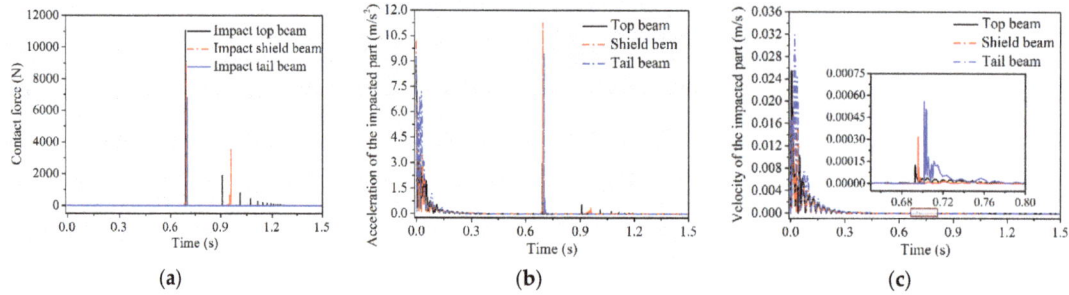

Figure 5. (**a**) Contact force; (**b**) Centroid acceleration; (**c**) Centroid velocity; Impact contact response of the different parts of the hydraulic support.

The props and tail beam jacks are replaced by equivalent springs. In the process of particle free falling, the hydraulic support produces vibrated self-stable equilibrium under the combined action of gravity and the equivalent spring. According to the acceleration and velocity curve of the center of mass of the impinged part of the hydraulic support in Figure 5, the utmost velocity of the center of mass of the tail beam reaches the maximum of the three in the process of self-stable equilibrium, the utmost velocity of the center of mass of the shield beam is the minimum, while the utmost acceleration of the center of mass of the shield beam is the maximum, and that of the center of mass of the tail beam is the minimum. The collision and contact process between the particles and the support is very short and the contact force generated during the process is very large. The impact contact force curve between coal and the different positions of the hydraulic support shows that continuous collision between the coal particle and the top beam happened more than 10 times, and coal particles impacted with the shield beam twice and then separated, eventually coming into contact with the tail beam with rebound impact, but the particle just impacted with the tail beam once and then separated. The collision frequency mainly depends on the support structure, posture and the length of the upper surface metal plate on the various parts and the residual collision velocity of the particle.

When the particles collide with the top beam, the shield beam and the tail beam, these three parts have reached self-stable equilibrium. After self-stabilization, the top beam only deflected at a minimal angle, and the impact between the particles and the top beam was approximately vertical. Therefore, the utmost contact force between coal particles and the top beam was the largest during the initial impact contact. The deflection angle of the tail beam is greater than that of the shield beam, which is supported by the combined "rigid" structure of the top beam, front and rear connecting rods, while one end of the tail beam is in the flexible state of equivalent spring support, so the utmost contact force between particle and the shield beam is greater than that between particle and the tail beam, i.e., $F_{top\ beam} > F_{shield\ bea} > F_{tail\ beam}$. Under the contact force, the relation of the maximum velocity and acceleration of the center of mass between the impinged parts is $v_{top\ beam} < v_{shield\ beam} < v_{tail\ beam}$, $a_{top\ beam} < a_{tail\ beam} < a_{shield\ beam}$. During the collision between the particles and the hydraulic support, the system energy is greatly consumed, leading to a small degree of the system response in the second collision. Compared with the first collision, the contact force and the acceleration of the center of mass of the impacted parts in the second collision between coal particles and the top beam and the shield beam are reduced by more than 82%.

5. Study on the Impact Contact Response and Response Differences between Coal Gangue and the Hydraulic Support

Hydraulic support is the multi-body parallel equipment, and there are interactions between the different parts. The difference in the properties of coal and gangue determines the difference in the impact contact response between coal gangue and the metal plate of the hydraulic support, which will lead to the difference in the impact contact response between coal gangue and the main parts of the hydraulic support, and lead to the difference in the contact response between the associated parts and the parts' joints. This section will determine the difference rule of the impact contact response between the coal gangue and the hydraulic support through the contact response analysis of the direct contact parts, indirect associated parts and parts connection units after coal gangue impact.

5.1. Difference in Contact Response of the Direct Contact Parts

When coal gangue impacts on the top beam, the shield beam and the tail beam, respectively, the contact force between coal gangue and the direct contact parts as well as the acceleration and velocity changing curves of the direct impact parts are shown in Figures 6–8. It can be seen that each impact contact process between coal gangue and the different positions of the hydraulic support is completed within a very short period of time. The contact force generated by the coal particle impacting is smaller than the gangue, and the vibration acceleration and velocity of the center of mass acquired by the impacted parts after coal particle impacting is also lower than the gangue. When colliding with the shield beam at the axial midpoint of the shield beam at the same height, the coal particle separated from the shield beam after two collisions and finally impacted on the tail beam, while the gangue separated from the shield beam after only once collision with the shield beam and then came into contact with the tail beam, that is, the impact frequency of coal gangue when impacting the shield beam was different.

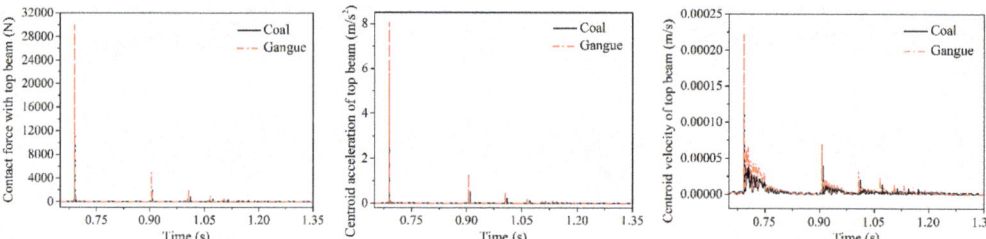

Figure 6. Impact contact response difference in the top beam.

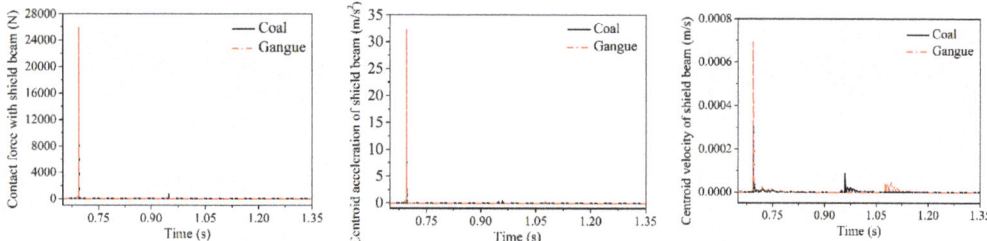

Figure 7. Impact contact response difference in the shield beam.

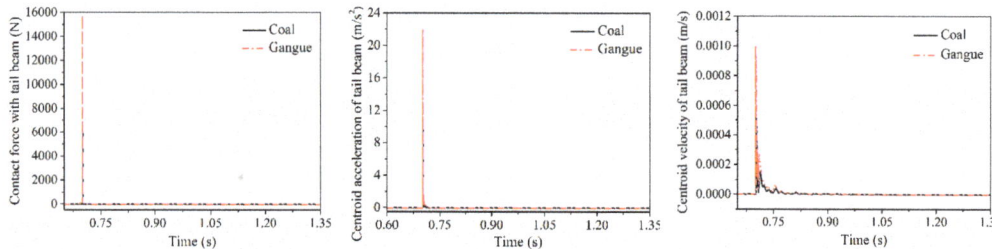

Figure 8. Impact contact response difference in the tail beam.

After coal gangue impacted on the axial midpoint of the shield beam and separated with it, the impact contact force when the particle impact on the tail beam, velocity and acceleration curve of the tail beam are shown in Figure 9, respectively. The impact contact time between the gangue and the tail beam lags behind that of coal, but the contact force and the amplitude of the vibration velocity and acceleration of the tail beam centroid produced by the impact contact action were greater than that of coal.

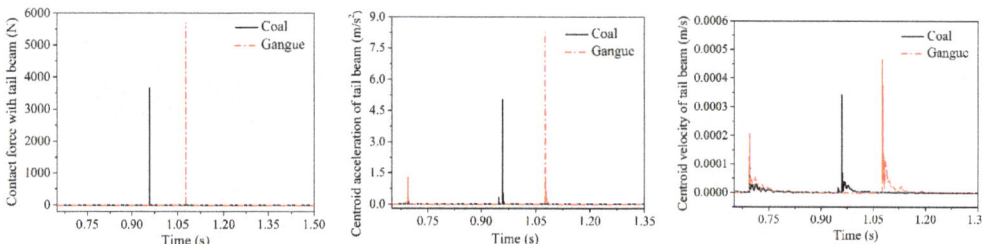

Figure 9. Impact contact response difference in the tail beam when the shield beam is impact directly.

To further clarify the difference in contact response of the direct contacted parts caused by the impact of coal gangue, the extreme values of each contact response in the process of initial collision contact were extracted, respectively, as shown in Table 2.

Table 2. Initial impact limit contact response.

Impact Position	Top Beam			Shield Beam			Tail Beam		
	Contact Force (N)	Velocity (m/s)	Acceleration (m/s^2)	Contact Force (N)	Velocity (m/s)	Acceleration (m/s^2)	Contact Force (N)	Velocity (m/s)	Acceleration (m/s^2)
Coal	11,063.6	1.2×10^{-4}	3.0	9114.9	3.2×10^{-4}	11.3	6792.0	5.6×10^{-4}	9.5
Gangue	30,020.1	2.2×10^{-4}	8.1	25,999.5	6.9×10^{-4}	32.3	15,773.9	1.0×10^{-3}	22.2
Difference value	18,956.5	9.8×10^{-4}	5.1	16,884.6	3.8×10^{-4}	21.0	8982.0	4.4×10^{-4}	12.7
Difference ratio	1.7	0.8×	1.7	1.9	1.2	1.9	1.3	0.8	1.3

During the impact with the top beam, the shield beam and the tail beam, the maximum difference in the contact force generated by the impact between coal gangue particles and the top beam is the greatest, the maximum difference in the vibration acceleration of the impacted parts generated by the impact of the shield beam is the greatest, and the maximum difference in the vibration velocity of the impacted parts generated by the tail beam is the greatest. The contact forces generated by the impact of gangue are all more than 1.3 times higher than those generated by the impact of coal. The vibration velocity and acceleration of the center of mass obtained by the impinged parts after gangue impact is 0.8 times and 1.3 times higher than that of the impinged parts caused by coal impact, respectively. It can be seen that the contact response caused by the coal gangue when impact with the hydraulic support is obviously different, so it is feasible to identify coal and gangue based on the impact contact response. By comparing the difference ratio of coal gangue impact contact response, the difference ratio of the contact force, the vibration velocity and acceleration acquired by the centroid of the shield beam when coal gangue impacts on the shield beam are all the largest, reaching up to 1.9 times, 1.2 times and more than 1.9 times higher, respectively. The difference in the impacted parts contact response caused by the impact of coal gangue when impact on the tail beam is the smallest. Among the contact force, the centroid velocity and acceleration of the impinged parts, the difference ratio between the centroid acceleration of the impinged part and the contact force after coal gangue impact is similar, which is much higher than that of the centroid velocity of the impinged part.

After coal gangue impacted the shield beam and separated with it, the ultimate responses when the coal gangue final impacts on the tail beam such as the contact force, the vibration velocity and acceleration of the center of mass of the tail beam is shown in Table 3. Among them, the ultimate contact force was reduced by more than 59% compared with the contact force when impacting on the shield beam. The difference ratio of the contact force after coal gangue impact can reach above 0.6, the maximum difference ratio of the centroid acceleration of the tail beam can reach 0.7. Although the difference ratio of the centroid velocity of the tail beam is minimum, it can also reach 0.4. When the coal gangue separated from the shield beam and impacted on the tail beam, the impact contact response between coal and gangue is still significantly different.

Table 3. Ultimate contact response when coal gangue rebound impacting the tail beam.

Contact Response	Contact Force (N)	Velocity (m/s)	Acceleration (m/s^2)
Coal	3656.9	3.4×10^{-4}	5.0
Gangue	5791.9	4.7×10^{-4}	8.3
Difference value	2135.0	1.3×10^{-4}	3.3
Difference ratio	0.6	0.4	0.7

5.2. Contact Response Difference of the Indirect Contact and Associated Part

Contact action between coal gangue and the hydraulic support will also be transmitted through the force transfer effect among various parts of the support, thus causing the indirect response of the related parts. In the process of impact contact between coal gangue and the different parts of the hydraulic support, three kinds of associated responses are existed, which are the vibration response of the shield beam and the tail beam jack after the tail beam is impacted by coal gangue, the associated vibration response of the top beam and the tail beam after the shield beam is impacted by coal gangue, and the vibration response of the prop and the shield beam after the top beam is impacted by coal gangue. Figures 10–12 show the centroid velocity response curve of the shield beam after the tail beam is impacted by coal gangue, the centroid velocity response curve of the tail beam after the shield beam is impacted by coal gangue, and the equivalent spring vibration response curve of the tail beam jack after the tail beam is impacted by coal gangue, respectively. It can be seen from the figures that after the impact of coal gangue, there are also significant differences in the correlation responses of non-directly contacting parts. After the different parts of the hydraulic support is impacted by gangue, the contact responses of the associated parts such as the centroid velocity of the shield beam, the centroid velocity of the tail beam and the equivalent spring force amplitude of the tail beam jack are significantly greater than that of coal.

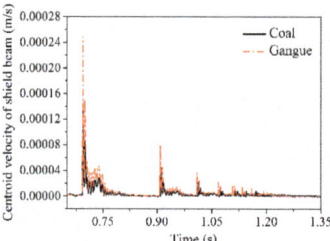

Figure 10. Velocity of shield beam.

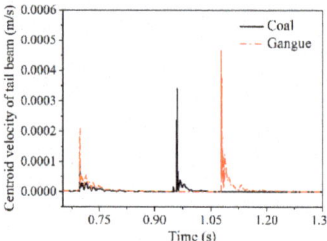

Figure 11. Velocity of tail beam.

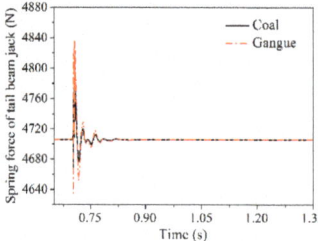

Figure 12. Spring force of tail beam.

The ultimate contact response of the indirect contact and associated parts of the top beam, the shield beam and the tail beam, respectively, were extracted, as shown in Table 4. The difference ratios of the vibration contact response of the associated parts caused by the impact of coal and gangue are all greater than 0.8. The difference ratio of the centroid vibration acceleration of the associated part is much greater than the difference ratio of the centroid vibration velocity. The sensitivity of the associated parts to the contact response when the shield beam and the tail beam are impacted are both greater than that of the top beam.

Table 4. Ultimate contact response of the associated parts.

Part	Top Beam		Shield Beam		Shield Beam		Tail Beam	
	Shield Beam		Top Beam		Tail Beam		Shield Beam	
Associated Part	Velocity (m/s)	Acceleration (m/s^2)	Velocity (m/s)	Acceleration (m/s^2)	Velocity (m/s)	Acceleration (m/s^2)	Velocity (m/s)	Acceleration (m/s^2)
Coal	1.4×10^{-4}	0.4	2.3×10^{-5}	0.4	9.8×10^{-5}	0.5	6.1×10^{-5}	0.4
Gangue	2.5×10^{-4}	0.9	4.7×10^{-5}	1.2	2.1×10^{-4}	1.3	1.1×10^{-4}	0.9
Difference value	1.1×10^{-4}	0.5	2.4×10^{-5}	0.8	1.1×10^{-4}	0.9	5.0×10^{-5}	0.5
Difference ratio	0.8	1.2	1.0	2.0	1.1	1.8	0.8	1.3

The ultimate variation of the equivalent springs for the prop and the tail beam jack when impacted by coal gangue, respectively, were extracted, as shown in Table 5. The response force of each spring after coal impact is lower than that of gangue, and the difference ratio is more than 0.8, that is, there is an obvious difference in the contact response of the hydraulic system. When the top beam and the tail beam are impacted, the front and rear props and the tail beam jacks are directly stressed, respectively. The equivalent spring force, and its difference for the tail beam jack when the tail beam is impacted, is larger than the equivalent spring force and the equivalent spring force difference in the front and rear props. When three different parts were impacted, indirect contact response was observed at the tail beam jack, among which, the difference of the spring response force for the tail beam jack caused by the impacting of coal gangue was more significant when the tail beam was directly impacted.

Table 5. Ultimate variation of the equivalent spring force.

Direct Impact Part	Top Beam			Shield Beam		Tail Beam (N)
	Front Row Prop (N)	Back Row Prop (N)	Tail Beam (N)	Associated Tail Beam (N)	Impacting Tail Beam (N)	
Coal	61.4	62.9	15.4	7.5	15.0	72.1
Gangue	110.8	113.4	27.9	14.9	54.8	131.4
Difference value	49.4	50.6	12.5	7.4	39.8	59.3
Difference ratio	0.8	0.8	0.8	1.0	2.7	0.8

5.3. Contact Response Differences of the Force Transmission Hinge Points

The top beam and the shield beam as well as the shield beam and the tail beam are connected by pin shafts. The connected positions hinge point 1 and hinge point 2, respectively, were defined, and the force transfer between the top beam, the shield beam and the tail beam is realized by hinge point 1 and hinge point 2. The ultimate variation values of the force at the two hinge points are extracted, respectively, as shown in Table 6.

Table 6. Force at the hinge points.

Impact Part	Top Beam		Shield Beam		Tail Beam	
	Hinge Point 1 (N)	Hinge Point 2 (N)	Hinge Point 1 (N)	Hinge Point 2 (N)	Hinge Point 1 (N)	Hinge Point 2 (N)
Coal	109.5	43.7	631.5	190.5	42.2	293.8
Gangue	205.3	75.4	1876.3	435.6	78.1	539.0
Difference value	95.8	31.7	1244.8	245.1	35.9	245.2
Difference ratio	0.9	0.7	2.0	1.3	0.9	0.8

When gangue impacts on the hydraulic support, the force at the hinge point is greater than that of coal, and the difference ratio is above 0.7. Compared with the contact force, the variation value of the active force at the hinge point after coal gangue impact is very small, and the active force at the hinge point 2 when coal gangue impacts on the top beam and the active force at the hinge point 1 when coal gangue impacts on the tail beam is small. By comparing the forces of hinge point 1 when coal gangue impacts on the top beam, the forces of hinge point 1 and hinge point 2 when coal gangue impacts on the shield beam, and the forces of hinge point 2 when coal gangue impacts on the tail beam, the difference in the active force for hinge point 1 when coal gangue impacts on the shield beam is the largest, while the difference in the active force for hinge point 1 when coal gangue impacts on the top beam is the smallest. Additionally, the difference ratios of the active force for two hinge points when coal gangue impacts the shield beam are both the largest.

6. Conclusions

In order to further clarify the contact action law and contact difference characteristics between the coal or gangue particle and the hydraulic support, and then to further lay a foundation for studying the interaction characteristics between coal gangue particles and the hydraulic support in the drawing stage of top coal caving mining, the impact contact behavior between the single coal/gangue particle and the different parts of hydraulic support is taken as the research object in this paper. Based on the construction of the accurate equivalent spring stiffness mathematical model of the hydraulic cylinder, the rigid–flexible coupling impact contact dynamic simulation model between single coal gangue particle and the hydraulic support was established by griding treatment to coal gangue particles and the main force transmission parts of the hydraulic support, and the simulation study on impact contact behavior between coal gangue and hydraulic support is carried out. The following conclusions are drawn:

(1) When the same coal particle impact on the top beam, shield beam and tail beam, respectively, from the same height of 0.8 m in the free-falling way, the coal particles collide with the top beam continuously more than 10 times, collide with the shield beam twice, then separate and eventually come into contact with the tail beam with rebound impact, and then collide with the tail beam only once and then separate. It follows that the change in impacted component will lead to the change in collision times between the particle and the hydraulic support.

(2) When the same coal particle impact on the top beam, shield beam and tail beam, respectively, from the same height of 0.8 m, in the free-falling way, the change in the impacted component will lead to the change in the contact responses value, and the relationship between the contact responses between the coal particle and the impacted component is $F_{\text{top beam}} > F_{\text{shield beam}} > F_{\text{tail beam}}$, $v_{\text{top beam}} < v_{\text{shield beam}} < v_{\text{tail beam}}$, $a_{\text{top beam}} < a_{\text{tail beam}} < a_{\text{shield beam}}$. It can be seen that the size relationship between the contact responses produced by the same particle impacting different parts of the same support is not the same.

(3) When the coal and gangue particles with the same radius impact at the same component of the hydraulic support with the same height, all the contact responses amplitude of the direct contact component, the indirect contact associated compo-

nents and the force transmission hinge points when the gangue impact are larger than that of coal.

(4) When the coal and gangue particles with the same radius impact at the top beam with the same height, the contact response difference ratios of contact force, velocity and acceleration are above 0.8, and the difference ratios of contact force and acceleration are even above 1.7. When the single particle impact at the shield beam, the contact response difference ratios of contact force, velocity and acceleration are above 1.2, and the difference ratios of contact force and acceleration are even above 1.9. When the particles impact at the tail beam, the contact response difference ratios of contact force, velocity and acceleration are above 0.8, and the difference ratios of contact force and acceleration are even above 1.3. Therefore, when coal or gangue particles with the same size impact the same part of the hydraulic support, the contact response caused by coal and gangue is obviously different.

(5) When the coal and gangue particles with the same radius impact at the same component of the hydraulic support with the same height, the associated responses of the non-directly contacting parts such as the velocity, acceleration and spring force were also significantly different, with the difference ratios above 0.8. The active forces at the hinge points when gangue impacts on the hydraulic support are greater than that of the coal, which the difference ratio is above 0.7, the differences were also significant.

(6) When the impact occurs with the hydraulic support, the direct contact response, the indirect contact associated response and the contact response of the force transmission hinge points caused by coal or gangue are all significantly different. Therefore, it is feasible to identify coal and gangue based on the impact contact response.

(7) When the coal and gangue particles with the same radius impact at the tail beam with the same height, not only the contact force, velocity and acceleration of the tail beam are obviously different, but the response force difference ratio of the tail beam jacks supporting the tail beam and the force difference ratio of the hinge point between the shield beam and the tail beam are both more than 0.8. Therefore, when conduct the coal gangue recognition technology research, the vibration responses of the tail beam with the contact responses of the tail beam jack and the connecting pin of the tail beam can be used as coal gangue recognition parameters.

The research content of this paper reveals the difference rule of contact response induced by the difference in impacted location and the impacting material properties during the contact between the coal/gangue particles and the hydraulic support, thereby providing theoretical support for coal gangue recognition in top coal caving based on the contact response difference.

Author Contributions: Conceptualization, Y.Y.; methodology, Y.Y.; investigation, Y.Y.; writing—original draft preparation, Y.Y.; writing—review and editing, Y.Y.; translation and typesetting, Y.Z.; project administration, Q.Z. (Qingliang Zeng), L.W. and Q.Z. (Qiang Zhang); funding acquisition, Q.Z. (Qingliang Zeng) and Y.Y. All authors have read and agreed to the published version of the manuscript.

Funding: This work was supported by National Natural Science Foundation of China (Grant No. 51974170), Special funds for Climbing Project of Taishan Scholars and Open Fundation of Shandong Provincial Key Laboratory of Mining Mechanical Engineering (Grant No. 2022KLMM310).

Data Availability Statement: The data used to support the findings of this study are included within the article.

Conflicts of Interest: The authors declare no conflict of interest.

References

1. Wang, J.C.; Wei, W.J.; Zhang, J.W.; Mishra, B.; Li, A. Numerical investigation on the caving mechanism with different standard deviations of top coal block size in LTCC. *J. Min. Sci. Technol.* **2020**, *30*, 583–591. [CrossRef]
2. Wang, J.A.; Yang, L.; Li, F.; Wang, C. Force chains in top coal caving mining. *Int. J. Rock Mech. Min. Sci.* **2020**, *127*, 104218. [CrossRef]

3. Yang, S.L.; Zhang, J.W.; Chen, Y.; Song, Z.Y. Effect of upward angle on the drawing mechanism in longwall top-coal caving mining. *Int. J. Rock Mech. Min. Sci.* **2016**, *85*, 92–101. [CrossRef]
4. Zhang, N.B.; Liu, C.Y.; Yang, P.J. Flow of top coal and roof rock and loss of top coal in fully mechanized top coal caving mining of extra thick coal seams. *Arab. J. Geosci.* **2016**, *9*, 465. [CrossRef]
5. Alehossein, H.; Poulsen, B.A. Stress analysis of longwall top coal caving. *Int. J. Rock Mech. Min. Sci.* **2010**, *47*, 30–41. [CrossRef]
6. Song, Z.Y.; Konietzky, H.; Herbst, M. Drawing mechanism of fractured top coal in longwall top coal caving. *Int. J. Rock Mech. Min. Sci.* **2020**, *130*, 104329. [CrossRef]
7. Zhang, Q.L.; Yue, J.C.; Liu, C.; Feng, C.; Li, H.M. Study of automated top-coal caving in extra-thick coal seams using the continuum-discontinuum element method. *Int. J. Rock Mech. Min. Sci.* **2019**, *122*, 104033. [CrossRef]
8. Zhang, D.S.; Ren, H.W.; He, M.; Bian, J.; Li, T.J.; Ma, Q. Experimental study on supporting status of internal and external loading of two-legs shielded hydraulic support. *Coal Sci. Technol.* **2019**, *47*, 135–142.
9. Wang, X.W.; Yang, Z.J.; Feng, J.L.; Liu, H.J. Stress analysis and stability analysis on doubly-telescopic prop of hydraulic support. *Eng. Fail. Anal.* **2013**, *32*, 274–282. [CrossRef]
10. Wang, G.F.; Pang, Y.H.; Li, M.Z.; Ma, Y.; Liu, X.H. Hydraulic support and coal wall coupling relationship in ultra large height mining face. *J. China Coal Soc.* **2017**, *42*, 518–526.
11. Ma, Y.Y.; Xie, L.Y.; Qin, X.F. Strength and Reliability Analysis of Hydraulic Support. *Adv. Mater. Res.* **2012**, *544*, 18–23.
12. Sun, H.B.; Jiang, J.Q.; Ma, Q. Research on hydraulic-powered roof supports test problems. *J. Coal Sci. Eng.* **2011**, *12*, 201–206. [CrossRef]
13. Xie, S.R.; Wang, L.; Chen, D.D.; Wang, E.; Li, H.; He, S.S. Spatial Load-bearing Characteristics of Four-pillar Chock-shield Support. *Adv. Eng. Sci.* **2020**, *52*, 56–65.
14. Wang, G.F.; Hu, X.P.; Liu, X.H.; Yu, X.; Liu, W.C.; Lv, Y.; Zheng, Z. Adaptability analysis of four-leg hydraulic support for underhand working face with large mining height of kilometer deep mine. *J. China Coal Soc.* **2020**, *45*, 865–875.
15. Ren, H.W.; Zhang, D.S.; Gong, S.X.; Zhou, K.; Xi, C.Y.; He, M.; Li, T.J. Dynamic impact experiment and response characteristics analysis for 1:2 reduced-scale model of hydraulic support. *Int. J. Min. Sci. Technol.* **2021**, *31*, 347–356. [CrossRef]
16. Gao, J.W. Analysis of Force Transmission of Hydraulic Support under Impact Load. *Mech. Manag. Dev.* **2021**, *9*, 138–140.
17. Zhai, G.D.; Liang, Z.H.; Qu, J.G.; Wang, Y.D.; Chang, J.B. Study on Hydraulic System Characteristics of Hydraulic Support Column under Impact Load. *Coal Eng.* **2019**, *51*, 131–135.
18. Luo, A.M. Analysis of Dynamic Characteristics of Hydraulic Support Column Based on Impact Load. *Coal Mine Mach.* **2021**, *42*, 88–91.
19. Yang, Y.; Zeng, Q.L.; Wan, L.R. Dynamic response analysis of the vertical elastic impact of the spherical rock on the metal plate. *Int. J. Solids Struct.* **2019**, *158*, 287–302. [CrossRef]
20. Yang, Y.; Zeng, Q.L. Influence Analysis of the Elastic Supporting to the Dynamic Response When the Spherical Rock Elastic Impacting the Metal Plate and to the Coal Gangue Impact Differences. *IEEE Access* **2019**, *7*, 143347–143366. [CrossRef]
21. Wan, L.R.; Chen, B.; Yang, Y.; Zeng, Q.L. Dynamic response of single coal-rock impacting tail beam of top coal caving hydraulic support. *J. China Coal Soc.* **2019**, *44*, 2905–2913.
22. Wan, L.R.; Li, Z.; Yang, Y.; Li, R. Analysis on the Difference of Impact Response between Single Coal-Rock Particle and the Box Structure-Based Tail Beam. *Shock. Vib.* **2021**, *2021*, 6688964. [CrossRef]
23. Yang, Y.; Wan, L.R.; Xin, Z.Y. Dynamic Response Analysis of the Coal Gangue-like Elastic Rock Sphere Impact on the Massless Tail Beam Based on Contact-Structure Theory and FEM. *Shock. Vib.* **2019**, *2019*, 6030542. [CrossRef]
24. Zeng, Q.L.; Xin, Z.Y.; Yang, Y.; Chen, B.; Wan, L.R. Stress analysis of hinge point in hydraulic support of coal gangue granular impact in caving based on Abaqus. *J. Shandong Univ. Sci. Technol. Nat. Sci.* **2019**, *38*, 35–42.
25. Liang, L.C.; Tian, J.J.; Zheng, H.; Jiao, S.J. A study on force transmission in a hydraulic support under impact loading on its canopy beam. *J. China Coal Soc.* **2015**, *40*, 2522–2527.
26. Liang, L.C.; Ren, H.W.; Zheng, H. Analysis on mechanical-hydraulic coupling rigidity characteristics of hydraulic powered support. *Coal Sci. Technol.* **2018**, *46*, 141–147.
27. Wan, L.R.; Liu, P.; Meng, Z.S.; Lu, Y.J. Analysis of the influence of impact load on shield beam of hydraulic support. *J. China Coal Soc.* **2017**, *42*, 2462–2467.
28. Liu, X.K.; Zhao, Z.H.; Zhao, R. Study on Dynamic Features of Leg Applied to Hydraulic Powered Support Under Bumping Load. *Coal Sci. Technol.* **2012**, *40*, 66–70.
29. Yang, Y.; Xin, Z.Y.; Zeng, Q.L.; Liu, Z.H. Simulation Research on the Influence of the Clearance to the Impact Contact Characteristics between Coal Gangue and the Clearance-Contained Tail Beam Structure. *Adv. Mater. Sci. Eng.* **2021**, *2021*, 6627395. [CrossRef]
30. Dou, D.Y.; Zhou, D.Y.; Yang, J.G.; Zhang, Y. Coal and gangue recognition under four operating conditions by using image analysis and ReliefSVM. *Int. J. Coal Prep. Util.* **2020**, *40*, 473–482. [CrossRef]
31. Lai, W.H.; Zhou, M.R.; Hu, F.; Bian, K.; Song, H.P. A study of Multispectral Technology and Two-dimension Autoencoder for Coal and Gangue Recognition. *IEEE Access* **2020**, *8*, 61834–61843. [CrossRef]
32. Liu, W.; He, K.; Liu, C.Y.; Gao, Q.; Yan, Y.H. Coal-gangue interface detection based on Hilbert spectral analysis of vibrations due to rock impacts on a longwall mining machine. *Proc. Inst. Mech. Eng. Part C J. Mech. Eng. Sci.* **2015**, *229*, 1523–1531. [CrossRef]
33. Song, Q.J.; Jiang, H.Y.; Song, Q.H.; Zhao, X.G.; Wu, X.X. Combination of minimum enclosing balls classifier with SVM in coal-rock recognition. *PLoS ONE* **2017**, *12*, e0184834.

34. Pu, Y.Y.; Apel, D.B.; Szmigiel, A.; Chen, J. Image Recognition of Coal and Coal Gangue Using a Convolutional Neural Network and Transfer Learning. *Energies* **2019**, *12*, 1735. [CrossRef]
35. Hou, W. Identification of coal and gangue by feed-forward neural network based on data analysis. *Int. J. Coal Prep. Util.* **2019**, *39*, 33–43. [CrossRef]
36. Zhang, N.B.; Liu, C.Y. Radiation characteristics of natural gamma-ray from coal and gangue for recognition in top coal caving. *Sci. Rep.* **2018**, *8*, 190. [CrossRef] [PubMed]
37. Alfarzaeai, M.S.; Niu, Q.; Zhao, J.Q.; Eshaq, R.M.A.; Hu, E.Y. Coal/Gangue Recognition Using Convolutional Neural Networks and Thermal Images. *IEEE Access* **2020**, *8*, 76780–76789. [CrossRef]
38. Zhang, J.W.; Han, X.; Cheng, D.L. Improving coal/gangue recognition efficiency based on liquid intervention with infrared imager at low emissivity. *Measurement* **2022**, *189*, 110445. [CrossRef]
39. Yang, Y.; Zeng, Q.L. Impact-slip experiments and systematic study of coal gangue "category" recognition technology Part I: Impact-slip experiments between coal gangue mixture and top coal caving hydraulic support and the study of coal gangue "category" recognition technology. *Powder Technol.* **2021**, *392*, 224–240. [CrossRef]
40. Yan, P.C.; Sun, Q.S.; Yin, N.N.; Hua, L.L.; Shang, S.H.; Zhang, C.Y. Detection of coal and gangue based on improved YOLOv5.1 which embedded scSE module. *Measurement* **2022**, *188*, 110530. [CrossRef]
41. Yuan, Y.; Wang, J.W.; Zhu, D.S.; Wang, J.C.; Wang, T.H.; Yang, K.H. Feature extraction and classification method of coal gangue acoustic signal during top coal caving. *J. Min. Sci. Technol.* **2021**, *6*, 711–720. [CrossRef]
42. Wang, J.C.; Pan, W.D.; Zhang, G.Y.; Yang, S.L.; Yang, K.H.; Li, L.H. Principles and applications of image-based recognition of withdrawn coal and intelligent control of drawing opening in longwall top coal caving face. *J. China Coal Soc.* **2022**, *47*, 87–101.
43. Liu, Y. Study on Particle Dynamics of Impact Separation for Coal and Gangue Underground. Doctoral Dissertation, China University of Mining and Technology, Xuzhou, China, 2011.
44. Johnson, K.L. *Contact Mechanics*; Cambridge University Press: Cambridge, UK, 1985.
45. Thornton, C.; Cummins, S.J.; Cleary, P.W. On elastic-plastic normal contact force models, with and without adhesion. *Powder Technol.* **2017**, *315*, 339–346. [CrossRef]
46. Brake, M.R.W. An analytical elastic plastic contact model with strain hardening and frictional effects for normal and oblique impacts. *Int. J. Solids Struct.* **2015**, *62*, 104–123. [CrossRef]
47. Hou, Y.L.; Wang, Y.; Jing, G.N.; Deng, Y.J.; Zeng, D.X.; Qiu, X.S. Chaos phenomenon and stability analysis of RU-RPR parallel mechanism with clearance and friction. *Adv. Mech. Eng.* **2018**, *10*, 1687814017746253. [CrossRef]
48. Zheng, E.L.; Zhu, R.; Zhu, S.H.; Lu, X.J. A study on dynamics of flexible multi-link mechanism including joints with clearance and lubrication for ultra-precision presses. *Nonlinear Dyn.* **2016**, *83*, 137–159. [CrossRef]
49. Minamoto, H.; Kawamura, S. Moderately high speed impact of two identical spheres. *Int. J. Impact Eng.* **2011**, *38*, 123–129. [CrossRef]
50. Khemili, I.; Romdhane, L. Dynamic analysis of a flexible slider-crank mechanism with clearance. *Eur. J. Mech. A Solids* **2008**, *27*, 882–898. [CrossRef]
51. Abdallah, M.A.B.; Khemili, I.; Aifaoui, N. Numerical investigation of a flexible slider-crank mechanism with multijoints with clearance. *Multibody Syst. Dyn.* **2016**, *38*, 173–199. [CrossRef]
52. Yang, Y.; Wan, L.R. Study on the Vibroimpact Response of the Particle Elastic Impact on the Metal Plate. *Shock. Vib.* **2019**, *2019*, 6325472.

Article

Deformation, Failure, and Acoustic Emission Characteristics under Different Lithological Confining Pressures

Shuo Wu [1,2], Guangpeng Qin [1,2,*] and Jing Cao [3,*]

1. College of Resource, Shandong University of Science and Technology, Tai'an 271000, China; skdyjsws@163.com
2. National Engineering Laboratory for Coalmine Backfilling Mining, Shandong University of Science and Technology, Tai'an 271000, China
3. College of Finance and Economics, Shandong University of Science and Technology, Tai'an 271000, China
* Correspondence: skd992807@sdust.edu.cn (G.Q.); skd993798@sdust.edu.cn (J.C.)

Abstract: When a temporary support is used to control new surrounding rock in a deep mining roadway, the new surrounding rock is supported by the working resistance of the temporary support. In this study, the influence of deep well boring roadway deformation and rock failure characteristics under different surrounding pressure was investigated. In this paper, for each confining pressure, we experimentally identified the stress-strain, strength, and acoustic emission characteristics of the rocks. The results show that: the surrounding pressure has a significant effect on the damage deformation characteristics of the rock, and the change of the surrounding pressure directly affects the strength, damage form and elastic modulus of the rock; the strength limit of the rock increases with the surrounding pressure, and the damage form of the rock gradually changes to ductile damage with increase of the surrounding pressure; the elastic modulus of the rock increases non-linearly with the increase of the surrounding pressure. The acoustic emission signal of a rock can be divided into three stages: calm, sudden increase, and destruction. The acoustic emission ringing count rate increases suddenly and reaches a peak before the main fracture. Therefore, a sudden increase in the acoustic emission value can be considered a precursor to rock destruction.

Keywords: rock mechanics; compression test; acoustic emission; ambient pressure effect; rupture precursor

1. Introduction

With the gradual depletion of shallow resources, there is an increasing need for deep energy mining. As the most important pillar industry in China, coal mining continues to deepen [1]. Rock underground engineering is generally buried at great depths, and the crossing formations, ground stress, and transmission of the underground structure are complex. Under a state of high ground stress, deep interaction relationships are enhanced. Rock disturbances caused by excavation of the upper rock body become more complex, and disasters are more frequent. Studies have shown that the mechanical response mechanisms of rock change with depth [2]. Deng et al. [3,4] conducted an indoor triaxial compression test on marble from the Jinping Grade II Hydropower Station in southwest China, and used maximum entropy to determine the strength characteristic parameters of the probability density function. Zhang et al. [5] studied the mechanical properties of rocks under the dual action of freezing and melting cycles and surrounding pressure. Mukang et al. [6] studied the relationship between sandstone acoustic emissions and compression deformation by establishing a fine numerical model. L. Gao et al. [7] studied the energy evolution characteristics, based on uniaxial unloading tests, of five different types of rocks, determined their damage rupture thresholds, used energy release and dissipation rates to describe the changes in rock volume and unit strain energy, and concluded that the evolution characteristics of strain energy rates could be easily identified by the crack expansion thresholds. Cheng Hongming et al. [8] found that the energy parameters at the closure stress had a power relationship with the enclosing pressure, while the energy parameters at the initiation stress, damage stress and peak stress had a linear relationship with the enclosing pressure, and the difference between the input energy and elastic strain energy corresponding to the characteristic value points decreased gradually with increase of the enclosing pressure, through graded cyclic loading and unloading tests on sandstones with different enclosing pressures. Ji et al. [9] conducted conventional triaxial compression acoustic emission tests with different circumference pressures to identify rupture precursors for granite. Gong [10] used the acoustic emission frequency analysis algorithm to study precursory signals of rock rupture with instantaneous frequency. Zhang et al. [11] conducted uniaxial compression acoustic emission tests of coal gangue and identified the main frequency and information entropy theory. Wang et al. [12] studied the deformation and strength of three kinds of sandstone under the influence of ambient pressure, using triaxial compression tests. Wang et al. [13] investigated the influence of ambient pressure on shale acoustic emission characteristics through conventional triaxial compression acoustic emission tests. With the continued development of rock engineering, it is particularly important to study rock mechanics and acoustic emission properties under different ambient pressures [13]. The change pattern of acoustic emission signals during rock rupture evolution was studied to reveal its rupture evolution mechanism and explore its rupture precursors [14–17]. Ye Wanjun et al. [18] studied the changes of fine microstructure and macro-mechanical properties of paleosols under the action of dry and wet cycles. Li Shu-Lin et al. [19] studied the acoustic emission properties before peak intensity of rocks under incremental cycling with unloading. Jiang Jingdong et al. [20] studied the mechanical properties and energy characteristics of mudstone under different water-bearing states. Zhang Yanbo et al. [15,21] studied the acoustic emission spectral characteristics of the rock fracture process by conducting uniaxial compression acoustic emission experiments on water-saturated granite, and extracting the primary and secondary frequencies of acoustic emission signals using fast Fourier transform. Fine sandstone and mudstone are often encountered during coal mining; however, little is known about their associated surrounding pressure effect, acoustic emission characteristics, or rupture precursors. Therefore, in this paper, in order to avoid the fact that the acoustic emission characteristics of single lithology rocks have a chance effect of surrounding pressure and its main rupture precursor information, we adopted a comparative study of different lithology rocks to investigate in depth the mechanical and acoustic emission characteristics of rocks under different surrounding pressure.

Pressure and deformation of the support required for deep rock excavation come from the deformation or rupture of the surrounding rock in the affected area of the excavation roadway. After selecting the appropriate roadway rock support, the surrounding rock is in a three-way force state. With the continuous advancement of the road machine, the surrounding pressure of the roadway rock changes; therefore, the morphology of the surrounding rock and its change status have an important influence on the support characteristics.

2. Engineering Background

We used a rock sample from the Fucun coal mine, operated by the Zao Mining Group, to analyze mechanical phenomena after the excavation of the roadway and quarry. The downhole of the roadway is located in the southeast of the Dongshi mining area. The roadway is surrounded by the Huancheng fault to the east, the 1009 working face to the west, the Magizhi 1 fault to the south, and the 1008 transport lane under level 3 to the north. Adjacent excavations include the 1009 working face of level 3 to the north of the roadway, the east 8 mining area in the south, and the east 10 mining area in the west. The East Sephuancheng fault is adjacent to the mine. The upper part is the 1009 working face of level 3. The coal seam excavated in this roadway is number 3 of the Shanxi Group; this seam has a stable coal thickness of 2.3–4.5 m (average of 2.5 m) and a simple structure. Three lower coal seams form the original structure of coal; the coal body structure is a strip, while the endogenous fissure is more developed. Closure of the adjacent mined working surface and the excavated roadway did not impact the coal seam structure. The test standard used in the single-axis test part of this paper was DZ/T 0276.18-2015, and the test standard used in the three-axis test part was DZ/T 0276.20-2015. The surrounding rock and lithological characteristics of the 1009 working face of the Fucun coal mine are shown in Table 1 and the research route is shown in Figure 1.

Table 1. Comprehensive histogram of working face conditions.

Thickness (m)	Depth (m)	Formation
1.85	52.66	Medium sandstone
9.0	61.66	Fine sandstone
10	71.66	Medium sandstone
5	76.66	Siltstone
0.3	76.96	Mudstone
5.5	82.46	coal seam
0.3	82.76	Mudstone
5.12	87.88	Siltstone
0.15	88.03	Mudstone

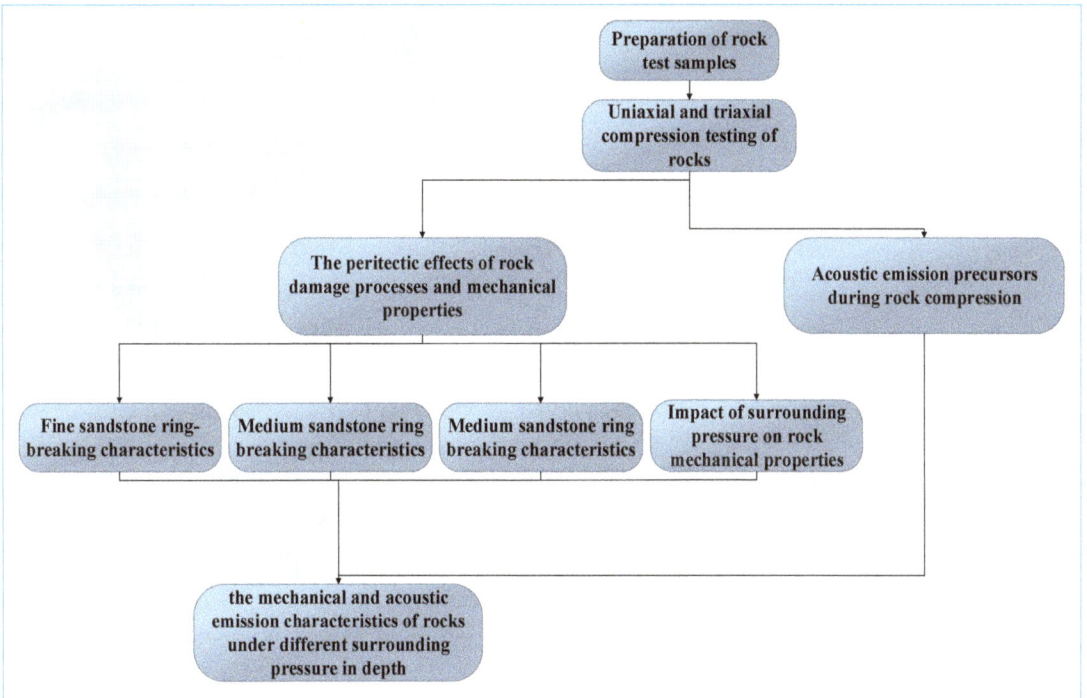

Figure 1. Research Route.

3. Test Protocol

3.1. Preparation of Rock Test Samples

The rocks selected for this experiment were sampled in strict accordance with the unified standards to avoid the influence of specimen anisotropy on the experimental results. The rocks were cored by a rock coring machine, and then the specimens were cut and polished to the standard size by a smoothing machine. We used Φ50 mm × 100 mm standard test coal samples [22], which were polished with error controlled at ±0.02 mm [23] (Figure 2).

Figure 2. Standard test samples.

3.2. Test Equipment

To test the rock samples, we used the axial displacement loading method with a simultaneous acoustic emission device to collect the acoustic emission data of the triaxial compression test. The loading equipment was automatically controlled by a full digital computer, which could adopt various load methods (e.g., force, displacement, and axial strain), and could conduct high-speed data collection. It had the advantages of high test accuracy and stable performance.

Uniaxial compression tests were carried out using a microcomputer-controlled o-hydraulic servo universal test machine. The displacement sensor measured the axial relative displacement parameters, and the axial load parameter of the rock sample was measured by a 100-KN load sensor. The data were automatically converted into the corresponding strain and stress for the data acquisition system terminal, and the radial strain of the rock test piece was used to ensure complete data.

Triaxial compression tests were conducted using a Rock600-50 rock three-axial multi-field coupling mechanical test system (Rock600-50 Triaxial and Multi Field Coupling Rock Mechanical Test System; Figure 3) equipped with an acoustic emission positioning test system (Figure 4). The test surrounding pressure was selected according to the lithology (medium sandstone 7 or 11 MPa; fine sandstone, 4, 8, or 12 MPa; mudstone, 1 or 2 MPa).

Figure 3. Rock600-50 test system.

Figure 4. Specimen and sensor apparatus.

4. Rock Failure and Mechanical Characteristics under Different Surrounding Pressures

4.1. Fine Sandstone Ring-Breaking Characteristics

The fine sandstone damage characteristics were closely related to the surrounding pressure (Figure 5). When the surrounding pressure was 0 (i.e., a uniaxial compression state), the fine sandstone surface first formed small cracks. With increasing pressure, fracture expansion was followed by transverse crack formation. Finally, owing to loss of carrying capacity, the rock sample failed.

When the surrounding pressure was 4 MPa, the initial crack appeared in the middle of the sample, followed by small cracks on the rock surface. After rock failure, small cracks remained on the rock surface, but there was no obvious transverse crack formation. Compared with 0 MPa, the sample remained relatively complete at the end of the experiment.

At an ambient pressure of 8 MPa, the initial rock crack appeared at the top of the sample; as the axial load increased, microcracks were constantly produced and expanded, and eventually ran through the entire sample resulting in shear damage. Ultimately, the rock showed only small lateral cracks, mainly because the presence of ambient pressure limited the lateral expansion of the fine sandstone.

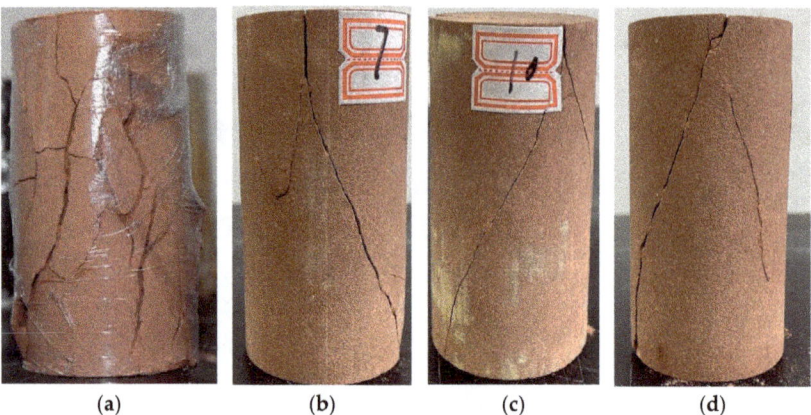

Figure 5. Fine sandstone damage under different confining pressures. Confining pressures of (**a**) 0 MPa, (**b**) 4 MPa, (**c**) 8 MPa, and (**d**) 12 MPa.

Under a circumference pressure of 12 MPa, there was an obvious limitation on transverse crack formation. The development angle of rock fissures gradually reduced with increasing surrounding pressure, and there were almost no transverse fissures. The rock mainly suffered from shear damage, and remained mostly complete after completion of the experiment.

In summary, the level of rock damage decreased with increasing surrounding pressure.

4.2. Medium Sandstone Ring Breaking Characteristics

The fracture surface of the damage of medium sandstone under different circumferential pressure varied greatly, as shown in Figure 6. Under low circumferential pressure of 0 MPa, the damage of medium sandstone was shown in the formation of many macroscopic fractures, and the damage of uniaxial compression (circumferential pressure = 0 MPa) was serious; the fracture surface was large, and the fracture extension angle was large and close to upright in the case of no lateral constraint or the existence of smaller constraint. In addition to the main fracture, there were also secondary macroscopic fractures and transverse fractures developed, and the main fractures and secondary fractures showed macroscopic cross fractures, which made the rock fragmentation very serious.

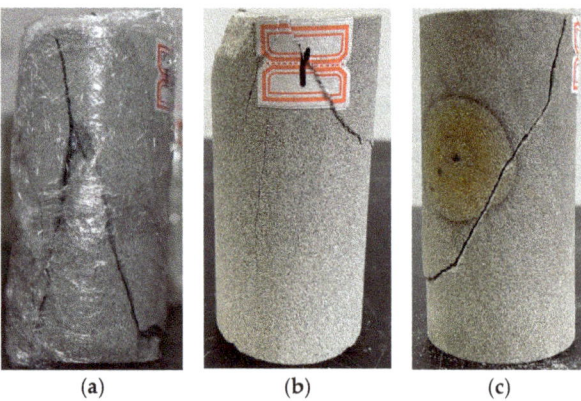

Figure 6. Medium sandstone damage under different confining pressures. Confining pressures of (**a**) 0 MPa, (**b**) 7 MPa, and (**c**) 11 MPa.

With increase of circumferential pressure = 7 MPa, the smaller fractures inside the rock were closed, the development of small fractures reduced, the secondary fractures reduced and were only reflected in subtle places on the rock surface. The crushing of the rock was reduced, compared with the uniaxial state.

The fracture development macroscopically showed a single major fracture and almost no secondary fracture. The dip angle of the fracture surface decreased, the damage angle became smaller, and the rock crushing degree was single.

4.3. Mudstone Ring Breaking Characteristics

As shown in Figure 7.

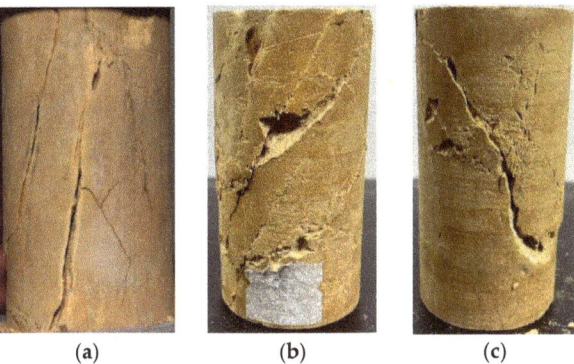

(a) (b) (c)

Figure 7. Mudstone damage under different confining pressures. Confining pressures of (**a**) 0 MPa, (**b**) 1 MPa, and (**c**) 2 MPa.

In the 0 MPa state, the mudstone had splitting damage, i.e., the damage surface of the mudstone was parallel to the direction of the main compressive stress. The fracture slope was close to vertical, there were more secondary fractures, more minor crushing, more broken blocks, and the degree of damage was serious, mainly because the surrounding pressure was 0, which could not limit the development of transverse fractures in the mudstone.

In the state of 1 MPa, there were still more fractures in the mudstone, but the damage surface was inclined to the direction of the main compressive stress, and the damage of the mudstone changed from splitting damage to shear damage. The degree of fine fragmentation of the mudstone was reduced, compared with that in the state of 0 MPa. However, because of the small circumferential pressure, the development of transverse fractures was greater, and there were still more broken pieces of the rock, although it was reduced compared with the circumferential pressure of 0.

In the state of 2 MPa, the dip angle of the rock rupture fissure increased, under the action of axial stress, a shear slip occurred in the mudstone, and the mudstone was destroyed into monolithic. The enclosing pressure increased, the mudstone was compacted, the degree of fine crushing decreased, the development of a transverse fissure was limited, the degree of fine crushing of rock was greatly reduced, and the larger crushing was almost absent.

4.4. Impact of Surrounding Pressure on Rock Mechanical Properties

The stress-strain curves of the rock specimens under uniaxial compression are shown in Figure 8. Similar to that of conventional rocks, the uniaxial compression process could be divided into three stages: elasticity, yield, and destruction. In the elastic phase, stress and strain were basically linear, in line with Huke's law, and had obvious elastic deformation characteristics. In the yield phase, the strain no longer increased linearly with stress, the curve was concave, the rocks began to lose their ability to resist deformation, and underwent irreversible deformation, and gradually changed from elastic to plastic.

According to the test results, it can be seen that the stress-strain results of the three lithologies, fine sandstone, medium sandstone and mudstone, changed in the same trend with increase of the surrounding pressure. The axial stress and axial deformation of the rocks increased and the plastic strain and residual strain also increased significantly. In the uniaxial compression state, the peak compressive strength of the rock was the lowest and the brittle damage was obvious.

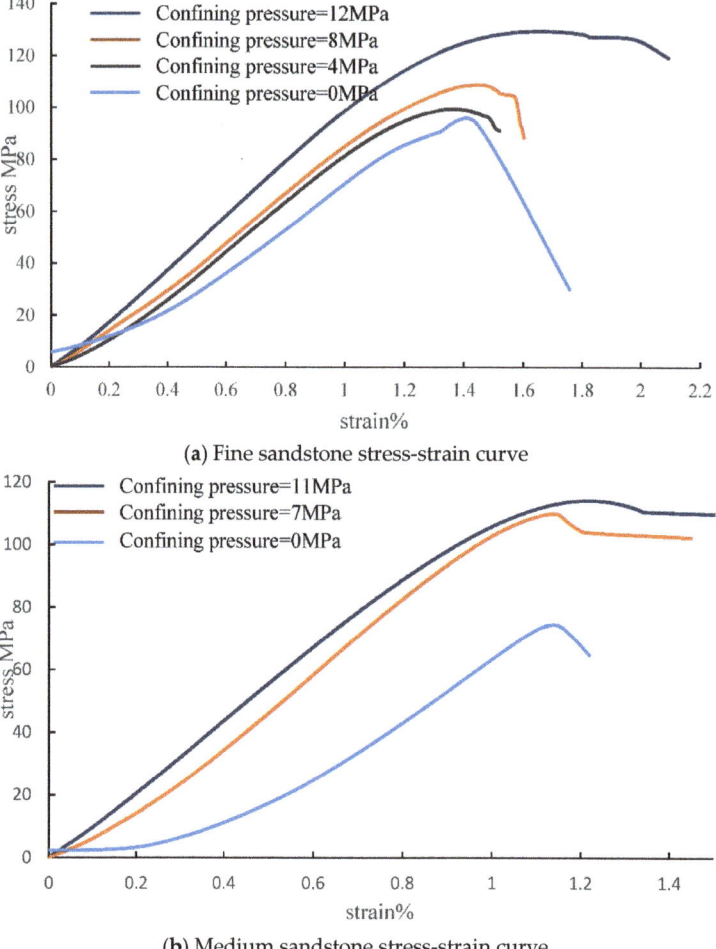

(a) Fine sandstone stress-strain curve

(b) Medium sandstone stress-strain curve

Figure 8. *Cont.*

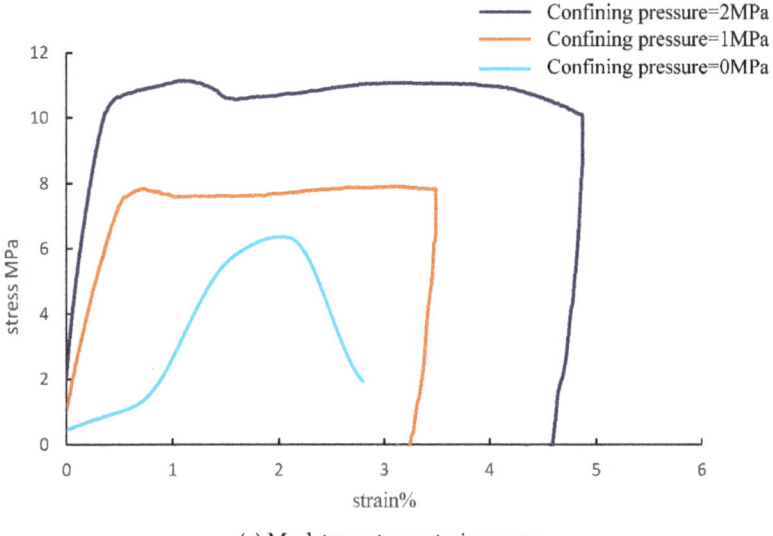

(c) Mudstone stress-strain curve

Figure 8. Stress-strain curves of different rock types. (**a**) Fine sandstone, (**b**) medium sandstone, and (**c**) mudstone.

At the same time, the peak compressive strength of the rock was closely related to the enclosing pressure and increased with increase of the enclosing pressure, and the compressive strength in the uniaxial state was significantly lower than that in the triaxial state. When the enclosing pressure increased, the strain value of the rock increased by nearly 1.5 times, and the rock needed a longer period of compressive deformation before it was destroyed. The peak compressive strength increased significantly, and the strain before destruction also increased, and the linear elastic phase of the rock accounted for the proportion of the pre-peak axial stress-strain curve segment. The proportion of the rock linear elastic phase in the pre-peak axial stress-strain curve segment gradually increased, the elastic modulus gradually increased, and the increase of the surrounding pressure significantly improved the strain capacity of the rock.

4.5. Summary of the Chapter

Comparing the characteristics of uniaxial and triaxial compression damage, it can be seen that all three different lithologies of the rock were severely damaged by uniaxial compression, and the damage of the rock was reduced by increase of the surrounding pressure, while the mechanical properties of the rock were also enhanced with increase of the surrounding pressure. In other words, the damage process and mechanical properties of the rocks are less dependent on the lithology of the rocks.

In addition to the main rupture fracture, there were also many secondary fractures in the uniaxial state of the rock, and the cross damage of the main and secondary rupture fractures made the rock sample break to a serious degree, and the extension and expansion angle of the rock rupture fracture was close to vertical. In the triaxial experiment, the fracture surface of the rock was close to parallel to the main stress direction with a large dip angle in the low circumferential pressure state. The dip angle of the fracture surface of the rock gradually decreased as the circumferential pressure increased, which made the smaller fractures inside the rock compress and close. Comparing the damage patterns of rocks with different surrounding pressure states under the same lithology, with increase of surrounding pressure, the development of transverse fracture extension gradually decreases and the degree of rock damage decreases, which shows that the existence of surrounding pressure restricts the development of transverse fractures in rocks.

Comparing the mechanical characteristics of uniaxial and triaxial compression, we can see that the uniaxial peak compressive strength is the smallest in all three lithologies. The linear elastic stage accounts for the smallest proportion of the pre-peak axial stress–strain curve segment, and the pre-damage strain and peak compressive strength of the rock increases with increase of the enclosing pressure. The reason for this is that in the triaxial compression state, the existence of the enclosing pressure limits the lateral deformation capacity of the rock, and the damage mode of the rock is a single shear damage.

5. Acoustic Emission Precursors during Rock Compression

Rock is a natural granular material containing native cracks and pores; under stress, there is crack expansion of internal fractures, pore closure, and plastic deformation, all of which are acoustic emission sources [24]. Generally, the primary pore fracture compression closure phase is very short. In this phase, due to the rock internal microfracture closure, there is resulting small amplitude acoustic emission generation. In the line elastic deformation phase, rock does not produce damaging damage, the ring count is calm. In fracture initiation and the stable expansion phase, the load exceeds the elastic limit of the rock, and the internal fracture begins to rupture and expand, causing the ring count to jump and increase. Each significant increase indicates that there are fissures sprouting or rupture in the specimen. In the non-stable extension stage of fracture, the ringing count increases sharply and intensively, which is due to the rapid fracture expansion when the main rupture occurs, and the original fracture penetration, thus developing into a fracture network, leading to the formation of macro rupture [25].

Acoustic emission tests were conducted using a displacement load; that is, the data changed with loading time. An acoustic emission device was used to analyze the relationship between acoustic emission and rock deformation characteristics. The main stress-time-bell count ratio curves for different ambient pressure levels for medium sandstone, fine sandstone, and mudstone are shown in Figure 9, Figure 10, and Figure 11, respectively.

As shown in Figure 9, the development of the ringing count rate-time in the main stress-time-bell count rate curve can be divided into three stages: destruction, surge, and calm. When the surround pressure was 7 MPa, the end point of the calm phase was near the peak stress of 90%. When the circumference pressure increased to 11 MPa, the end point occurred before the peak stress, consistent with the elastic stage of the stress-strain curve. During the internal rock crack initiation stage, acoustic emission activity was low, with only a few cracks and small bell count rate.

During the surge phase there was a rapid expansion of internal cracks, consistent with the yield phase of the stress-strain curve; as the stress increased, internal cracks constantly expanded and propagated, and the acoustic emission events increased accordingly (i.e., bell counting rate increased to the maximum value). At a pressure of ~7 MPa, the surge phase occurred after the peak rock stress. As the surrounding pressure increased to 11 MPa, the surge phase occurred at 70–80% of peak stress (i.e., prior to the peak). Therefore, the beginning of the bell counting rate also indicated the peak stress (i.e., the point at which damage occurred). In other words, a sudden increase in the bell count rate could be used as a precursor of rock damage.

The rock destruction stages for the different surrounding pressures corresponded to the destruction stages of the stress-time curve. Medium sandstone destruction occurred immediately after reaching the peak stress; compared with the other rock types, it had fewer acoustic emission events and a smaller bell counting rate.

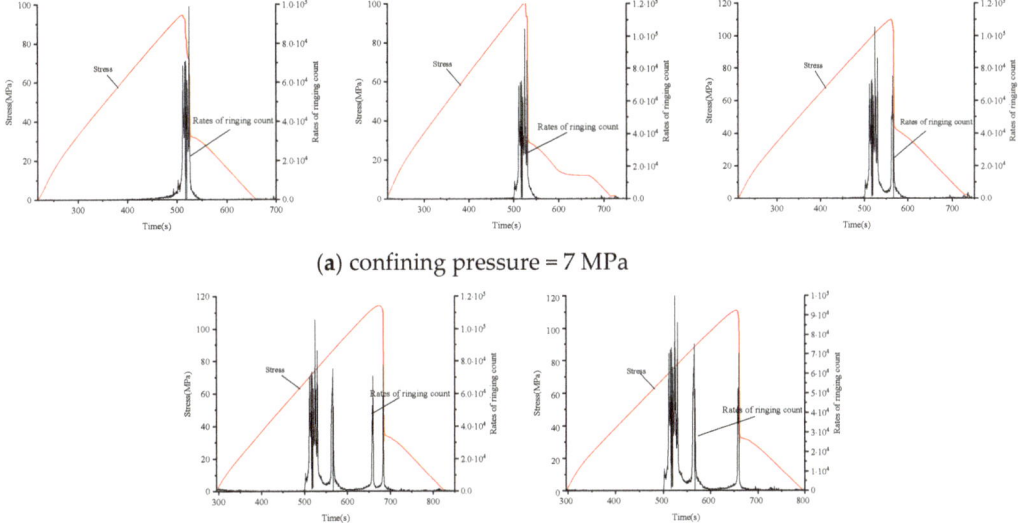

Figure 9. Count rate, stress, and time relationship of medium sandstone failure under different confining pressures. Confining pressure of (**a**) 7 MPa and (**b**) 11 MPa.

For the fine sandstone, we observed a decreasing rock bell count rate during all three stages with increasing circumference pressure (Figure 10). The calm stages for surrounding pressures of 4, 8, and 12 MPa all occurred before the peak stress, because the surrounding pressure removed the small crack pressure inside the rock.

In the surge phase, the rock began to undergo irreversible deformation, consistent with the yield phase of the stress-strain curve, and cracks began to form; rock damage occurred after reaching peak stress. At the same time, acoustic-emission events became active; the bell count rate increased substantially, compared with the calm phase. For a surrounding pressure of 4 MPa, the surge phase began in the destruction phase, while for a surrounding pressure of 8 MPa it began near the peak stress. In contrast, for a surrounding pressure of 12 MPa, the surge phase occurred close to the peak stress. The sudden increase in the count rate of fine sandstone acoustic emission ringing under high ambient pressure could be used as a precursor to rock damage.

The damage stage occurred over a short period after the peak stress decreased rapidly. Owing to the surrounding pressure, samples did not fail immediately. First, a large number of internal cracks began to expand until they ran all through the rock; while, at the same time, the acoustic and emission events were large.

For the mudstone, the calm phase was distributed near the peak stress (Figure 11); the rock produced a small bell count rate before the peak. The spike phases for all surrounding pressures occurred at 70–100% peak stress, consistent with the yield phase of the stress-strain curve. The damage stage occurred after the peak stress, after which cracks developed until the mudstone failed completely and the bell count rate fell.

Increased surrounding pressure significantly reduced the bell count rate of the mudstone; moreover, as the timings of the three stages were similar for surrounding pressures of 1 and 2 MPa, the bell count rates were also similar. Increasing bell count rate indicated the impending failure of the mudstone.

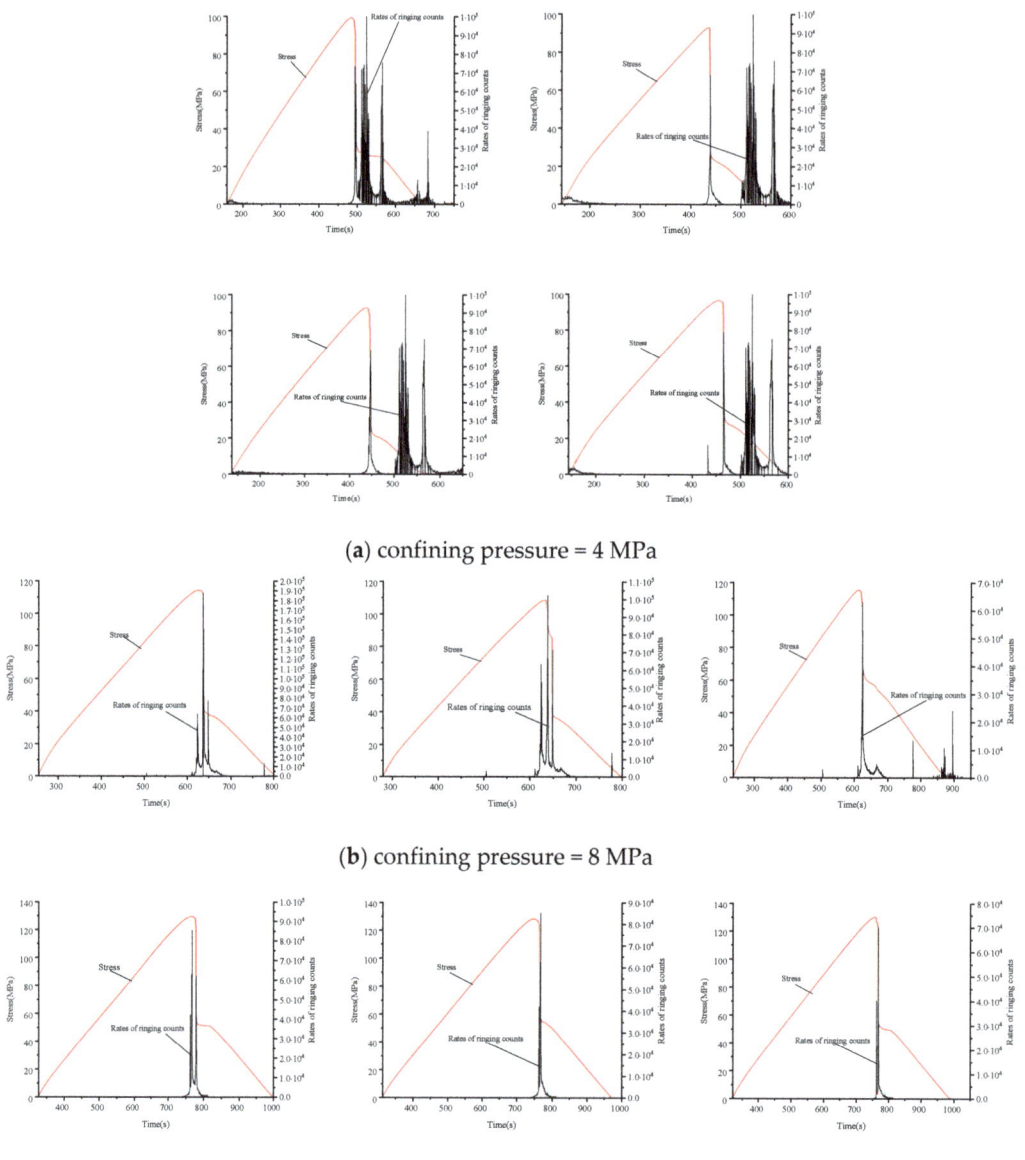

Figure 10. Count rate, stress, and time relationship of fine sandstone failure under different confining pressures. Confining pressure of (**a**) 4 MPa, (**b**) 8 MPa, and (**c**) 12 MPa.

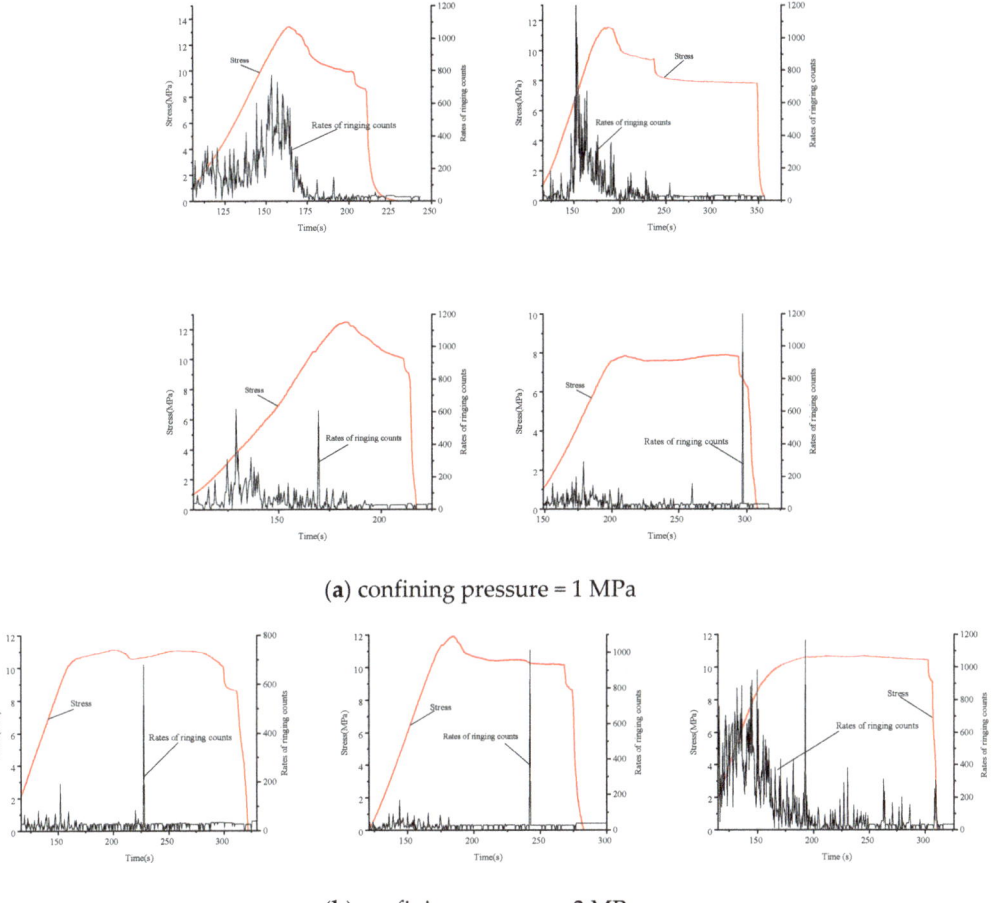

Figure 11. Count rate, stress, and time relationship of mudstone failure under different confining pressures. Confining pressure of (**a**) 1 MPa and (**b**) 2 MPa.

6. Conclusions

In this study, we performed triaxial and uniaxial compression experiments to analyze the deformation and damage characteristics of fine sandstone, medium sandstone and mudstone under different surrounding pressure. The results show that:

The surrounding pressure effects of rock damage processes and mechanical properties are less correlated with rock lithology. The presence of confining pressure leads to the closure of smaller internal cracks, limiting the lateral deformation capacity of the rock, which has a significant impact on the deformation and damage characteristics of the rock. The damage mode of rocks with higher confining pressure is single shear damage. Increasing the confining pressure reduces the development of transverse cracks and increases the compressive strength of the rock.

The acoustic emission ringing count rate reflects the development of fractures within the rock. As the surrounding pressure increases, the acoustic emission ringing count rate decreases as small cracks close. When the acoustic emission ring count rate is small, crack production decreases. The increase of the surrounding pressure makes the fine fractures in the rock destruction process tight and closed, so the acoustic emission ringing count rate of the rock also decreases with the increase of the rock surrounding pressure, and it also further proves the limitation effect of the surrounding pressure on rock fracture development.

A relationship exists between the acoustic emission pattern and the mechanical properties of the rock. A high-frequency band of acoustic emission ringing rate appears in the yielding phase of the stress-strain curve, and in triaxial tests, the ringing rate of the rock increases abruptly before and after the stress peak. In summary, surge in the ringing count rate is significant and can be used as a precursor to rock damage.

Author Contributions: Writing—original draft preparation, S.W.; writing—review and editing, substantial contributions to the conception or design of the work, G.Q., J.C. All authors have read and agreed to the published version of the manuscript.

Funding: This study was sponsored by the Shandong Provincial Key R&D Plan of China (Grant No. 2019SDZY034-1), and the Engineering Laboratory of Deep Mine Rockburst Disaster Assessment Open Project (LMYK2020007), and the National Natural Science Foundation of China (Grant No. 51504145, 51804182). The authors are grateful for their support.

Institutional Review Board Statement: Not applicable.

Informed Consent Statement: Informed consent was obtained from all subjects involved in the study.

Data Availability Statement: The data used to support the findings of this study are available from the corresponding author upon request.

Conflicts of Interest: The authors declare no conflict of interest.

References

1. Xie, H.; Zhou, H.; Xue, D. Research and consideration on deep coal mining and critical mining depth. *J. China Coal Soc.* **2012**, *37*, 535–542.
2. Jiang, Y.D.; Pan, Y.S.; Jiang, F.X.; Dou, L.M.; Ju, Y. State of the art review on mechanism and prevention of coal bumps in China. *J. China Coal Soc.* **2014**, *39*, 205–213.
3. Deng, J.; Li, S.; Jiang, Q.; Chen, B. Probabilistic analysis of shear strength of intact rock in triaxial compression:a case study of J INPING II project. *Tunn. Undergr. Space Technol.* **2021**, *111*, 103833. [CrossRef]
4. Zhang, J.C.; Lin, Z.N.; Dong, B.; Guo, R.X. Triaxial compression testing at constant and reducing confining pressure for the mechanical characterization of a specific type of sandstone. *Rock Mech. Rock Eng.* **2021**, *41*, 1999–2012. [CrossRef]
5. Zhang, H.M.; Xia, H.J.; Yang, G.S.; Zhang, M.J.; Peng, C.; Ye, W.J.; Shen, Y.J. Experimental research of influences of freeze-thaw cycles and confining pressure on physical-mechanical characteristics of rocks. *Rock Mech. Rock Eng.* **2018**, *43*, 441–448.
6. Mu, K.; Li, T.; Yu, J. Visual simulation of the relationship between sandstone acoustic emission and compression deformation under peripressure effect. *J. Rock Mech. Eng.* **2014**, *33*, 2786–2793.
7. Gao, L.; Gao, F.; Zhang, Z.; Xing, Y. Research on the energy evolution characteristics and the failure intensity of rocks. *Int. J. Min. Sci. Technol.* **2020**, *30*, 705–713. [CrossRef]
8. Cheng, H.; Yang, X.; Liu, J. Confining pressure effect on energy parameters of sandstones based on damage evolution. *J. Min. Sci. Technol.* **2020**, *5*, 249–256. (In Chinese)
9. Ji, H.; Lu, X. Characteristics acoustic emission and rock fracture precursors of granite under conventional triaxial compression. *Chin. J. Rock Mech. Eng.* **2015**, *34*, 694–702.
10. Gong, Y.X.; He, M.C.; Wang, Z.H.; Yin, Y. Research on time-frequency analysis algorithm and instantaneous frequency precursors for acoustic emission data from rock failure experiment. *Chin. J. Rock Mech. Eng.* **2013**, *32*, 787–799.
11. Zhang, Y.B.; Liang, P.; Liu, X.; Liu, S.; Tian, B. Experimental study on precursor of rock burst based on acoustic emission signal dominant-frequency and entropy. *Chin. J. Rock Mech. Eng.* **2015**, *34*, 2959–2967.
12. Wang, Y.F.; Su, H.; Wang, L.P.; Jiao, H.Z.; Li, Z. Study on the difference of deformation and strength characteristics of three kinds of sandstone. *J. China Coal Soc.* **2020**, *45*, 1367–1374.
13. Wang, Z.; Gu, Y.; Li, Q. Research on confining pressure effect on acoustic emission of shale and its main fracture precursor information. *Chin. J. Undergr. Space Eng.* **2018**, *14*, 78–85.
14. Lai, X.P.; Zhang, S.; Cui, F.; Wang, Z.; Xu, H.; Fang, X. Energy release law during the damage evolution of water-bearing coal and rock and pick-up of AE signals of key pregnancy disasters. *Chin. J. Rock Mech. Eng.* **2020**, *39*, 433–444.
15. Zhang, Y.B.; Yang, Z.; Yao, X.L.; Tian, B.; Liu, X.; Liang, P. Experimental study on rock burst acoustic emission signal and fracture characteristics in granite roadway. *J. China Coal Soc.* **2018**, *43*, 95–104.
16. Zhao, H.; Liang, Z.; Liu, X. Acoustic Emission Evolution Characteristics of Rock Under Three-point Bending Tests. *J. Water Resour. Archit. Eng.* **2020**, *18*, 6–12.

17. Xia, D.; Yang, T.H.; Xu, T.; Wang, P.; Zhao, Y. Experimental study on AE properties during the damage process of water-saturated rock specimens based on time effect. *J. China Coal Soc.* **2015**, *40*, 337–345.
18. Ye, W.J.; Wu, Y.T.; Yang, G.S.; Jing, H.; Chang, S.; Chen, M. Study on microstructure and macro-mechanical properties of pale sol under dry-wet cycles. *Chin. J. Rock Mech. Eng.* **2019**, *38*, 2126–2137. (In Chinese)
19. Li, S.L.; Zhou, M.J.; Gao, Z.P.; Chen, D.X.; Zhang, J.L.; Hu, J.Y. Experimental study on acoustic emission characteristics before the peak strength of rocks under incrementally cyclic loading-unloading methods. *Chin. J. Rock Mech. Eng.* **2019**, *38*, 724–735. (In Chinese)
20. Jiang, J.D.; Chen, S.S.; Xu, J.; Liu, Q.S. Mechanical properties and energy characteristics of mudstone under different containing moisture states. *J. China Coal Soc.* **2018**, *43*, 2217–2224. (In Chinese)
21. Yao, X.; Zhang, Y.; Liu, X.; Liang, P.; Sun, L. Optimization method for key characteristic signal of acoustic emission in rock fracture. *Rock Soil Mech.* **2018**, *39*, 375–384. (In Chinese)
22. Teng, J.; Tang, J.; Wang, J.; Zhang, Y. The evolution law of the damage of bedded composite rock and its fractal characteristics. *Chin. J. Rock Mech. Eng.* **2018**, *37*, 3263–3278.
23. Liu, X.H.; Zhang, R.; Liu, J. Dynamic test study of coal rock under different strain rates. *J. China Coal Soc.* **2012**, *37*, 1528–1534.
24. Mei, Z.; Hu, Y.; Bao, C. Numerical Simulation of Mechanical and Acoustic Emission Characteristics in Rock Compression Under Different Confining Pressures. *J. Shaoxing Univ.* **2017**, *37*, 15–19.
25. Yang, Y.J.; Wang, D.C.; Guo, M.F.; Li, B. Study of rock damage characteristics based on acoustic emission tests un-der triaxial compression. *Chin. J. Rock Mech. Eng.* **2014**, *33*, 98–104.

Article

Inversion Method of the Young's Modulus Field and Poisson's Ratio Field for Rock and Its Test Application

Yanchun Yin [1,2], Guangyan Liu [1], Tongbin Zhao [2], Qinwei Ma [1,*], Lu Wang [1] and Yubao Zhang [2]

1. School of Aerospace Engineering, Beijing Institute of Technology, Beijing 100081, China
2. College of Energy and Mining Engineering, Shandong University of Science and Technology, Qingdao 266590, China
* Correspondence: maqw@bit.edu.cn

Abstract: As one typical heterogeneous material, the heterogeneity of rock micro parameters has an important effect on its macro mechanical behavior. The study of the heterogeneity of micro parameters is more important to reveal the root cause of deformation and failure. However, as a typical heterogeneous material, the current testing and inversion method is not suitable for micro parameters measurement for the rock. Aiming at obtaining the distribution of micro Young's modulus and micro Poisson's ratio of the rock, based on the digital image correlation method (DIC) and finite element method (FEM), this paper proposed a parameter field inversion method, namely the DF-PF inversion method. Its inversion accuracy is verified using numerical simulation and laboratory uniaxial compression test. Considering the influences of heterogeneity, stress state and dimension difference, the average inversion error of Young's modulus field and Poisson's ratio field are below 10%, and the proportion of elements with an error of less than 15% accounts for more than 86% in the whole specimen model. Compared with the conventional measuring method, the error of macro Young's modulus and macro Poisson's ratio calculated by the DF-PF inversion method is less than 2.8% and 9.07%, respectively. Based on the statistical analysis of Young's modulus field and Poisson's ratio field, the parameter homogeneity and quantitative function relation between the micro parameter and the principal strain can also be obtained in laboratory tests. The DF-PF inversion method provides a new effective method of testing Young's modulus field and Poisson's ratio field of the rocks under complex stress states.

Keywords: rock; Young's modulus; Poisson's ratio; parameters field; inversion method

1. Introduction

Rock is a typical heterogeneous material. Its microstructure and micro parameter distribution are the main controlling factors leading to strain localization, stress concentration, nonlinear damage evolution, and so on [1–4]. The study of rock heterogeneity is very important for revealing the mechanism of rock deformation and failure and even guiding the stability control of rock mass engineering. Therefore, in the field of rock mechanics, the microstructure characterization and micromechanical properties research of rocks have always been a hot issue.

At present, rock heterogeneity is mainly studied by numerical simulation methods. Weibull function [5,6] and digital image processing technology (DIP) [7,8] are often used to characterize the heterogeneity of rock microparameters and microstructures. By using the Weibull function, Chen et al. [9] and Pan et al. [10] studied the effect of rock parameter homogeneity on its macro mechanical behavior. Based on the micrographs of rocks, Shah et al. [11] built a microstructure-based numerical model and evaluated the microscale failure response of various weathering grade sandstones. These studies have a certain significance for revealing the mechanical properties of the rock. However, in these kinds of numerical simulation methods, the microparameters are hard to quantitatively assigned.

Even if it is characterized by Weibull function, it is based on artificial assumptions, and is inconsistent with the real distribution of the rock. Therefore, the accurate assignment of micro parameters in microstructure numerical models is important to research.

The laboratory test is another commonly used method to study rock microproperties. The commonly used test methods include computerized tomography (CT) [12], scanning electron microscope (SEM) [13], digital image correlation method (DIC) [14–17], and others. CT and SEM are always used to obtain the microstructures and microfractures distribution of the rock. Using DIC, the rock's deformation field and strain localization characteristics can be analyzed. The laboratory test methods mainly focus on the research of rock microstructures and lack effective micro parameters testing methods.

Combining numerical simulation methods and laboratory test methods, a variety of parameter inversion methods for materials have been proposed, such as the virtual field method [18] and finite element model updating method (FEMU) [19]. These kinds of methods can obtain material mechanical parameters under complex stress. By using weight FEMU and DIC, Mathieu et al. [20] estimated the parameters of the pure titanium sample under tensile loading. Ogierman et al. [21] proposed a novel two-step optimization procedure inversion method of the elastic properties of the composite constituents, which improved the solution efficiency. Based on the digital speckle correlation method (DSCM) and finite element method (FEM), Song et al. [22,23] proposed the DSCM-FEM inversion method, which can obtain the Young's modulus and Poisson's ratio of rock soil materials.

The above inversion methods can obtain the macro Young's modulus and other material parameters under a certain stress state, which provide ideas for measuring microparameters. Liu et al. [24,25] proposed a double iterative inversion method based on DIC and FEM. They obtained Young's modulus field (i.e., elements micro Young's modulus distribution on specimen surface) and damage variable field of graphite material. In this method, it is assumed that the damage variable of the micro Young's modulus of the whole model conforms to the same equation, which is suitable for the inversion of Young's modulus field of homogeneous materials such as graphite. However, for rock materials affected by the diagenetic process, in-situ stress environment, engineering construction disturbance and specimen processing, the rock before the test has certain initial damage, and the microparameters and macroparameters have obvious discreteness [26–28]. The parameter damage of elements of the specimen is difficult to be described by the same quantitative equation. It is verified that this method cannot obtain the Young's modulus field of heterogeneous rock.

As a typical heterogeneous material, the current parameter inversion method is not suitable for micro parameters inversion for the rock. In order to effectively obtain the micro parameters of the rock, this paper proposed a parameter field inversion method for rocks based on DIC and FEM, namely DF-PF inversion method. This method can perform simultaneous inversion of Young's modulus field and Poisson's ratio field (elements' micro Poisson's ratio distribution on specimen surface). The accuracy of the inversion results is verified by a numerical simulation test, and the inversion test of laboratory rock uniaxial compression test was carried out. The tests reveal that the DF-PF inversion method can obtain Young's modulus field and Poisson's ratio field of rocks under a complex stress state.

2. DF-PF Inversion Method

One key link in the parameter inversion process based on DIC and FEM is the matching between experimental and simulated strain fields. The strain field of rock is not only related to Young's modulus, but Poisson's ratio is also an important factor, and micro Poisson's ratio of elements is also heterogeneous. Therefore, to ensure the accuracy of inversion results, Poisson's ratio field should be carried out simultaneously during the inversion of rock Young's modulus field.

Based on the high-efficiency double iterative inversion method proposed by the authors' previous work [24], this paper optimizes its objective function and iterative process. It puts forward the DF-PF inversion method, which can realize the synchronous inversion of rock Young's modulus field and Poisson's ratio field, as shown in Figure 1. The specific methods are as follows:

(1) Carry out the laboratory test of rock specimen, and obtain the real strain field on the specimen surface by DIC method.
(2) Then, FEM is used to establish the numerical model, and the geometric dimensions and boundary conditions of the numerical model are consistent with those in the laboratory test. Assign each element's micro Young's modulus E_i and micro Poisson's ratio μ_i separately, and their initial value can be set as the results measured by ISRM suggested method.
(3) Export the stress field obtained by FEM, and calculate the strain field by Hooke's law. Then, an objective function Q_i is established as the squared difference between the strains measured with the DIC and the strains calculated by FEM. Each element i establishes an independent objective function Q_i, in which micro Young's modulus E_i and micro Poisson's ratio μ_i are taken as the inversion variables. The idea is to iteratively minimize the objective function with respect to Young's modulus E_i and micro Poisson's ratio μ_i. The objective function Q_i is:

$$Q_i = \left(\frac{\sigma_{xi}}{E_i} - \frac{\mu_i \sigma_{yi}}{E_i} - \varepsilon_{xi}^{DIC}\right)^2 + \left(\frac{\sigma_{yi}}{E_i} - \frac{\mu_i \sigma_{xi}}{E_i} - \varepsilon_{yi}^{DIC}\right)^2 + \left(\frac{2(1+\mu_i)\tau_i}{E_i} - \gamma_i^{DIC}\right)^2 \quad (1)$$

where $\sigma_{xi}, \sigma_{yi}, \tau_i$ is the horizontal stress, vertical stress and shear stress of element i obtained by FEM, and $\varepsilon_{xi}^{DIC}, \varepsilon_{yi}^{DIC}, \gamma_i^{DIC}$ is the horizontal strain, vertical strain and shear strain of element i obtained by the DIC test.

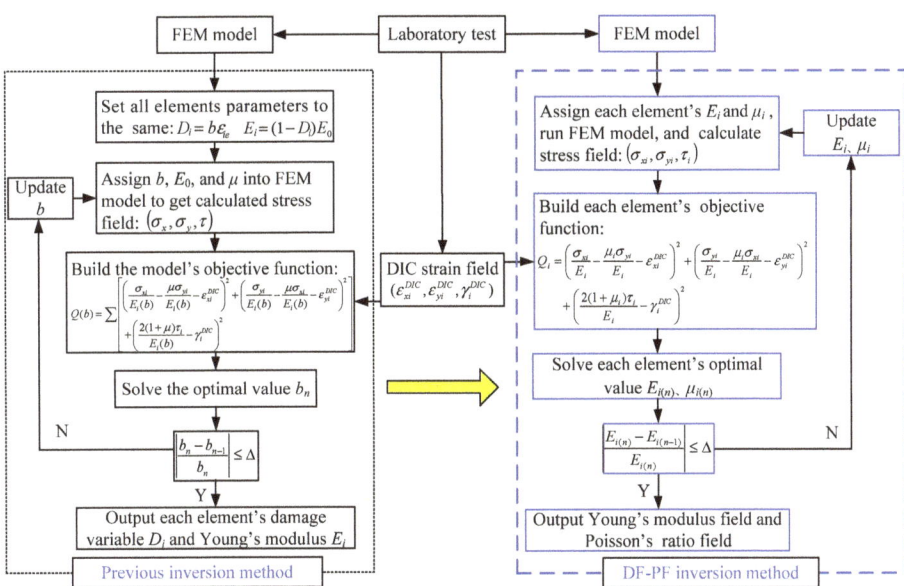

Figure 1. Flow charts of the DF-PF inversion method.

To minimize the objective function Q_i by iteratively seeking for optimal parameters value, the Nelder–Mead simplex method is used [24]. For an optimization problem with two parameters, the Nelder–Mead simplex method is a classical and successful optimization method for unconstrained optimization problems without requiring gradient information.

(4) Solve the objective function Q_i with the Nelder–Mead simplex method, and the optimal micro Young's modulus $E_{i(n)}$ and micro Poisson's ratio $\mu_{i(n)}$ of each element in the current iterative step n can be obtained.

(5) Input the new micro Young's modulus and micro Poisson's ratio of each element into the FEM model and repeat steps (2), (3) and (4). When the difference in the value of each element's micro Young's modulus between the two iteration steps is less than the allowable error Δ, stop the iteration. Then output the micro Young's modulus and micro Poisson's ratio of each element, and draw Young's modulus field and Poisson's ratio field.

In the DF-PF inversion method, the damage variable D is removed, and Young's modulus field and Poisson's ratio field obtained by inversion are the results considering damage evolution. Thus, this method can be used for micromechanical parameter inversion of the rock both in elastic and plastic deformation states.

3. Verification of the Method

3.1. Verification Thought

Due to the obvious discreteness of rocks, the distribution of micro Young's modulus and micro Poisson's rock ratio cannot be determined quantitatively in laboratory tests. It is difficult to verify the accuracy of the DF-PF inversion method by laboratory test method. Therefore, this paper verifies the inversion method using the numerical simulation method. Firstly, establish one heterogeneous rock specimen model through Abaqus, and take the strain field on one surface as DIC strain field data, which surrogate the laboratory test results. Thus, the micro parameters of the rock are known. Then, the DF-PF inversion method is used to inverse Young's modulus field and Poisson's ratio field on the rock surface. Finally, the true value of micro parameters of the specimen is compared with the inversion results to verify the accuracy of the results. The specific method is shown in Figure 2.

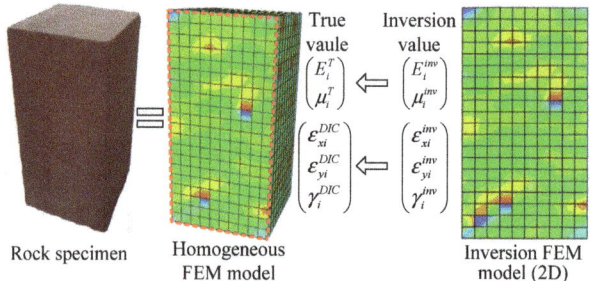

Figure 2. Verification thought of the inversion method.

In the rock specimen model simulating the laboratory test, micro Young's modulus and micro Poisson's ratio of elements of the rock model were heterogeneous, and its homogeneity was set by the shape parameter m in the Weibull function [9,10]. The greater the homogeneity m is, the more homogeneous the micromechanical parameters of rock are. Meanwhile, the damage variable D was added to the model, which is related to the strain of the element [29]. The micro Young's modulus and micro Poisson's ratio of each element was changed with increasing the strain, as shown in Equation (2). Parameters' heterogeneous distribution and damage evolution were realized by writing a user material subroutine (UMAT).

$$\begin{cases} E_i = E_{i0}(1-D) \\ \mu_i = \mu_{i0}(1-D) \\ D = ae^{b\varepsilon_e} + c \end{cases} \qquad (2)$$

where E_{i0} and μ_{i0} are the initial value of micro Young's modulus and micro Poisson's ratio of element i; ε_e is the equivalent strain; a, b, and c is the constant.

3.2. Verification Schemes

In this section, five factors influencing the inversion accuracy were studied, which were micro Young's modulus homogeneity m_E, micro Poisson's ratio m_μ, stress level σ, stress state and model dimension difference (i.e., the laboratory test is a 3D-dimensional model, and the inversion adopts a 2D-dimensional model). Five verification schemes were mainly designed, as shown in Table 1.

Table 1. Verification schemes of the DF-PF inversion method.

Scheme	Influence Factor	Test Type	m	σ/MPa
I	m_E and σ	2D uniaxial compression test	2, 4, 6, 8	5, 10, 15
II	Stress state	2D diametral compression test	2, 4, 6	10
III	Model dimensional difference	3D uniaxial compression test	2, 4, 6	10
IV	m_E, and m_μ	2D and 3D uniaxial compression test	2	10

(1) In scheme I, a 2D uniaxial compression test was carried out to study the influences of micro Young's modulus homogeneity and stress level. The size of the rock specimen model was 50 mm × 100 mm and was divided into 800 quadrilateral elements. The initial macro Young's modulus and macro Poisson's ratio were set to 10 GPa and 0.25. Micro Young's modulus was set as heterogeneity, while micro Poisson's ratio was homogeneous.

(2) In scheme II, the 2D diametral compression test of the ring specimen was carried out to study the influence of the stress state. Compared with the uniaxial compression test, the element is in a complex stress state in the diametral compression test. The ring's outer and inner diameters were 100 mm and 60 mm, respectively, divided into 608 quadrilateral elements.

(3) Scheme III mainly studied the influence of the dimensional difference between the test and inversion models on the inversion results. The test model was a 3D uniaxial compression specimen, a cuboid of 50 mm × 50 mm × 100 mm and divided into 16,000 hexahedral elements. The inversion model was still the 2D uniaxial compression specimen, the same as in scheme I.

(4) Scheme IV mainly studied the inversion accuracy when micro Poisson's ratio and micro Young's modulus were both heterogeneous, and it was researched by 2D uniaxial compression test and 3D uniaxial compression test.

In the verification process, the following indexes were mainly used to evaluate the accuracy of the results:

(1) $e_{\varepsilon x}$, $e_{\varepsilon y}$, and e_γ: the mean value of the relative error of horizontal strain, vertical strain and shear strain between the DIC strain field and inversion strain field.

(2) $R_{\varepsilon x}$, $R_{\varepsilon y}$, and R_γ: the correlation coefficient of horizontal, vertical, and shear strain between DIC strain field and inversion strain field. The more the correlation coefficient tends to 1, the better the correlation.

(3) e_E, and S_E: the relative error and its standard deviation between the true value and the inversion value of Young's modulus field.

(4) e_μ, and S_μ: the relative error and its standard deviation between the true value and the inversion value of Poisson's ratio field.

(5) R_i: the proportion of the elements in the whole specimen.

3.3. Results and Analysis

The results of the four schemes show that the correlation coefficient between the DIC strain field and the inversion strain field of all tests was greater than 0.93, and the relative error was less than 2%. The strain data almost completely coincide, which shows high inversion accuracy of the strain field. The inversion results of the specimen with homogeneity $m_E = 2$ and an axial stress $\sigma = 15$ MPa in scheme I were taken as an example. Micro Young's modulus inversion results are shown in Figure 3. The maximum relative error e_E of elements' micro Young's modulus between the true value and the inversion value is less than 10%, the mean value is 1.89%, and the standard deviation S_μ is 1.68%.

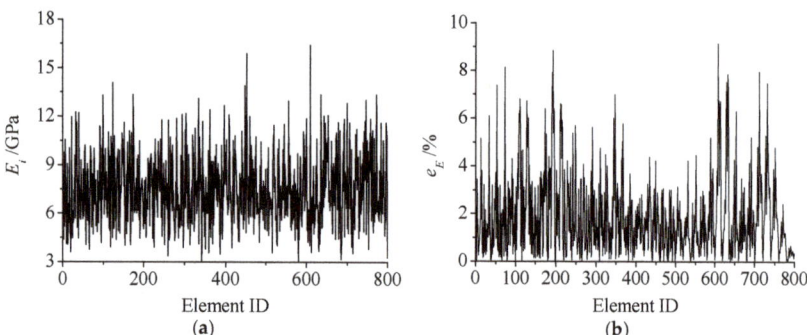

Figure 3. (a) Inversion value and (b) relative error of micro Young's modulus of the specimen with homogeneity $m_E = 2$.

The inversion error of Young's modulus field of scheme I is shown in Figure 4. The stress level only significantly impacts the inversion results of rocks with small homogeneity (such as 2 and 4). With the decrease in homogeneity, the mean value and standard deviation of the relative error of the inversion results gradually increase. However, the mean value is still less than 2.5%, which shows that the inversion accuracy meets the requirements.

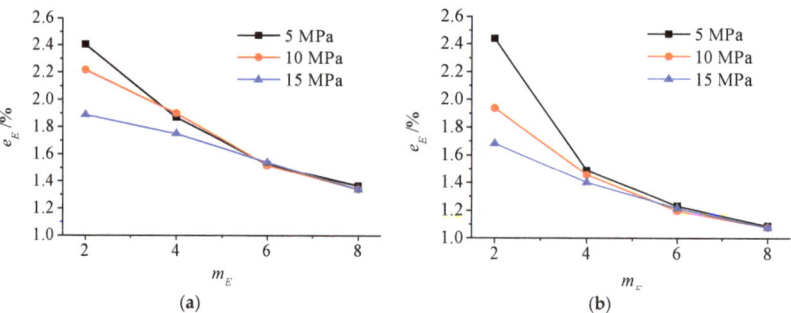

Figure 4. (a) The mean value and (b) standard deviation of relative error of Young's modulus field of Scheme I.

Young's modulus field inversion results of schemes II and III were listed in Table 2. The mean value and standard deviation of the relative error of scheme II are relatively small, indicating that the complexity of the specimen stress state has little impact on the inversion accuracy of micro Young's modulus. The relative error of scheme III is larger than that of scheme II. For example, when the homogeneity $m_E = 2$, the average relative error is 6.61%, but the micro Young's modulus error of most elements is still small, and the proportion R_i of elements with an error less than 15% is 91.75%, as shown in Figure 5. Considering the dimensional difference between the laboratory test and the inversion model, the deformation field of the outer surface of the rock specimen is not only related to the elements on the surface. It is also affected by the associated internal elements, resulting in a larger inversion error of micro Young's modulus of some elements. However, this part accounts for a small proportion, and the inversion accuracy of Young's modulus field of the specimen can still meet the test requirements.

Table 2. Inversion error of Young's modulus field of schemes II and III.

Influence Factor	e_E			S_E		
m_E	2	4	6	2	4	6
Scheme II	1.87%	0.88%	0.82%	1.80%	1.23%	0.92%
Scheme III	6.61%	3.47%	2.60%	5.25%	2.97%	2.21%

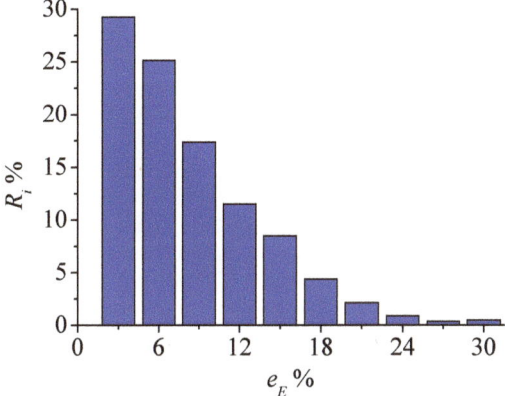

Figure 5. Statistics of the inversion error of Young's modulus field of scheme III.

When considering Poisson's ratio heterogeneity m_μ, the inversion results of Young's modulus field and Poisson's ratio field are listed in Table 3. Compared with the homogeneous Poisson's ratio, its heterogeneous distribution has a certain impact on the inversion result. The maximum value of the relative error of Young's modulus field is 9.89%, while the error of most elements is still small. The proportion of elements with an error of less than 10% accounts for more than 72%, and the proportion of elements with an error of less than 15% accounts for more than 86%, as shown in Figure 6a. For Poisson's ratio field, the average relative error of the 2D uniaxial compression test is 6.18%, the proportion of elements with error less than 10% is more than 87%, and the proportion of elements with error less than 15% is more than 94%, as shown in Figure 6b.

Table 3. Inversion error of Young's modulus field and Poisson ratio field of Scheme IV.

Test Model	e_E	e_μ
2D uniaxial compression test	8.67%	6.18%
3D uniaxial compression test	9.89%	5.83%

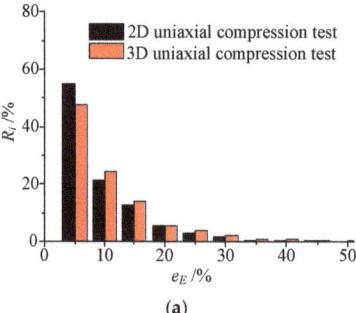 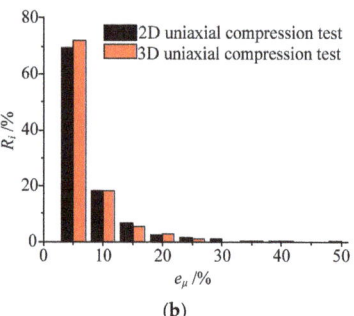

(a)   (b)

Figure 6. Statistics of the inversion error of (**a**) Young's modulus field and (**b**) Poisson's ratio field when Poisson's ratio is heterogeneous.

4. Application in Laboratory Test

4.1. Test Scheme

In order to verify the effectiveness of the DF-PF inversion method in a laboratory test, aluminum and sandstone specimens were selected for the uniaxial compression test, and the specimen was cuboids of 50 mm × 50 mm × 100 mm. The loading device was RLJW-2000 rock mechanics testing machine, developed by Shandong University of Science and Technology (Qingdao, China). The loading speed was 0.1 mm/min. During the test, the axial deformation of the specimen was monitored by using LVDT displacement sensor, produced by Changchun Testing Machine Co. Ltd (Changchun, China). DIC test was carried out simultaneously, and four groups of optical extensometers were arranged [17]. The sampling frequency of the speckle image was 10 frames/s. The specific test system is shown in Figure 7.

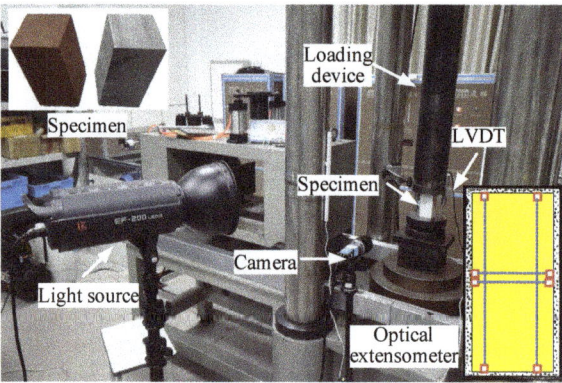

Figure 7. Testing system.

4.2. Inversion Results

Since the inversion results of the strain field, Young's modulus field and Poisson's ratio field of aluminum and sandstone are similar, the sandstone with low homogeneity was analyzed as an example. The inversion results of the sandstone parameter field under axial stress of 26 MPa (99%σ_c) are shown in Figure 8.

For axial strain ε_y field and transverse strain ε_x field, the average relative error between the test results and the inversion results is less than 2%, and the correlation coefficient is greater than 0.99. However, the inversion error of the shear strain γ field is large. Its correlation coefficient is 0.537, and the mean value of the relative error is 49.37%. However, its absolute error is small, and the mean value is only 1.47×10^{-6}. Since the mean value of shear strain field γ is 2.97 × 10^{-6}, its relative error is relatively larger. Overall, the average inversion error of the three strain fields is 17.13%, and the correlation coefficient is 0.842. The strain localization region in the DIC results and the inversion results are similar, as shown in Figure 8a. DIC strain field matches well with inversion strain field for rocks with large discreteness.

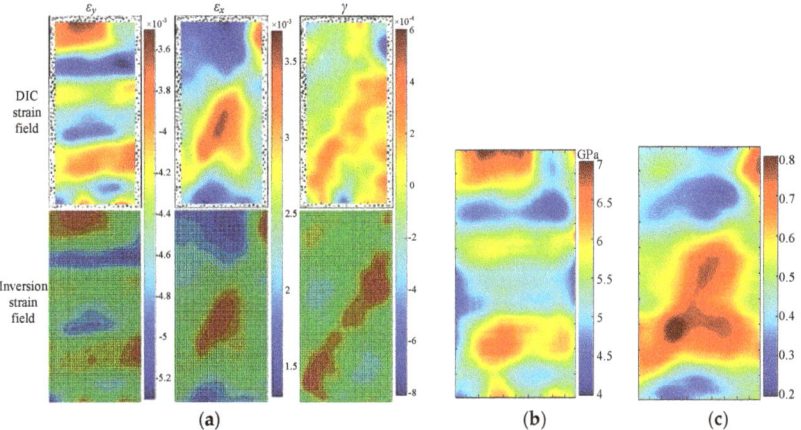

Figure 8. (a) Strain field; (b) Young's modulus field and (c) Poisson field of the sandstone under axial stress of 26 MPa.

Young's modulus and Poisson's ratio fields show obvious localization characteristics. In the uniaxial compression test, the axial strain is related to Young's modulus, and the transverse strain is related to Poisson's ratio. The inversion results show that the distribution patterns of Young's modulus field and axial strain ε_y field, Poisson's ratio field and transverse strain ε_x field are similar, as shown in Figure 8b,c. But they are not exactly the same due to rock heterogeneity. At the same time, affected by the loading end effect, the Poisson's ratio in the middle of the specimen is significantly greater than that in the upper and lower ends.

The statistical distribution of micro Young's modulus of sandstone and aluminum is shown in Figure 9. The distribution form of micro Young's modulus is well fitted with the Weibull function [10], and the homogeneity of aluminum (m_E = 12.86) is greater than that of sandstone (m_E = 10.32). The Poisson's ratio also presents a similar law.

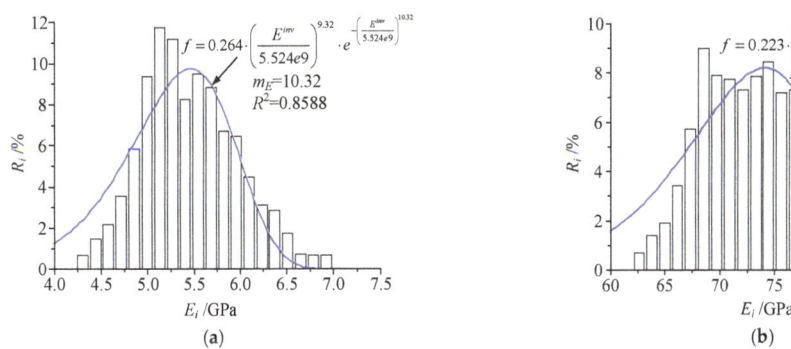

Figure 9. Statistical distribution of micro Young's modulus of (a) sandstone and (b) aluminum.

The above analysis shows that the inversion results of Young's modulus field and Poisson's ratio field obtained by the DF-PF inversion method are reasonable and reliable in the laboratory.

4.3. Comparison with Conventional Test

Based on Young's modulus field and Poisson's ratio field obtained by the DF-PF inversion method, the mean value of the whole field or local field data can be used to calculate the macro Young's modulus and Poisson's ratio of the specimen. Therefore, the accuracy of the inversion method can be further verified by comparing the macro parameters calculated by the DF-PF inversion method with the conventional test results. In the laboratory test, the macro Poisson's ratio was measured and calculated by DIC optical extensometer [17], and the macro Young's modulus was measured by a displacement sensor and DIC optical extensometer. In the DF-PF inversion method, the mean value of the whole field data of Young's modulus field was taken as the macro Young's modulus. The mean value of the Poisson's ratio field data in the area surrounded by the two groups of transverse DIC optical extensometers (as shown in Figure 7) was taken as the macro Poisson's ratio. The comparison and relative error of macro mechanical parameters obtained by the three methods are listed in Table 4. For Poisson's ratio, with the increase in stress level, the Poisson ratio increases gradually. The Poisson's ratio of sandstone under axial stress of 99%σ_c exceeds 0.5, indicating that the sandstone has already entered the plastic deformation stage, which is consistent with the research conclusions of the literature [30,31].

Table 4. Comparison of the inversion results and the conventional testing results.

Specimen	Stress/MPa	Macro Young's Modulus/GPa			Macro Poisson's Ratio	
		DF-PF	LVDT (Error)	DIC (Error)	DF-PF	DIC (Error)
Sandstone	18(69%σ_c)	6.16	5.89 (4.58%)	5.93 (3.88%)	0.314	0.344 (8.72%)
	22(84%σ_c)	5.85	6.09 (3.94%)	5.88 (0.51%)	0.436	0.463 (5.83%)
	26(99%σ_c)	5.52	6.00 (8.00%)	5.45 (1.28%)	0.604	0.681 (11.3%)
	Mean value	5.84	5.99 (2.50%)	5.75 (1.57%)	0.451	0.496 (9.07%)
Aluminum	160	71.74	71.54 (0.28%)	70.59 (1.63%)	0.139	0.152 (8.55%)
	180	71.49	72.55 (1.46%)	70.00 (2.12%)	0.154	0.167 (7.78%)
	200	74.60	73.53 (1.46%)	72.73 (2.57%)	0.193	0.210 (8.10%)
	Mean value	72.61	72.54 (0.10%)	71.11 (2.11%)	0.162	0.176 (7.96%)

Compared with the results of the laboratory sensor method and DIC optical extensometer method, the average error of macro Young's modulus obtained by the DF-PF inversion method is less than 2.5%. The average error of macro Poisson's ratio is less than 9.07%, which can meet the accuracy requirements of rock mechanical parameters with large discreteness. At the same time, the error of mechanical parameters of aluminum is less than that of sandstone, which is the same as the simulation test conclusion. That is, the more homogeneous the specimen material is, the smaller the inversion error is.

4.4. Micro Parameter Evolution Analysis

In the construction of a rock constitutive model, the evolution equation of Young's modulus is the key to the consistency between the constitutive model and test. It is often determined by function fitting with a stress-strain curve combined with a theoretical model. In the DF-PF inversion method, the values of micro Young's modulus and micro Poisson's ratio under different DIC strains of the elements were obtained, and the function fitting of the evolution equations of Young's modulus and Poisson's ratio were also carried out. It is verified that the micro Young's modulus and the maximum principal strain $|\varepsilon_{max}^{DIC}|$ fit well, and the relationship between them is an exponential function, which is consistent with the heterogeneous damage constitutive model of rock [32]. The fitting between micro Poisson's ratio and principal strain ratio $|\varepsilon_{min}^{DIC}/\varepsilon_{max}^{DIC}|$ is good, and they have a linear relationship, as shown in Figure 10.

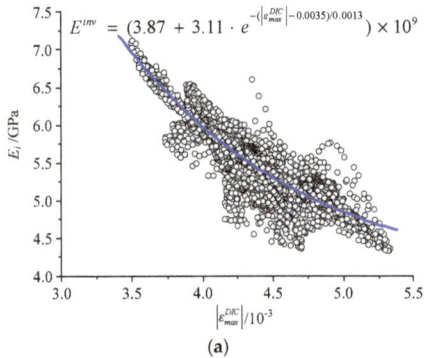 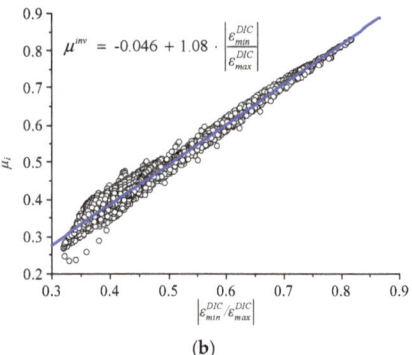

Figure 10. Evolution curves of (**a**) micro Young's modulus and (**b**) Poisson ratio.

5. Discussion

For typical heterogeneous materials like rocks, the mechanical parameters are very discrete. Previous works showed that the 95% confidence interval for rock mechanical parameters is (1 ± 20%) or higher [33,34]. For the DF-PF inversion method, the average error of Young's modulus field and Poisson's ratio field is below 10% in the numerical simulation test. The error of macro parameters is less than 9.07% in a laboratory test. It is verified that the DF-PF inversion method can meet the application requirements of rock parameter field inversion.

Double iterative inversion method [24] is a high-efficiency Young's modulus field and damage field inversion method. But this method is only suitable for homogeneous materials, not for rocks. Compared with the double iterative inversion method, the DF-PF inversion method is an effective inversion method for rocks while retaining high computational efficiency. Since this method is proposed based on DIC and FEM, the DF-PF inversion method is suitable for rocks in elastic and plastic deformation stages. This method will not be applicable when the rock generates obvious macro fractures.

Further, the DF-PF method can be used with the microstructure simulation method. In the microstructure simulation model, the size and distribution of mineral structures can be consistent with real rocks based on DIP [1,7,8]. But the micro parameters of different minerals are hard to assign accurately, and a trial and error method is generally used to determine the appropriate parameter values. For this problem, the DF-PF inversion method can provide accurate micro parameters for the microstructure simulation model.

6. Conclusions

In order to obtain the distribution of micro Young's modulus and micro Poisson's ratio of the rock, based on DIC and FEM, a parameters field inversion method named DF-PF inversion method is proposed, and its inversion accuracy is verified. The main outcome of the study can be summarized as follows:

(1) DF-PF inversion method provides a new effective method for the inversion of Young's modulus field and Poisson's ratio field for rocks in elastic and plastic deformation stages without obvious macro fractures. The average relative error is less than 10%.
(2) The Young's modulus field and Poisson's ratio field are obtained in the laboratory test. The parameters field shows obvious localization characteristics, which provides abundant data for the research of the nonlinear damage evolution of rocks under different stress states.
(3) Compared with the conventional measuring method, the error of macro Young's modulus and macro Poisson's ratio calculated by the DF-PF inversion method is less than 2.8% and 9.07%, respectively. The DF-PF inversion method can be used to measure rock macro mechanical parameters.

(4) The relationship between micro Young's modulus and maximum principal strain is an exponential function, and the relationship between micro Poisson's ratio and principal strain ratio is linear. It provides a new method for the determination of damage equation in the rock heterogeneous damage constitutive model.

Author Contributions: Conceptualization, Y.Y. and T.Z.; Formal analysis, Q.M. and L.W. and Y.Z.; Methodology, Y.Y., G.L. and L.W.; Software, G.L. and Q.M.; Writing—original draft, Q.M.; Writing—review & editing, Y.Y., T.Z. and Y.Z. All authors have read and agreed to the published version of the manuscript.

Funding: The study was financially supported by the Major Program of Shandong Provincial Natural Science Foundation (No. ZR2019ZD13), National Science Foundation of China (No. 52074167), and China Postdoctoral Foundation (No. 2019M660024).

Institutional Review Board Statement: Not applicable.

Informed Consent Statement: Informed consent was obtained from all subjects involved in the study.

Data Availability Statement: The data used to support the findings of this study are available from the corresponding authors upon request.

Conflicts of Interest: The authors declare no conflict of interest.

References

1. Zhang, Y.B.; Zhao, T.B.; Yin, Y.C.; Qiu, Y. Numerical Research on Energy Evolution in Granite under Different Confining Pressures Using Otsu's Digital Image Processing and PFC2D. *Symmetry* **2019**, *11*, 131. [CrossRef]
2. Carpinteri, A.; Corrado, M.; Lacidogna, G. Heterogeneous materials in compression: Correlations between absorbed, released and acoustic emission energies. *Eng. Fail. Anal.* **2013**, *33*, 236–250. [CrossRef]
3. Birck, G.; Iturrioz, I.; Lacidogna, G.; Carpinteri, A. Damage process in heterogeneous materials analyzed by a lattice model simulation. *Eng. Fail. Anal.* **2016**, *70*, 157–176. [CrossRef]
4. Tan, X.; Konietzky, H.; Chen, W. Numerical Simulation of Heterogeneous Rock Using Discrete Element Model Based on Digital Image Processing. *Rock Mech. Rock Eng.* **2016**, *49*, 4957–4964. [CrossRef]
5. Lin, P.; Ma, T.H.; Liang, Z.Z.; Tang, C.A.; Wang, R. Failure and overall stability analysis on high arch dam based on DFPA code. *Eng. Fail. Anal.* **2014**, *45*, 164–184. [CrossRef]
6. Sanchidrian, J.A.; Ouchterlony, F.; Segarra, P.; Moser, P. Size distribution functions for rock fragments. *Int. J. Rock Mech. Min. Sci.* **2014**, *71*, 381–394. [CrossRef]
7. Chen, S.; Yue, Z.Q.; Kwan, A.K.H. Actual microstructure-based numerical method for mesomechanics of concrete. *Comput. Concr.* **2014**, *12*, 1–18. [CrossRef]
8. Chen, S.; Yue, Z.Q.; Tham, L.G. Digital image-based numerical modeling method for prediction of inhomogeneous rock failure. *Int. J. Rock Mech. Min. Sci.* **2004**, *41*, 939–957. [CrossRef]
9. Chen, S.; Qiao, C.S.; Ye, Q.; Khan, M.U. Comparative study on three-dimensional statistical damage constitutive modified model of rock based on power function and Weibull distribution. *Environ. Earth Sci.* **2018**, *77*, 108. [CrossRef]
10. Pan, X.H.; Guo, W.; Wu, S.F.; Chu, J. An experimental approach for determination of the Weibull homogeneity index of rock or rock-like materials. *Acta Geotech.* **2020**, *15*, 375–391. [CrossRef]
11. Shah, K.S.; Hashim, M.H.B.M.; Rehman, H.; Ariffin, K.S.B. Evaluating microscale failure response of various weathering grade sandstones based on micro-scale observation and micro-structural modelling subjected to wet and dry cycles. *J. Min. Environ.* **2022**, *13*, 341–355.
12. Yin, D.H.; Xu, Q.J. Comparison of sandstone damage measurements based on non-destructive testing. *Materials* **2020**, *13*, 5154. [CrossRef]
13. Tian, G.L.; Deng, H.W.; Xiao, Y.G.; Yu, S.T. Experimental study of multi-angle effects of micron-silica fume on micro-pore structure and macroscopic mechanical properties of rock-like material based on NMR and SEM. *Materials* **2022**, *15*, 3388. [CrossRef] [PubMed]
14. Ma, Q.W.; Sandali, Y.; Zhang, R.N.; Ma, F.Y.; Wang, H.T.; Ma, S.P.; Shi, Q.F. Characterization of elastic modulus of granular materials in a new designed uniaxial oedometric system. *Chin. Phys. Lett.* **2016**, *33*, 038101. [CrossRef]
15. Zhao, T.B.; Yin, Y.C.; Tan, Y.L.; Song, Y.M. Deformation tests and failure process analysis of anchorage structure. *Int. J. Min. Sci. Technol.* **2015**, *25*, 237–242. [CrossRef]
16. Pan, B. Digital image correlation for surface deformation measurement: Historical developments, recent advances and future goals. *Meas. Sci. Technol.* **2018**, *29*, 082001. [CrossRef]
17. Munoz, H.; Taheri, A.; Chanda, E.K. Pre-peak and post-peak rock strain characteristics during uniaxial compression by 3D digital image correlation. *Rock Mech. Rock Eng.* **2016**, *49*, 2541–2554. [CrossRef]

18. Pierron, F.; Sutton, M.A.; Tiwari, V. Ultra high speed DIC and virtual fields method analysis of a three point bending impact test on an aluminium bar. *Exp. Mech.* **2011**, *51*, 537–563. [CrossRef]
19. Wan, H.P.; Ren, W.X. Parameter selection in finite-element-model updating by global sensitivity analysis using Gaussian process metamodel. *J. Struct. Eng.* **2015**, *141*, 04014164. [CrossRef]
20. Mathieu, F.; Leclerc, H.; Hild, F.; Roux, S. Estimation of elastoplastic parameters via weighted FEMU and integrated-DIC. *Exp. Mech.* **2015**, *55*, 105–119. [CrossRef]
21. Ogierman, W.; Kokot, G. Analysis of strain field heterogeneity at microstructure level and inverse identification of composite constituents by means of digital image correlation. *Materials* **2020**, *13*, 287. [CrossRef] [PubMed]
22. Song, Y.M.; Ling, X.K.; Zhang, J.Z.; Zhu, C.L.; Ren, H.; Yuan, D.S. Experimental study of DSCM-FEM inversion of mechanical parameters of rock and soil materials. *Rock Soil Mech.* **2021**, *42*, 2855–2864. (In Chinese) [CrossRef]
23. Wu, J.N.; Xing, T.Z.; Song, Y.M. Rock mechanical parameter inversion based on DSCM-FEMU. *Chin. J. Undergr. Space Eng.* **2021**, *17*, 350–355, 364. (In Chinese)
24. Liu, G.Y.; Wang, L.; Yi, Y.N.; Sun, L.B.; Shi, L.; Jiang, H.; Ma, S.P. Inverse identification of tensile and compressive damage properties of graphite material based on a single four-point bending test. *J. Nucl. Mater.* **2018**, *509*, 445–453. [CrossRef]
25. Liu, G.Y.; Wang, L.; Yi, Y.N.; Sun, L.B.; Shi, L.; Ma, S.P. Inverse identification of graphite damage properties under complex stress states. *Mater. Des.* **2019**, *183*, 108135. [CrossRef]
26. Xie, H.P.; Li, C.; He, Z.Q.; Li, C.B.; Lu, Y.Q.; Zhang, R.; Gao, M.Z.; Gao, F. Experimental study on rock mechanical behavior retaining the in situ geological conditions at different depths. *Int. J. Rock Mech. Min. Sci.* **2021**, *138*, 104548. [CrossRef]
27. Zhou, X.P.; Cheng, H.; Feng, Y.F. An experimental study of crack coalescence behavior in rock-like materials containing multiple flaws under uniaxial compression. *Rock Mech. Rock Eng.* **2014**, *47*, 1961–1986. [CrossRef]
28. Zhao, T.B.; Fang, K.; Wang, L.; Zou, J.; Wei, M. Estimation of elastic modulus of rock using modified point-load test. *Geotech. Test. J.* **2017**, *40*, 329–334. [CrossRef]
29. Liu, X.S.; Ning, J.G.; Tan, Y.L.; Gu, Q.H. Damage constitutive model based on energy dissipation for intact rock subjected to cyclic loading. *Int. J. Rock Mech. Min. Sci.* **2016**, *85*, 27–32. [CrossRef]
30. Ji, M.; Guo, H.J. Elastic-plastic threshold and rational unloading level of rocks. *Appl. Sci.* **2019**, *9*, 3164. [CrossRef]
31. Yang, S.Q.; Yang, J.; Xu, P. Analysis on pre-peak deformation and energy dissipation characteristics of sandstone under triaxial cyclic loading. *Geomech. Geophys. Geo-Energy Geo-Resour.* **2020**, *6*, 24. [CrossRef]
32. Wen, Z.J.; Tian, L.; Jiang, Y.J.; Zuo, Y.J.; Meng, F.B.; Dong, Y.; Lin, G.; Yang, T.; Lv, D.W. Research on damage constitutive model of inhomogeneous rocks based on strain energy density. *Chin. J. Rock Mech. Eng.* **2019**, *38*, 1332–1343. (In Chinese) [CrossRef]
33. Jiang, Q.; Zhong, S.; Cui, J.; Feng, X.T.; Song, L.B. Statistical characterization of the mechanical parameters of intact rock under triaxial compression: An experimental proof of the Jinping marble. *Rock Mech. Rock Eng.* **2016**, *49*, 4631–4646. [CrossRef]
34. Pepe, G.; Cevasco, A.; Gaggero, L.; Berardi, R. Variability of intact rock mechanical properties for some metamorphic rock types and its implications on the number of test specimens. *Bull. Eng. Geol. Environ.* **2017**, *76*, 629–644. [CrossRef]

Article

Damage Evolution Characteristics of Back-Filling Concrete in Gob-Side Entry Retaining Subjected to Cyclical Loading

Xicai Gao [1,2,*], Shuai Liu [1,2], Cheng Zhao [1,2], Jianhui Yin [3] and Kai Fan [4]

1. State Key Laboratory of Coal Resources in Western, Xi'an University of Science and Technology, Xi'an 710054, China
2. Key Laboratory of Western Mine Exploitation and Hazard Prevention, Ministry of Education, Xi'an University of Science and Technology, Xi'an 710054, China
3. Shaanxi Coal and Chemical Technology Institute Co., Ltd., Xi'an 710065, China
4. Sichuan Chuanmei Huarong Energy Co., Ltd., Panzhihua 617000, China
* Correspondence: gxcai07@163.com

Citation: Gao, X.; Liu, S.; Zhao, C.; Yin, J.; Fan, K. Damage Evolution Characteristics of Back-Filling Concrete in Gob-Side Entry Retaining Subjected to Cyclical Loading. *Materials* 2022, 15, 5772. https://doi.org/10.3390/ma15165772

Academic Editors: Xuesheng Liu, Yunliang Tan, Xuebin Li and Yunhao Wu

Received: 14 July 2022
Accepted: 17 August 2022
Published: 21 August 2022

Publisher's Note: MDPI stays neutral with regard to jurisdictional claims in published maps and institutional affiliations.

Copyright: © 2022 by the authors. Licensee MDPI, Basel, Switzerland. This article is an open access article distributed under the terms and conditions of the Creative Commons Attribution (CC BY) license (https://creativecommons.org/licenses/by/4.0/).

Abstract: The back-filling body in the gob-side entry retaining is subject to continuous disturbance due to repeated mining. In this study, uniaxial and cyclical loading tests of back-filling concrete samples were carried out under laboratory conditions to study damage evolution characteristics with respect to microscopic hydration, deformation properties, and energy evolution. The results showed that, due to the difference in the gradation of coarse and fine aggregates, the cemented structure was relatively loose, and the primary failure modes under cyclical loading were tensile and shearing failure, which significantly decreased its strength. With an increasing number of loadings, a hysteresis loop appeared for the axial strain, and the area showed a pattern of decrease–stabilization–increase. This trend, to a certain extent, reflected the evolution of the cracks in the back-filling concrete samples. The axial, radial, and volumetric plastic strain curves of the back-filling concrete samples showed a "U" shape. The plastic strain changed in three stages, i.e., a rapid decrease, stabilization, and a rapid increase. A damage parameter was defined according to the plastic strain increment to accurately characterize the staged failure of the samples. The plastic strain and energy dissipation of the samples were precursors to sample failure. Prior to the failure of the back-filling samples, the amount and speed of change of both the plastic strain and energy parameters increased significantly. Understanding the characteristics of plastic strain, damage evolution, and energy dissipation rate of the back-filling samples are of great reference value for realizing real-time monitoring of back-filling concrete in the gob-side entry retaining and providing early warning of failure.

Keywords: gob-side entry retaining; cyclical loading; back-filling concrete; damage evolution; energy dissipation

1. Introduction

Gob-side entry retaining technology is one of the key directions of development in pillar-free coal mining. Concrete is the most common roadside back-filling material due to its characteristics of fast curing, high load-bearing capacity, and simple process [1]. The back-filling body in the gob-side entry retaining is subject to repeated loads such as mining disturbance and roof breakage; thus, it undergoes a cyclical loading and unloading process. Due to damage accumulation and energy dissipation, stiffness is degraded and stress redistributes, reducing the load-bearing performance of the concrete material and bringing great difficulty to the stability control of the surrounding rock in the gob-side entry retaining [2–5].

Many studies have been carried out on the stress state and deformation mechanism of the back-filling body in the gob-side entry retaining under the influence of mining activities. Huang et al. [6] monitored the pressure of the roadside back-filling body in the gob-side entry retaining and analyzed the deformation and stress characteristics of the roadside backfilling body. Kan et al. [7] established the continuous laminate model under different roof conditions and proposed an equation to calculate the supporting resistance of the roadside back-filling body in the gob-side entry retaining. Feng et al. [8] reported that the stress and deformation of the roadside backfill body for gob-side entry of fully mechanized caving in thick coal seams were directly related to the breaking of the working face roof. The stress first increased and then stabilized with the advancement of the working face. Meng et al. [9] analyzed the roadside backfill body in gob-side entry retaining under combined static and dynamic loading and discussed the influence of the roof cutting angles on the behavior of the roadside backfill body using discrete element methods. Fan et al. [10] proposed an innovative BCR-GER approach for the complex geostress environment of deep coal seams based on the mechanical analysis of the roadside backfill body and the surrounding rock. The results from the above studies regarding the stress distribution and deformation of the back-filling body provide reference points for the design of roadway support in the gob-side entry retaining. However, implementing supporting protection in the deep gob-side entry retaining in recent years has shown that the back-filling body is often subject to repeated disturbance of roof subsidence in the current section and advancement of the working face in the lower section, causing nonlinear deformation, severe rock pressure, and failure of the back-filling body. Yet, few studies have focused on the damage or instability mechanism of the back-filling body in the gob-side entry retaining under cyclical loading conditions.

A large number of studies test different rock materials under cyclical loading conditions. For instance, Deng et al. [11] carried out a uniaxial cyclical loading test for sandstone. Considering the influence of residual deformation and the hysteresis effect, the authors modified the calculation method of energy parameters during the cyclical loading process. The results showed that rapid changes in energy parameters and residual strain were able to predict failure in the sandstone samples. Using digital image correlation technology, Yang et al. [12] defined a non-uniform deformation index and analyzed the evolution of non-uniform deformation and localized characteristic parameters during cyclical loading. Li et al. [13] carried out uniaxial cyclical loading testing of red sandstone and obtained the staged evolution characteristics of axial deformation, elastic energy, and dissipated energy. Wu et al. [14] analyzed the evolution of rock energy under different graded cyclical loading and unloading modes and established an evolution equation between dissipated energy, the cyclic stress level, and the number of loading cycles. This study laid a solid foundation for analyzing rock damage under cyclical loading based on energy dissipation. Yu et al. [15] analyzed the volumetric strain and volume expansion characteristics of marble fractures using variable amplitude cyclical loading tests under different confining pressures. These authors found that with increased loading times, the initial volumetric strain of marble fractures increased, and there was an exponential relationship between the increment of volumetric strain in a single loading cycle and stress amplitude. Tang et al. [16] studied local micro-cracks and nonlinear energy evolution of limestone under variable amplitude cyclical loading. The results indicated that both the ratio of elastic energy to total energy and the ratio of dissipated energy to total energy show staged characteristics. Variation in dissipated energy was related to the distribution of prefabricated cracks, the spatial location of micro-cracks, and the expansion rate. In addition, through triaxial cyclical loading tests of granite, Miao et al. [17] studied the evolution characteristics of dissipation energy, friction energy dissipation, and crushing energy dissipation under cyclical loading conditions. These authors found that the rock damage variable based on crushing energy dissipation could reasonably describe the damage evolution process of granite under cyclical loading.

In terms of the mechanical properties of concrete, Breccolotti et al. [18] proposed a constitutive model capable of accurately describing the damage accumulation of plain concrete subjected to cyclic uniaxial compressive loading. Park et al. [19] analyzed the microstructure and elastic–plastic characteristics of concrete subjected to cyclical loading using a nonlinear resonant ultrasonic method and quantitatively characterized the nonlinearity variation and load damage of concrete. Song et al. [20] studied the energy dissipation characteristics of concrete subjected to uniaxial cyclical loading based on dissipated energy and found that energy dissipation and damage evolution were stress-path-dependent. Hu et al. [21] analyzed the structural performance of recycled aggregate concrete subjected to cyclical loading, identified the characteristic points pertaining to the hysteresis loop, and proposed a simplified constitutive equation. Hutagi et al. [22] studied the stress–strain characteristics of geopolymer concrete (GPC) under cyclical loading; the analytical equations of the envelope curve, the common point curve, and the stable point were proposed. From the above studies, it is notable that there is little published work on the prediction of damage accumulation, progressive failure, and instability of back-filling concrete subjected to cyclical loading.

There are significant differences in rock deformation and energy evolution under cyclical loading compared to conventional uniaxial loading. Therefore, this article selects the back-filling concrete material as the research object, considering the repeated mining action, designs a continuous cyclic loading test, explores and characterizes the back-filling concrete material fracture extension, damage accumulation effect, and energy dissipation mechanism of back-filling concrete subjected to cyclical loading, and reveals the deep gob-side entry retaining deformation failure mechanism. It provides a basis for the monitoring and stability analysis of back-filling body in the gob-side entry retaining, which has important practical engineering value.

2. Materials and Methods

2.1. Sample Preparation

The back-filling concrete was composed of Portland cement, river sand, crushed stone, and a high-efficiency water-reducing agent. The cementing material was Jidong 42.5 ordinary Portland cement, and the coarse aggregate was crushed stone with a particle size of 5–16 mm and a density of 2.72 kg/m^3. The fine aggregate was medium sand with a fineness modulus of 2.6–3.0 and a density of 2.6 g/cm^3. The high-efficiency water-reducing agent was a polycarboxylic acid-type water-reducing agent, with a surface density of 550–650 G/L, an active ingredient of greater than 99%, and a pH value (10% aqueous solution) between 7.0 and 8.0.

According to the formula for concrete in the gob-side entry retaining (cement: river sand (0–5 mm): crushed stone (5–15 mm): water = 1:0.82:1:0.38), the materials were mixed and stirred well before being poured into a mold. The mixture was vibrated sufficiently to reduce air bubbles, and then the surface was covered with fresh film to prevent the surface from drying and cracking, curing it with a mold. The concrete sample mold was placed in the standard curing box and the temperature was set to 20 ± 2 °C; while the relative humidity was 90% after curing for 28 days. According to the recommendations of the International Association of Rock Mechanics, the concrete samples were formed into a cylindrical shape with a diameter of 50 mm and a height of 100 mm, and the diameter error was controlled to within 0.3 mm (Samples No. J1–J20). Moreover, both ends of the samples were ground such that surface flatness was within 0.02 mm. In addition, an ultrasonic test device was used to test the wave velocity of the samples, and those with comparable wave velocities were chosen for testing. At a normal temperature, the average longitudinal wave velocity of the samples was 2147 m/s, and the average density was 2.17 g/cm^3.

2.2. Test Scheme

Based on the mechanical conditions of the back-filling body in the gob-side entry retaining, a cyclical loading test was set up in an indoor laboratory. The linear correlation between the compressive strength and the ultrasonic wave velocity was established based on the results of uniaxial loading (Figure 1). The loading path is shown in Figure 2. First, the compressive strength σ_c of the samples was first predicted based on the linear correlation ($\sigma_c = 0.9358v - 1982.6$) and the initial wave velocity, then the upper and lower limit of the cyclical load was determined ($\sigma_{\min} = 0.4\sigma_c$, $\sigma_{\max} = 0.8\sigma_c$), as shown in Table 1. The load was applied under force control and the loading process was as follows: (1) In the first stage, the sample was linearly loaded at a speed of 0.2 kN/s to $\sigma_0 = 0.6\sigma_c$; (2) in the second stage, sine wave cyclical loads were applied to the sample (amplitude: 0.4–$0.8\sigma_c$, frequency: 0.4 Hz) until sample failure. Ultrasonic testing was carried out continuously throughout the entire loading cycle. After data processing, the parameters related to the strength, deformation, and failure of the sample and the energy evolution under cyclical loading were obtained.

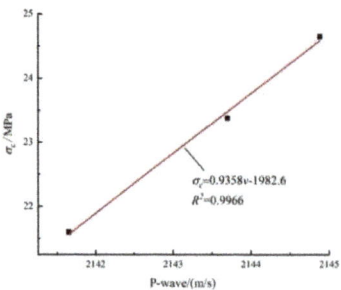

Figure 1. Relationship between uni-axial compressive strength and p-wave velocity.

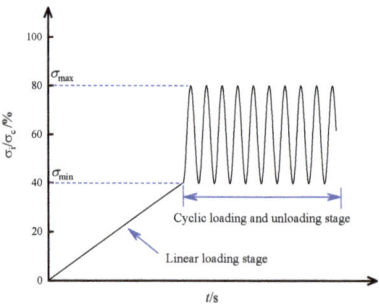

Figure 2. Cyclic loading method.

Table 1. Cyclic loading and unloading test parameters.

Sample	Prediction of Uniaxial Compressive Strength (MPa)	Cycle Loading Times	Failure Stress (MPa)	Intensity Decay Ratio (%)
J-4	24.83	55	17.01	31.49
J-13	23.83	64	17.00	28.66
J-14	24.83	48	17.57	29.24
J-20	23.82	50	18.82	20.99
J-11	24.82	90	18.81	24.21
J-17	24.83	108	18.52	25.41
J-18	26.82	90	19.98	25.50
J-19	25.69	95	19.97	22.27

2.3. Equipment

The MTS815 material testing machine was used via electro-hydraulic driven operation (Figure 3). The axial load capacity was 2300 kN, the upper limit of confining pressure was 140 MPa, and the accuracy of the pressure transducer was 0.001 kN. Deformation was measured with an axial extensometer and a circumferential extensometer. The measuring range of the extensometers was between 5 mm and 8 mm, respectively, and the resolutions were both 10^{-4} mm. Through the Flex Test GT digital controller, the MTS machine was able to realize load and displacement control modes.

Figure 3. Sample loading device and ultrasonic pressure transducer layout.

An ultrasonic system was used for damage accumulation monitoring (Figure 3). The transmitting and receiving transducers were arranged on the upper and lower ends of the samples, and a coupler was applied to ensure good contact between the sample and the transducer. The resonance frequency (HYN, 55 KHz) and damping meet the requirements for mechanical testing of concrete materials. A waveform generator actively transmitted a pulse signal (main frequency: 90 kHz; amplitude: 10 Vpp). After receiving the pulse signal, one end of the transducer generated instantaneous vibration. The vibration propagated in the sample and was then received by the transducer on the other end. The data acquisition frequency was 10 Mpoints/s, and the duration was 5 ms.

The Quanta450& IE250X-MAX50 scanning electron microscope was used to scan the microstructure of back-filling concrete samples. Block samples were taken from back-filling concrete samples of different ages and fixed with conductive adhesive. In order to enhance the electrical conductivity of back-filling concrete, gold was sprayed on its surface.

3. Results and Discussion

3.1. Deterioration of Load-Bearing Performance of Concrete Samples Subjected to Cyclical Loading

(1) Sample microstructures

The microscopic morphology of the back-filling concrete samples with different curing periods is shown in Figure 4 (magnification 5000×). At 7 days of curing, a small amount of hydration product calcium-silicate-hydrate (C-S-H) particles was seen around the aggregate, as shown in Figure 4a. Note that the longer the curing period of the back-filling concrete, the better the crystallinity of the hydration product. Figure 4b shows that the surface was rough, containing particles bound by C-S-H, and the structure of the concrete samples was relatively loose at 14 days. Increasing the curing period gradually increases the amount of hydration product, and a large number of links formed between C-S-H and the aggregates, while the internal pores were continuously filled, which enhanced the strength of the aggregate–cement interface, as shown in Figure 4c,d. Overall, a large number of micropores and cracks were formed in the concrete samples, which had strong heterogeneity.

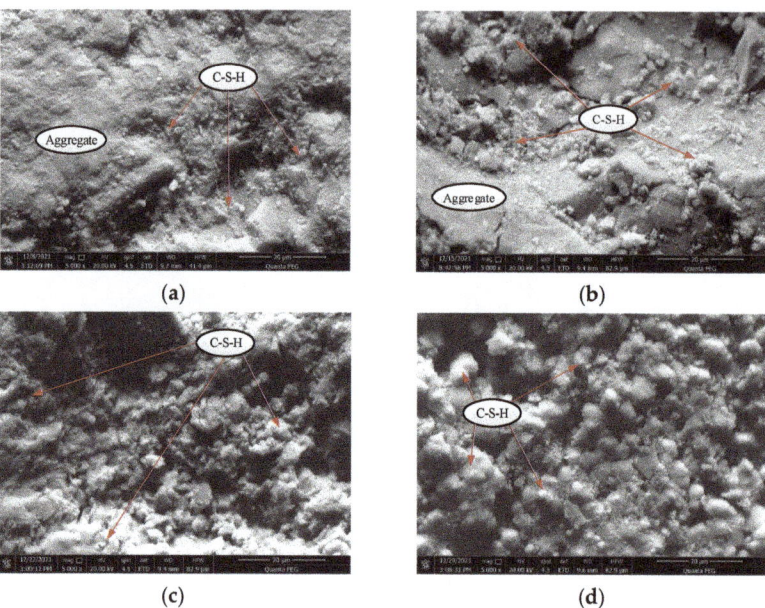

Figure 4. Scanning electron microscope of back-filling concrete at different ages. (**a**) 7 days; (**b**) 14 days; (**c**) 21 days; (**d**) 28 days.

(2) Macroscopic fracture characteristics

The fracture patterns of the samples under different loading paths are shown in Figure 5a,b. In Figure 5a, the red lines represent the main cracks and the thin black line represents the secondary cracks. When the loading path was changed from uniaxial loading to constant-amplitude cyclical loading, the main cracks were connected upon sample failure, accompanied by a large number of secondary cracks. The number and density of macroscopic cracks increased significantly in Figure 5b. Specifically, a large number of cracks appeared parallel to the principal stress direction. The shear failure zone showed a large number of fragments and localized exfoliation.

Figure 5. Fracture characteristics of back-filling concrete samples under different loading paths. (**a**) J-1 sample uniaxial loading path; (**b**) J-11 sample under cyclic loading path.

During the loading process, lateral deformation of the sample was large, and there was clear volumetric expansion. This was mainly because the load did not decrease to 0 kN during the unloading phase, and damage accumulated under cyclical loading. When the samples were damaged, a macroscopic crack penetrating the whole sample appeared, and the failure modes were mainly tensile and shearing failure. In addition, the peak ultimate failure load of the samples decreased significantly under cyclical loading. The intensity reduction ratio was 20.99–31.49% of that obtained during uniaxial compressive loading (Table 1), i.e., the load-bearing performance of the concrete sample declined significantly.

3.2. Deformation Characteristics of Concrete Samples Subjected to Cyclical Loading

(1) Stress–strain curve

As Table 1 shows, the total number of loading cycles of the J-4 sample was 55 times and the J-19 sample was 95 times in the two groups' samples, and J-4 and J-19 are selected as the representatives of both groups of samples. The stress–strain curve of the concrete samples subjected to cyclical loading is shown in Figure 6a,b, The failure stress of the concrete samples was relatively close, and the stress–strain curves showed hysteresis loops. With the increase in the loading cycle, the area of the hysteresis loop gradually increased. The reason for this was that each loading cycle led to new damage to the internal samples, and the damage continued to accumulate, resulting in a continuous increase in the axial strain on the sample. The lower load limit was above 0 kN, thus, the micro-cracks generated during cyclical loading were always in a compressive state, and the cumulative plastic strain gradually increased. Compared with J19 in Figure 6b, sample J4 exhibited more obvious ductility during fatigue failure as shown in Figure 6a.

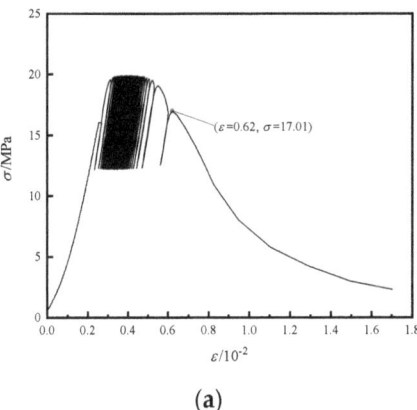
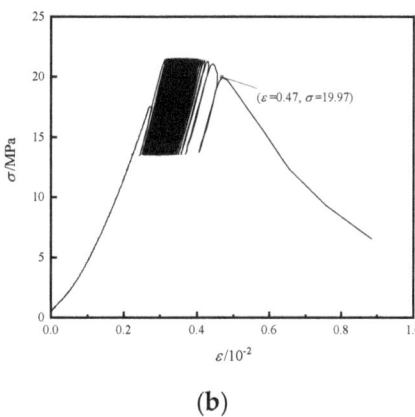

(a) (b)

Figure 6. Stress–strain curves of cyclic loading and unloading concrete samples. (**a**) J-4; (**b**) J-19.

With the increased loading cycle, irreversible deformation of the concrete samples increased and showed a nonlinear increasing trend. Macroscopically, the sample showed clear volumetric expansion and fatigue failure.

(2) Plastic hysteresis loop

Based on the maximum hysteresis loop area before failure, the area of the plastic hysteresis loop in each loading cycle was normalized. Taking the ratio of the total number of loading cycles to the number of cycles at fatigue failure as the x-axis, n_i/n represents the total number of loading cycles to the number of cycles at fatigue failure, and the variation of the area of the plastic hysteresis loop with the number of loading cycles is obtained (Figure 7). Figure 7a notes that as the number of loading cycles increased, the area of the plastic hysteresis loop at each time showed a staged pattern of decrease–stabilization–rapid increase and showed a "U" shape. The loading stress in the first loading cycle ($\sigma_{max} = 0.8\sigma_c$) exceeded the yield strength of the concrete material, and the sample entered the plastic deformation stage. There was a large amount of plastic deformation, indicated by the large area of the plastic hysteresis loop. After unloading, the elastic deformation recovered, and the area of the plastic hysteresis loop decreased significantly.

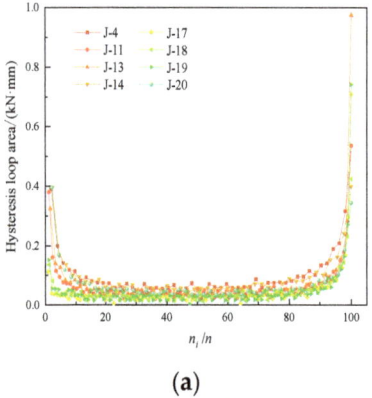
(a)

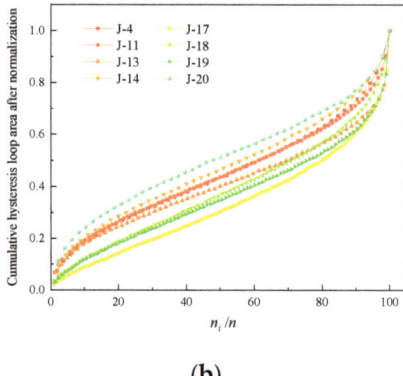
(b)

Figure 7. Variation rule of hysteresis loop area of cyclic loading and unloading. (**a**) Hysteresis loop area each cycle. (**b**) Normalized cumulative hysteresis loop area.

With the progression of cyclical loading, the cracks in the samples were in a constant compressive state, and the plastic strain continuously accumulated, although the sample was still in a stable condition. In the later stage of cyclical loading, the area of the plastic hysteresis loop in a single cycle gradually increased with increasing amplitude, indicating an increasing degree of fatigue damage. Once macroscopic cracks penetrated the sample, the load-bearing performance greatly declined.

Figure 7b shows that the area of each hysteresis curve was normalized, and the normalized cumulative hysteresis loop area showed a nonlinear increase with the increase in the number of cycles, and its growth rate showed a trend of "decrease–stable–increase".

(3) Variation of plastic strain

The elastic strain ε_e, plastic strain ε_p, and volumetric strain ε_v of the concrete samples subjected to cyclical loading were calculated to quantify the evolution of the plastic strain. The physical implications of elastic and plastic strains are shown in Figure 8, and the elastic and plastic strains are calculated as [23]:

$$\varepsilon_e = \varepsilon_{\max} - \varepsilon_{\min} \tag{1}$$

$$\varepsilon_p = \varepsilon_{\min} - \varepsilon'_{\min} = \varepsilon_{\min} - \frac{\sigma_{\min}}{E} \tag{2}$$

where ε_e and ε_p are the elastic strain and plastic strain, respectively; $\varepsilon_{\max}$ is the strain corresponding to the upper load limit, E is the elasticity modulus, $\varepsilon_{\min}$ represents the strain corresponding to the lower load limit, and $\varepsilon'_{\min}$ represents the elastic strain corresponding to the lower load limit.

Under uniaxial cyclical loading, assuming the axial strain is ε_1 and the radial strain is ε_3, the volumetric strain is calculated as [24]:

$$\varepsilon_v = \varepsilon_1 + 2\varepsilon_3 \tag{3}$$

where ε_v is the volumetric strain, ε_1 is the axial strain, and ε_3 is the radial strain.

Figure 9a–f show the relationship between the axial strain, radial strain, and volumetric plastic strain of some samples (J-4, J-11, J-13, J-14, J-17, J-18, J-19, and J-20) and the number of loading cycles.

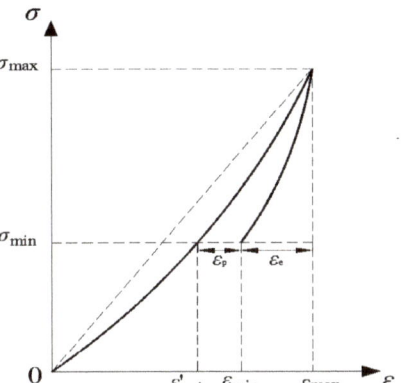

Figure 8. Schematic diagram of strain parameters.

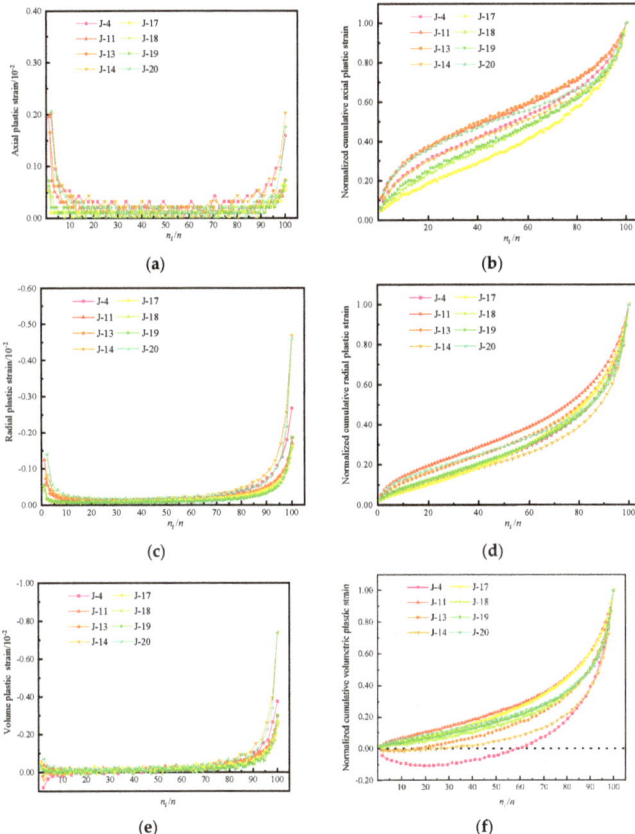

Figure 9. Variation curve of plastic strain with cyclic loading and unloading times. (**a**) Axial plastic strain; (**b**) normalized accumulative axial plastic strain; (**c**) radial plastic strain; (**d**) normalized accumulative radial plastic strain; (**e**) volumetric plastic strain; (**f**) normalized accumulative volumetric plastic strain.

Note from Figure 9 that the variation of the axial, radial, and volumetric strains with the number of loading cycles was essentially consistent, showing a "U" shape. The plastic strain presented a staged pattern of a rapid decrease–stabilization–a rapid increase. Thus, the variation of plastic strain showed clear stage characteristics. n is the total cycle number of each sample, the cycle number of the demarcation point between the rapid decrease and stabilization stages was set as n_1, and the cycle number of the demarcation point between the stabilization stage and the rapid increase stage was set as n_2. The n_1 and n_2 values of each sample are shown in Table 2.

Table 2. Node classification of plastic strain stage.

Samples	n	n_1	n_1/n	n_2	n_2/n
J-4	55	6	10.91%	45	81.82%
J-11	90	11	12.22%	76	84.44%
J-13	64	9	14.06%	53	82.81%
J-14	48	6	12.50%	39	81.25%
J-17	108	11	10.19%	92	85.19%
J-18	90	10	11.11%	72	80.00%
J-19	95	10	10.53%	80	84.21%
J-20	50	6	12.00%	41	82.00%
average	/	/	11.69%	/	82.72%

As one can see from Figure 9a,c, in the early stage, the loading stress was above the yield stress, which led to a large amount of plastic deformation. After unloading, the elastic deformation was recovered, and the axial and radial deformations decreased significantly. In the late stage, the internal cracks propagated due to repeated compression and the degree of fatigue damage increased, as indicated by the increasing axial and radial plastic deformation. Then, cracks penetrated the sample, and the plastic deformation increased dramatically to the maximum value, and the sample underwent shearing-tensile failure, and significant radial cracks developed.

With increasing loading cycles, the plastic strain gradually accumulated, yet the sample was still in a stable state. The accumulative velocity of axial plastic strain first increases and then decreases with the increase in the loading cycles as shown in Figure 9b,d. During the cyclical loading process, the volumetric strain showed a significant nonlinear increasing trend as shown in Figure 9e. In the initial stage, the volumetric plastic strain of the J-4, J-13, and J-14 samples had a positive increment, indicating that axial compressive deformation was much larger than radial expansion deformation. However, the positive increment was small and rapidly became negative, indicating that there were few initial cracks in the concrete sample and micro-cracks were initiated. Furthermore, the volumetric strain gradually changed from axial compression deformation to radial expansion deformation. For the J-11, J-17, J-18, J-19, and J-20 samples, the initial volumetric strain had negative increments, showing clear radial volumetric expansion as shown in Figure 9f. As cyclical loading continued, the internal micro-cracks were repeatedly squeezed and propagated, and the proportions of axial, radial, and volumetric plastic strains all gradually increased. In addition, the growth rate of the plastic strain increased sharply before sample failure (80.00–85.19% of the total number of cycles) as shown in Table 2. As a result, primary and secondary cracks were connected, and the concrete sample was damaged.

3.3. Energy Dissipation of Concrete Samples Subjected to Cyclical Loading

(1) Strain energy density

During the cyclical loading process, the samples exchanged energy with the environment. As the loading increased on the concrete samples, the accumulated strain energy was continuously transformed and dissipated, accompanied by the deformation and failure of the samples. In the stage before peak stress, energy conversion involved only elastic energy storage and energy dissipation. The total strain energy density, W, is the sum of the elastic strain energy density, W_e, and the energy dissipation density, W_d, which can be expressed as [25,26]:

$$W = W_e + W_d \tag{4}$$

$$W = \int_0^{\varepsilon_i} \sigma_i d\varepsilon_i \tag{5}$$

$$W_e = \int_{\varepsilon_p}^{\varepsilon_i} \sigma_i' d\varepsilon_i' \tag{6}$$

where σ_i and σ_i' are the loading and unloading stress of the i-th cycle, respectively; ε_i and ε_i' are the axial strain in the loaded and unloaded state in the i-th cycle, respectively.

(2) Energy evolution characteristics

Figure 10 shows that elastic strain energy first increased with the increasing loading cycles, then stabilized with small fluctuations, then decreased rapidly before sample failure. Furthermore, the total strain energy and dissipated energy of the concrete samples had similar trends with obvious stage characteristics under cyclical loading. The total strain energy and dissipated energy were high in the early stage due to the first cycle having higher loading and deformation; with increasing loading cycles, they decreased first and then stabilized and increased rapidly in the late stage. The slope change of curve was taken as the basis for stage. The turning point of dissipated energy rapidly decreases to stability with small fluctuations was taken as the stage segmentation point from stage I to II, such as Figure 10a (6,0.80) and 10b (5,0.36). The turning point of dissipated energy steadily develops to a rapid increase was taken as the stage segmentation point from stage II to III, such as Figure 10a (41,0.75) and 10b (81,0.38). And Failure of the concrete samples was accompanied by a large increase in dissipated energy.

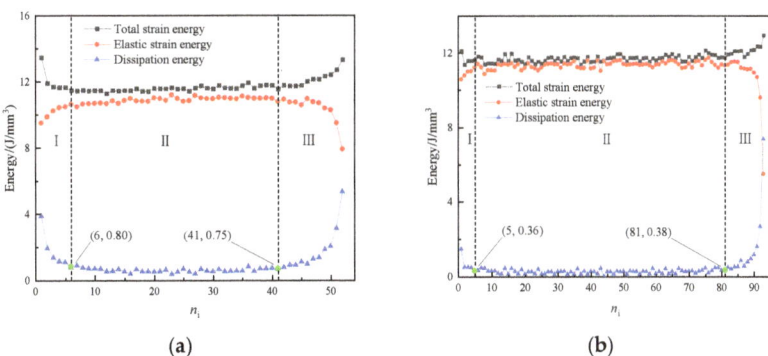

Figure 10. Energy evolution curve of back-filling concrete sample. (a) J-4; (b) J-19.

To accurately describe the damage evolution process of the concrete samples subjected to cyclical loading and to characterize the energy dissipation, the variation in the energy dissipation rate with the number of cycles for different samples is shown in Figure 11. The x-axis is the total number of loading cycles in Figure 11a, the x-axis is the ratio of the total number of loading cycles to the number of cycles at fatigue failure in Figure 11b, and the y-axis is the energy dissipation rate. The ratio of energy dissipation density (W_d) to total strain energy (W) was defined as the energy dissipation rate [26].

$$W_k = \frac{W_d}{W} \tag{7}$$

where W_k is the energy dissipation rate.

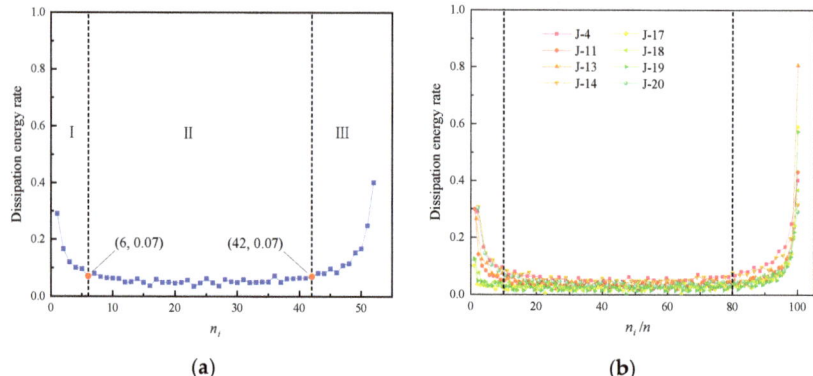

Figure 11. Variation of dissipation energy rate with the number of cycles. (**a**) Dissipation energy rate change curve of J-4 sample. (**b**) Stage division of dissipation energy rate of samples.

Note from Figure 11 that the energy dissipation rate shows a pattern of sharp decrease–steady fluctuation–rapid increase with increased loading cycles. Take J-4 sample for example, the slope change of energy dissipation rate curve was taken as the basis for three stages. The turning point of energy dissipation rate sharp decrease to steady fluctuation was taken as the point from stage I to II, such as Figure 11a (6,0.07). The turning point of energy dissipation rate steadily develops to a rapid increase was taken as the point from stage II to III, such as Figure 11a (42,0.07).

According to the slope of the energy dissipation rate curve, the energy dissipation process can be divided into three stages (I, II and III):

Initial dissipation stage I (0–10% of the cycle): In the early stage of cyclical loading, the loading stress in the first cycle exceeded the yield stress, resulting in large plastic deformation and crack development. Thus, energy dissipation was large, and the plastic strain energy rate accounted for 30% of the total strain energy. After the first cycle, the number of new micro-cracks gradually decreased, the plastic deformation increment decreased sharply, and the plastic strain energy decreased gradually. The energy dissipation rate decreased sharply, while the accumulated elastic strain energy increased gradually.

Stable dissipation stage II (10–80% of the cycle): As cyclical loading continued, the lower stress limit in the unloaded state did not return to zero, there was compaction of internal pores and cracks and propagation of micro-cracks, and plastic deformation continued increasing with a small increment. The energy dissipation rate was maintained at 0.02–0.08 at this stage, and there was a small fluctuation. Most of the energy was converted into elastic strain energy, and energy dissipation was small.

Rapid dissipation stage III (80–100% of the cycle): With increased loading cycles, the micro-cracks gradually propagated to form large cracks, and the large cracks gradually penetrated to form macro-cracks. Plastic deformation increased rapidly at this stage, plastic strain energy density increased, and the energy dissipation rate increased significantly until the sample failed.

As shown in Figure 11a, before the failure of the J-4 sample, the amount of change and the change in speed of both the plastic strain and energy parameters increased significantly, indicating that the plastic strain and energy dissipation rates could predict the failure of the concrete samples.

4. Conclusions

(1) Back-filling concrete samples were prepared using Portland cement, river sand, crushed stone, and a high-efficiency water-reducing agent. With an increased curing period, hydration products increased, pores and fractures decreased, and the strength of the aggregate cement interface increased continuously, which has the characteristics

of the fast lifting of the bearing strength and strong ductility. Due to the heterogeneity of the mesostructure, the failure modes were mainly tensile and shear failure, and the load-bearing performance decreased significantly after failure.

(2) Under cyclical loading, the stress and strain curves of the concrete samples show obviously staged characteristics. With increased loading cycles, the evolution of plastic hysteresis loops shows the characteristic of "sparse–dense–sparse", the area of the hysteresis loops showed a pattern of decrease–stabilization–rapid increase, and the larger the hysteresis loop area, the more obvious the volume expansion and the more obvious the bearing capacity reduction. Furthermore, the axial, radial, and volumetric plastic strain showed a "U" shape, and the plastic strain presented a staged pattern of a rapid decrease–stabilization–rapid increase, which directly reflects the regularity of fracture development and damage accumulation evolution in back-filling concrete samples under the influence of repeated mining.

(3) Variation in the dissipated energy of the concrete sample subjected to cyclical loading was closely related to micro-crack propagation and damage evolution. The elastic energy increases with the increase in cyclic loading and unloading times, and the energy dissipation process could be divided into three stages: The initial dissipation stage, the stable dissipation stage, and the rapid dissipation stage. Furthermore, the dissipated energy first decreased, then stabilized with small fluctuations, and then increased rapidly. The calculation method of the dissipative energy rate is introduced to accurately describe the phase-change law of dissipative energy accumulation, which provides a basis for predicting the failure precursor of back-filling concrete samples.

Author Contributions: Conceptualization, X.G. and C.Z.; methodology, C.Z.; software, S.L.; validation, J.Y., X.G. and K.F.; formal analysis, X.G.; investigation, C.Z.; resources, K.F.; data curation, S.L.; writing—original draft preparation, X.G.; writing—review and editing, X.G.; visualization, S.L.; supervision, C.Z.; project administration, J.Y.; funding acquisition, K.F. All authors have read and agreed to the published version of the manuscript.

Funding: This research was funded by the Natural Science Basic Research Program of Shaanxi (Program No. 2021JLM-10) and the Natural Science Basic Research Program of Shandong (Program No. 2019JZZY020326). The APC was funded by the Natural Science Basic Research Program of Shaanxi (Program No. 2021JLM-10).

Informed Consent Statement: Informed consent was obtained from all subjects involved in the study.

Data Availability Statement: The data used to support the findings of this study are included in the article.

Conflicts of Interest: The authors declare no conflict of interest.

References

1. Kang, H.P.; Zhang, X.; Wang, D.P.; Tian, J.Z.; Yi, Z.Y.; Jiang, H. Strata control technology and application of non-pillar coal mining. *J. China Coal Soc.* **2022**, *47*, 16–44. [CrossRef]
2. Chen, Y.; Bai, J.B.; Zhu, T.L.; Yan, S.; Zhao, S.H.; Li, X.C. Mechanisms of roadside support in gob-side entry retaining and its application. *Rock Soil Mech.* **2012**, *33*, 1427–1432. [CrossRef]
3. Yang, B.S.; Tang, X.S.; Ling, Z.Q.; Wang, S. Study on surrounding rock moving law of roadside filling's gob-side entry retaining in deep mining. *J. Saf. Sci. Technol.* **2012**, *8*, 58–64. [CrossRef]
4. Liu, Z.; Zhao, L.H.; Wu, X.B.; Hu, G.P.; Zhou, Q.Y. Damage model of concrete considering hysteretic effect under cyclic loading. *Adv. Eng. Sci.* **2020**, *52*, 117–123. [CrossRef]
5. Liu, H.Y.; Zhang, B.Y.; Li, X.L. Research on roof damage mechanism and control technology of gob-side entry retaining under close distance gob. *Eng. Fail. Anal.* **2022**, *138*, 106331. [CrossRef]
6. Huang, Y.L.; Zhang, J.X.; Ju, F. Technology of roadside packing in gob-side entry retaining and law of rock pressure. *J. Xi'an Univ. Sci. Technol.* **2009**, *29*, 515–520. [CrossRef]
7. Kan, J.G.; Zhang, N.; Li, B.Y.; Si, G.Y. Analysis of supporting resistance of back-filling wall for gob-side entry retaining under typical roof conditions. *Rock Soil Mech.* **2011**, *32*, 2778–2784. [CrossRef]

8. Feng, G.R.; Ren, Y.Q.; Wang, P.F.; Guo, J.; Qian, R.P.; Li, S.Y.; Sun, Q.; Hao, C.L. Stress distribution and deformation characteristics of roadside backfill body for gob-side entry of fully-mechanized caving in thick coal seam. *J. Min. Saf. Eng.* **2019**, *36*, 1109–1119. [CrossRef]
9. Meng, N.K.; Bai, J.B.; Chen, Y.; Wang, X.Y.; Wu, W.D.; Wu, B.W. Stability analysis of roadside backfill body at gob-side entry retaining under combined static and dynamic loading. *Eng. Fail. Anal.* **2021**, *127*, 105531. [CrossRef]
10. Fan, D.Y.; Liu, X.S.; Tan, Y.L.; Yan, L.; Song, S.L.; Ning, J.G. An innovative approach for gob-side entry retaining in deep coal mines: A case study. *Energy Sci. Eng.* **2019**, *7*, 2321–2335. [CrossRef]
11. Deng, H.F.; Hu, Y.; Li, J.L.; Wang, Z.; Zhang, X.J.; Hu, A.L. The evolution of sandstone energy dissipation under cyclic loading and unloading. *Chin. J. Rock Mech. Eng.* **2016**, *35*, 2869–2875. [CrossRef]
12. Yang, X.B.; Han, X.X.; Liu, E.L.; Zhang, Z.P.; Wang, T.J.; Zhang, L.H. Properties of non-uniform deformation evolution of rock under uniaxial cyclic loading and unloading. *J. China Coal Soc.* **2018**, *43*, 449–456. [CrossRef]
13. Li, J.T.; Xiao, F.; Ma, Y.P. Deformation damage and energy evolution of red sandstone under uni-axial cyclic loading and unloading. *J. Hunan Univ. Nat. Sci.* **2020**, *47*, 139–146. [CrossRef]
14. Wu, Z.H.; Song, Z.Y.; Tan, J.; Zhang, Y.Z.; Qi, Z.J. The evolution law of rock energy under different graded cyclic loading and unloading modes. *J. Min. Saf. Eng.* **2020**, *37*, 836–844+851. [CrossRef]
15. Yu, J.; Liu, Z.H.; Lin, L.H.; Huang, J.G.; Ren, W.B.; Zhou, L. Characteristics of dilatation of marble under variable amplitude cyclic loading and unloading. *Rock Soil Mech.* **2021**, *42*, 2934–2942. [CrossRef]
16. Tang, H.D.; Zhu, M.L.; Zhu, Z.D. Three-dimensional propagation of local micro-cracks and non-linear deterioration mechanism of limestone under variable amplitude cyclic loading. *Chin. J. Rock Mech. Eng.* **2021**, *40*, 1170–1185. [CrossRef]
17. Miao, S.J.; Liu, Z.J.; Zhao, X.G.; Huang, Z.J. Energy dissipation and damage characteristics of Beishan granite under cyclic loading and unloading. *Chin. J. Rock Mech. Eng.* **2021**, *40*, 928–938. [CrossRef]
18. Breccolotti, M.; Bonfigli, M.F.; D'Alessandro, A.; Materazzi, A.L. Constitutive modeling of plain concrete subjected to cyclic uniaxial compressive loading. *Constr. Build. Mater.* **2015**, *94*, 172–180. [CrossRef]
19. Park, S.J.; Kim, G.J.; Kwak, H.G. Characterization of stress-dependent ultrasonic nonlinearity variation in concrete under cyclic loading using nonlinear resonant ultrasonic method. *Constr. Build. Mater.* **2017**, *145*, 272–282. [CrossRef]
20. Song, Z.; Frühwirt, T.; Konietzky, H. Characteristics of dissipated energy of concrete subjected to cyclic loading. *Constr. Build. Mater.* **2018**, *168*, 47–60. [CrossRef]
21. Hu, X.; Lu, Q.; Xu, Z.; Zhang, W.; Cheng, S. Compressive stress-strain relation of recycled aggregate concrete under cyclic loading. *Constr. Build. Mater.* **2018**, *193*, 72–83. [CrossRef]
22. Hutagi, A.; Khadiranaikar, R.B.; Zende, A.A. Behavior of geopolymer concrete under cyclic loading. *Constr. Build. Mater.* **2020**, *246*, 118430. [CrossRef]
23. Hu, J.H.; Zeng, P.P.; Yang, D.J.; Xu, X. Analysis of damage deformation and mesoscopic structure of granite under deep cyclic loading. *Chin. J. Nonferrous Met.* **2022**, *32*, 1187–1198. [CrossRef]
24. Xie, H.P.; Peng, R.D.; Ju, Y. Energy dissipation of rock deformation and fracture. *Chin. J. Rock Mech. Eng.* **2004**, *23*, 3565–3570. [CrossRef]
25. Liu, X.H.; Hao, Q.J.; Hu, A.K.; Zheng, Y. Study on determination of uniaxial characteristic stress of coal rock under quasi-static strain rate. *Chin. J. Rock Mech. Eng.* **2020**, *39*, 2038–2046. [CrossRef]
26. Guo, H.J.; Sun, J.M.; Zhou, Z.G.Z. Energy evolution characteristics of red sandstone under cyclic load. *J. Min. Strat. Control Eng.* **2021**, *3*, 15–23. [CrossRef]

Experimental Study of Energy Evolution at a Discontinuity in Rock under Cyclic Loading and Unloading

Wei Zheng [1], Linlin Gu [1,*], Zhen Wang [2,*], Junnan Ma [3], Hujun Li [2] and Hang Zhou [1]

1. Department of Civil Engineering, Nanjing University of Science and Technology, Nanjing 210094, China
2. School of Mechanical Engineering, Nanjing University of Science and Technology, Nanjing 210094, China
3. Department of Civil Engineering, Nagoya Institute of Technology, Nagoya 466-8555, Japan
* Correspondence: linlin_gu@njust.edu.cn (L.G.); wangzhen2012@njust.edu.cn (Z.W.)

Abstract: Energy is often dissipated and released in the process of rock deformation and failure. To study the energy evolution of rock discontinuities under cyclic loading and unloading, cement mortar was used as rock material and a CSS-1950 rock biaxial rheological testing machine was used to conduct graded cyclic loading and unloading tests on Barton's standard profile line discontinuities with different joint roughness coefficients (JRCs). According to the deformation characteristics of the rock discontinuity sample, the change of internal energy is calculated and analyzed. The experimental results show that under the same cyclic stress, the samples harden with the increase in the number of cycles. With the increase of cyclic stress, the dissipated energy density of each stage gradually exceeds the elastic energy density and occupies a dominant position and increases rapidly as failure becomes imminent. In the process of increasing the shear stress step-by-step, the elastic energy ratio shows a downward trend, but the dissipated energy is contrary to it. The energy dissipation ratio can be used to characterize the internal damage of the sample under load. In the initial stage of fractional loading, the sample is in the extrusion compaction stage, and the energy dissipation ratio remains quasi-constant; then the fracture develops steadily, the damage inside the sample intensifies, and the energy dissipation ratio increases linearly (albeit at a low rate). When the energy storage limit is reached, the growth rate of energy dissipation ratio increases and changes when the stress level reaches a certain threshold. The increase of the roughness of rock discontinuity samples will improve their energy storage capacity to a certain extent.

Keywords: rock mechanics; rock discontinuity; energy evolution; energy dissipation ratio

1. Introduction

There are many different types of rock discontinuities in natural rock mass, such as joints, folds, and faults. The existence of these rock discontinuities causes discontinuity and anisotropy in mechanical properties of rock mass, which directly affects the stability of rock mass [1–3]. In underground caverns, dams, abutments, and other rock mass projects, cyclic loads are common such as those induced by blasting, earthquakes, and changes in water level. Under the long-term action of these cyclic loads, rock mass discontinuities may slip and deform, resulting in failure of the rock mass structure [4,5]. Invoking the second law of thermodynamics, energy conversion is a basic feature of material physical processes, the essence of material damage is energy to drive the state of instability [6], and the failure of rock is an irreversible process of energy dissipation [7,8]. Therefore, studying the damage and failure of rock discontinuity from an energy perspective can provide a new idea for preventing and treating rock engineering disasters.

In recent years, scholars have conducted a series of studies on the mechanisms underpinning the evolution of the internal energy of a rock mass. For example, Xie et al. [9,10] proposed that failure is due to the abrupt change of energy dissipation under certain conditions and defined a rock strength failure criterion. In combination with damage mechanics and energy conservation theory, Tu et al. [11] established a slope instability criterion using a strength-reduction method based on energy conversion. Peng et al. [12] studied the relationship between crack angle and energy inside the loaded rock, and the results showed that angle was positively correlated with energy storage capacity. Wang et al. [13] used numerical simulation software to calculate the energy storage of surrounding rock, expounded the relationship between rock failure and elastic energy, and confirmed that the simulated strain energy analysis method could be used for rockburst prediction. Wu et al. [14] studied the energy evolution of rock under different loading modes. Wang et al. [15] conducted a uniaxial cyclic charge test on the lower dry and saturated sandstone, and analyzed the strength and deformation characteristics of the rock, as well as the change and distribution of energy under the dry and saturated state. Jia et al. [16] studied the energy variation law of rock mass at different depths during mining, and the results showed that the increase of the depth of rock mass would lead to the increase of all types of energy inside it. Deng et al. [17] carried out dynamic uniaxial compression tests on rock samples under impact velocity and analyzed the energy dissipation law in the dynamic failure process of rock. Zhang et al. [18] studied the influence of confining pressure on the change of energy inside the rock in the triaxial test, and the results showed that the increase of confining pressure could improve the efficiency of energy accumulation to a certain extent. Munoz et al. [19,20] defined a new rock brittleness index to elucidate the energy accumulation and release of rock failure. Song et al. [21] performed uniaxial cyclic loading and unloading tests on coal and rock, collected electromagnetic radiation signals released in the test process, and established the correlation between electromagnetic radiation and dissipated energy.

The above research has greatly enriched the application of energy theory in rocks, but most scholars focus on intact rocks and have little research on rock discontinuity, and in the cyclic loading and unloading test, often with unloading in the next cycle to 0 MPa, by changing the upper limit of cyclic loading for different stress amplitudes. The lower limit of cyclic load is often not zero, and therefore in the present work cement mortar was used to represent the rock, and a CSS-1950 biaxial rheological testing machine was adopted to conduct graded cyclic loading and unloading tests on Barton's standard profile line discontinuities of different joint roughness coefficients (JRCs). The energy evolution of samples in the process of failure was revealed, which can provide a theoretical basis for studying rock damage and failure mechanisms from the perspective of energy.

2. Test Equipment and Specimens

2.1. Test Equipment

The CSS-1950 (Model of testing machine, CSS is creep shear strength) rock biaxial rheological testing machine was used in this test (Figure 1a), which is manufactured by the Changchun Institute of Testing Machines in Changchun, China. The test machine includes vertical and horizontal loading systems with maximum vertical and horizontal compression loads of 500 and 300 kN, respectively. Two linear variable displacement transducers (LVDT) were used with a measurement range of 0–10 mm and accuracy of 0.0001 mm to monitor vertical and horizontal deformation, as shown in Figure 1b. The test machine adopts servo-motor control, a pressure system for screw pressure, load-rate control, and a continuous working time of more than 1000 h.

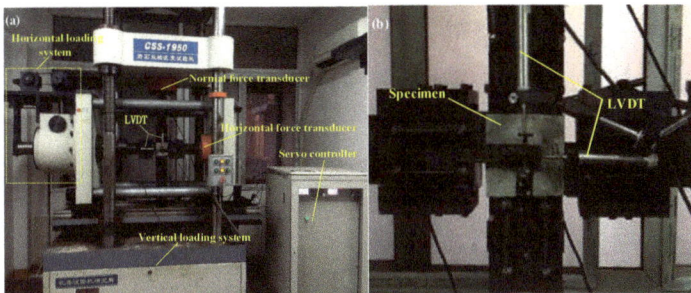

Figure 1. (**a**) CSS-1950 rock biaxial rheological testing machine; (**b**) specimen and monitoring system.

2.2. Sample Preparation

Due to the random composition and surface morphology of natural rock mass, it is difficult to prepare relatively uniform samples using natural rock mass, and it is impossible to quantify the roughness of the rock discontinuity, which results in difficulty when comparing test results; therefore, because cement mortar has good uniformity, it is often used as a kind of rock material to simulate rock discontinuities [22,23]. Therefore, in the present work, cement mortar was used as a similar material to prepare Barton's standard profile line discontinuities with different JRCs for testing.

In this test, the cement mortar samples named 1#, 4#, 8#, and 10# were respectively adopted to represent the features of structural planes with different roughness. Their Barton's standard profile line features are shown in Figure 2b. For the convenience of analysis, the JRC of the sample was taken as the intermediate value, that is, the cement mortar samples named 1#, 4#, 8#, and 10# were, respectively, 1, 7, 15, and 19 herein.

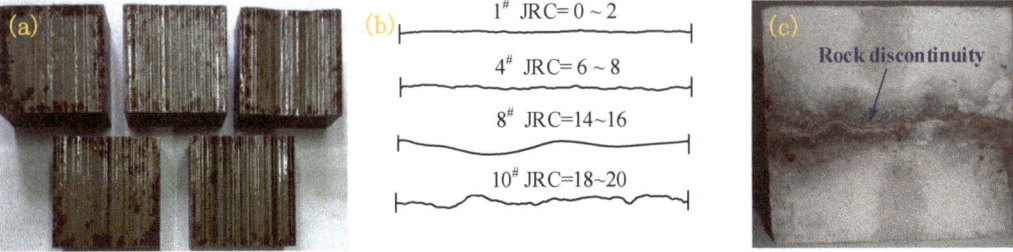

Figure 2. (**a**) Steel mold; (**b**) Barton's standard profile lines; (**c**) sample.

The steel mold used in this test was made by referring to Barton's standard profiles and using high precision (0.1 µm) computer control [24], as shown in Figure 2a. According to need, the upper and lower parts of the sample were prepared, and the two parts were combined to form a complete rock discontinuity sample (Figure 2c). Each specimen measured 100 × 100 × 100 mm.

The sample material was Portland cement with a compressive strength of 32.5 MPa, standard sand and water, and the mixing ratio was sand/cement/water of 4:2:1. After mixing evenly, the mold was filled. After filling, the sample was removed after the cement mortar was formed. After removal, the samples were placed in the curing chamber and stored in the standard curing chamber at the temperature and humidity of (20 ± 1) °C and 95%, respectively, for 28 days before testing.

According to the test requirements, a total of 29 specimens were prepared, including 5 intact samples and 24 rock discontinuity samples (the cement mortar samples named 1#, 4#, 8#, and 10# were prepared in 6 pieces each). In total, the intact samples were used for the uniaxial compression test, and the rock discontinuity samples were used for the direct shear test and graded cyclic shear test.

3. Test Procedure

3.1. Uniaxial Compression Test

To determine the normal stress value of shear test, uniaxial compression tests were conducted on five complete cement mortar specimens at a loading rate of 0.4 kN/s. The test results are shown in Table 1. Here, 10%, 20%, and 30% of the average compressive strengths were taken as the normal stresses of subsequent shear tests, which were 2.17, 4.35, and 6.52 MPa, respectively.

Table 1. Uniaxial compressive strength of each sample (σ_c is compressive strength).

Sample	Peak Stress/kN	σ_c/MPa
1	228.47	22.85
2	233.44	23.34
3	177.69	17.77
4	248.37	24.84
5	198.55	19.86
Average value	217.30	21.73

3.2. Direct Shear Test

To obtain the shear strength of rock discontinuities, the cement mortar samples named $1^\#$, $4^\#$, $8^\#$, and $10^\#$ (JRC = 1, 7, 15, 19) were selected and shear tests were conducted at a rate of 0.2 kN/s under the normal stress of 2.17, 4.35, and 6.52 MPa, respectively, until failure. The shear strength obtained was used as the basis for the classification of shear load grades in subsequent cyclic shear tests, and Barton's standard profile line number and normal stress were used to number the test results. For example, the test results of rock discontinuity No. 1 under the normal stress of 2.17 MPa were denoted "1–2.17", and the test data are listed in Table 2.

Table 2. Shear strength of each sample (τ_c is shear strength).

Sample	Peak Stress/kN	τ_c/MPa
1–2.17	7.5	1.5
1–4.35	15.2	3.0
1–6.52	18.5	3.7
4–2.17	10.1	2.0
4–4.35	15.2	3.0
4–6.52	22.5	4.5
8–2.17	12.2	2.4
8–4.35	20.0	4.0
8–6.52	25.0	5.0
10–2.17	12.5	2.5
10–4.35	22.1	4.4
10–6.52	27.5	5.5

3.3. Graded Cyclic Shear Tests

The cement mortar samples named $1^\#$, $4^\#$, $8^\#$, and $10^\#$ (JRC = 1, 7, 15, 19) were selected to conduct cyclic shear tests under normal stresses of 2.17, 4.35, and 6.52 MPa, respectively. The loading of samples is shown in Figure 3a. The normal stress was first added to a predetermined value, and the shear stress was applied after the normal deformation was stabilized. The shear stress was divided into multiple stages. The upper limit of the shear stress at the first stage was 30% of the shear strength, and each stage increased the stress by 10% of the shear strength. Under the same level of loading, the cyclic amplitude of shear stress was 10% of the shear strength. Under the same level of loading, 10 cycles were carried out with a loading rate of 0.2 kN/s. Continuous loading was conducted until failure. The actual loading stress is listed in Table 3.

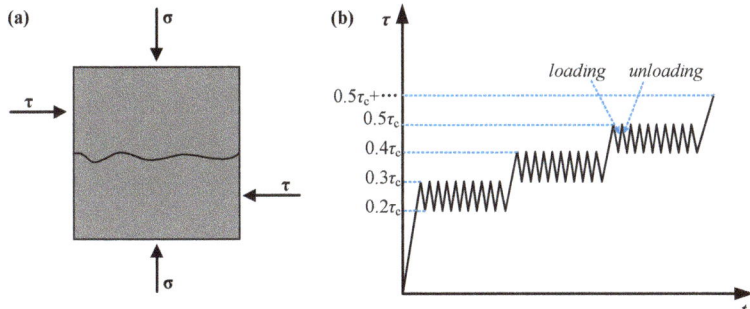

Figure 3. (a) Sample loading diagram; (b) schematic diagram of test loading path (σ is the normal stress and τ is the shear stress).

Table 3. Loading stress table.

Sample	Upper Limit of Loading Stress at First Stage/MPa	Amplitude/MPa
1–2.17	0.45	0.15
1–4.35	0.90	0.30
1–6.52	1.10	0.37
4–2.17	0.60	0.20
4–4.35	0.90	0.30
4–6.52	1.35	0.45
8–2.17	0.72	0.24
8–4.35	1.20	0.40
8–6.52	1.50	0.50
10–2.17	0.75	0.25
10–4.35	1.32	0.44
10–6.52	1.65	0.55

4. Results and Discussion

4.1. Deformation Characteristics of Specimen during Failure

The purpose of this research was to reveal the energy evolution of rock discontinuity under cyclic loading. Due to the limitation of word count, only one group of test data was studied, and the stress–strain relationship of other samples is similar, so it will be repeated here.

Figure 4 displays the whole process of the stress and displacement curve for sample 4–6.52. As shown in the figure, during each cycle, the unloading curve does not coincide with the original loading curve, and the unloading curve can form a completely closed annular area with the reloading curve, namely, the hysteresis loop [25]. The reason is that rock materials are not ideal elastomers, and there are a large number of internal defects such as pores and micro-cracks, which lead to the closure of the original cracks and the initiation of new cracks in the test process [26], resulting in some irreversible damage to the sample. The appearance of the hysteresis loop is not only an experimental phenomenon, but also a manifestation of energy dissipation, and the area of the hysteresis loop can be characterized as the energy dissipated by crack closure, expansion, and through-cracking failure of the loaded sample; the larger the area of the hysteresis loop, the more energy consumed and the more severe the damage to the sample.

The rock discontinuity sample is different from the intact rock sample, and in the first loading process of each grading cycle stage, the upper and lower parts of the rock discontinuity sample usually slip, resulting in large residual deformation. With the continuous increase of stress, this sliding deformation gradually increases, and the deformation curve of the sample is gradually sparse, indicating that the internal damage of the sample is increasingly aggravated with the increase of stress, and its ability to resist shear is gradually decreased, resulting in the gradual increase of irreversible deformation. However, under the same cyclic stress, the stress–strain curve changes from thinning to dense, and the hysteresis loop area decreases gradually, indicating that the specimen gradually compacts with the increase of cyclic number and the hardening degree increases.

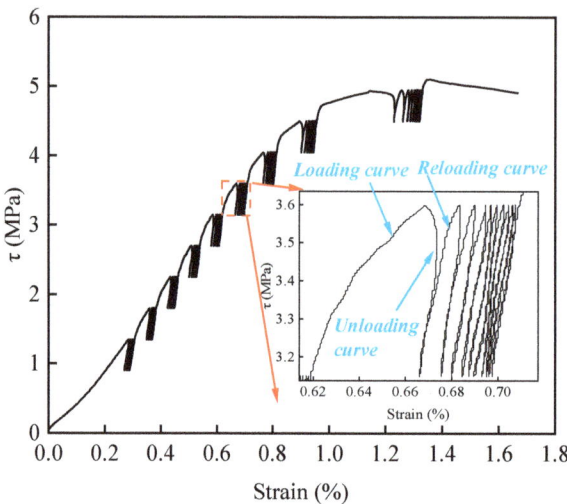

Figure 4. Stress–strain curves of specimen 4–6.52.

4.2. Energy Density Calculation Method

For the loaded sample, it often goes through the extrusion compaction stage, the stable development stage, and the unstable development stage of the crack before the final failure, and these processes are often accompanied by energy input, accumulation, dissipation, and release. The energy transformation of samples from deformation to failure is a dynamic process, which is represented by the transformation and balance among input energy, elastic energy, and dissipated energy. According to the first law of thermodynamics, the energy of substances in a thermodynamic system can be transformed and transferred, and the total amount of energy remains unchanged in the process of transformation and transfer [6]. The work performed by the surrounding to the rock mass can produce energy input into the rock mass, causing reversible deformation and irreversible deformation of the rock mass. The reversible deformation accumulates in the form of elastic energy. Irreversible deformation dissipates energy in the form of plastic deformation energy, internal friction of rock mass, and thermal radiation [27]. The sample in the test is deformed by external force loading and is a closed system. Therefore, the energy exchange between the specimen and the outside world is not considered [28,29]. According to the law of the conservation of energy,

$$U = U_e + U_d \tag{1}$$

where U is the total energy input from the outside; U_e denotes the elastic energy; and U_d is the dissipated energy, that is, the sum of the damage energy and plastic strain energy in rock mass.

At any time in the deformation process of the sample, there is a specific energy state corresponding to it [30], which is a function of stress, strain, and time. According to the stress–strain curve characteristics of samples, the elastic energy density and dissipation energy density of rock mass under cyclic loading were calculated. The relationship between the elastic energy density u_e and dissipation energy density u_d per unit of volume in the loading and unloading curve of samples under certain stress levels is shown in Figure 5. The total energy density u input by the outside world is the area formed by $OBAE$. The elastic energy density u_e denotes the area formed by *ACDE*. The dissipative energy density u_d is the total energy density u minus the elastic energy density u_e, which is the area formed by *OBACD*. The u_e and u_d are calculated as follows:

$$u_e = \int_{\varepsilon_1}^{\varepsilon_2} \sigma d\varepsilon \tag{2}$$

$$u_d = \int_{\varepsilon_0}^{\varepsilon_2} \sigma d\varepsilon - \int_{\varepsilon_1}^{\varepsilon_2} \sigma d\varepsilon \tag{3}$$

where σ is the stress at any point in the stress–strain curve; ε denotes the strain corresponding to σ; ε_0 refers to the strain corresponding to the initial stress σ_0 of the loading curve; ε_1 is the strain corresponding to the lower limit stress σ_0 of the unloading curve; and the strain corresponding to the loading of the upper limit stress σ_1 is represented by ε_2.

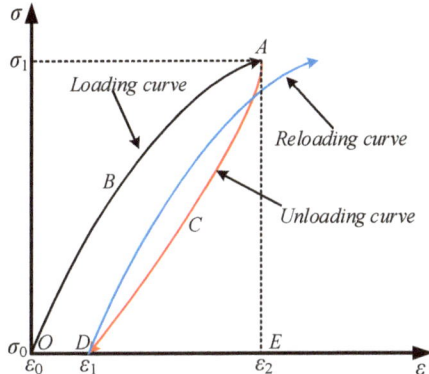

Figure 5. Schematic diagram of energy density calculation.

4.3. Energy Evolution Process in Samples

According to the above, during the first loading of each stage of the graded cycle, rock discontinuities usually produce large residual deformation, which makes it difficult for the unloading curve to form a hysteresis loop with the reloading curve, and the error in the calculation of the energy density is large; therefore, the second cycle was explored as if it were the first cycle. According to Equations (2) and (3), the energy density of the sample in each cycle of loading and unloading can be calculated. Taking specimen 4–6.52 as an example, the relationship between the energy density and cycle times under various cyclic stresses is shown in Figure 6.

As illustrated in Figure 6, the external input energy shows an overall downward trend with the increase in the number of cycles, but its change process is different under different cyclic stresses. When the stress is small (1.35 to 2.25 MPa), the external input energy decreases gradually with increasing number of cycles, and the decrease is significant in the first few cycles, small in the later stage, then tends to be stable. When the stress is large (2.70 to 4.95 MPa), the external input energy decreases rapidly with the increase in the number of cycles, and then decreases again after a small increase.

With the increase in the number of cycles, the elastic energy density shows an overall upward trend, and increases significantly in the early stage, and gradually stabilizes in the later stage, while the dissipated energy density gradually decreases, which is consistent with the phenomenon that the area of the hysteresis loop in the stress–strain curve decreases with increasing number of cycles. The results show that in the cyclic loading and unloading process associated with the same stress level, the micro-cracks inside the sample gradually close, the hardening of the sample increases, and the energy consumed by the friction inside the sample decreases.

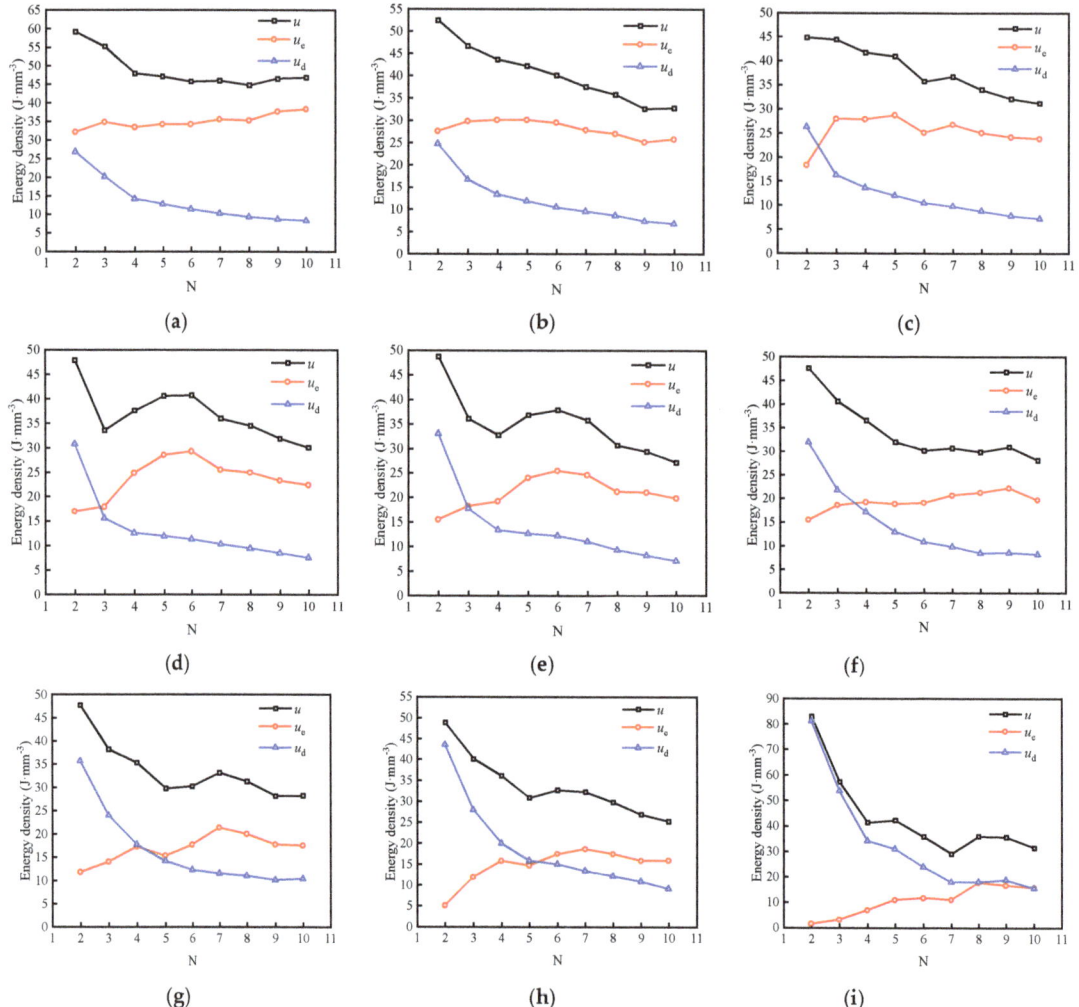

Figure 6. Variations in energy density of specimen 4–6.52 with the number of cycles N under various cyclic stresses: (**a**) 1.35 MPa; (**b**) 1.80 MPa; (**c**) 2.25 MPa; (**d**) 2.70 MPa; (**e**) 3.15 MPa; (**f**) 3.60 MPa; (**g**) 4.05 MPa; (**h**) 4.50 MPa; (**i**) 4.95 MPa.

With the continuous increase of cyclic stress, the dissipated energy density of each stage gradually exceeds the elastic energy density and gradually occupies a dominant position, indicating that the internal damage to the sample is gradually intensified.

The energy evolution of the other samples at each stress is akin to that of specimen 4–6.52, which is not repeated in this paper due to limited space.

According to the actual failure strength of the sample, the cyclic stress is normalized, and the energy density of the last nine cycles under all levels of cyclic stress is averaged to obtain the relationship between the average energy density and the stress level (the ratio of the upper limit of cyclic stress at all levels to the actual failure strength of the sample, as shown in Table 4), as shown in Figure 7. With the increase in stress, the energy absorbed by the sample from the outside firstly decreases greatly, and then decreases significantly. When the stress reaches a certain value, it shows a significant upward trend. The elastic energy density presents a non-linear decreasing trend with the increase of stress, and the rate of change decreases gradually thereafter. With the increase in stress, the dissipated energy density decreases slightly at first, then increases slowly at a low rate, and increases significantly as the specimen approaches failure.

Table 4. Stress level.

Sample	Stress Level
1–2.17	0.28, 0.37, 0.46, 0.55, 0.64, 0.74, 0.83, 0.92
1–4.35	0.27, 0.36, 0.46, 0.55, 0.64, 0.73, 0.82, 0.91
1–6.52	0.20, 0.27, 0.34, 0.41, 0.48, 0.55, 0.62, 0.69, 0.75, 0.82, 0.89
4–2.17	0.39, 0.52, 0.65, 0.79, 0.92
4–4.35	0.31, 0.41, 0.52, 0.62, 0.72, 0.83, 0.93
4–6.52	0.26, 0.35, 0.44, 0.53, 0.62, 0.71, 0.79, 0.88, 0.97
8–2.17	0.29, 0.39, 0.48, 0.58, 0.68, 0.78, 0.87, 0.97
8–4.35	0.31, 0.41, 0.51, 0.61, 0.71, 0.82, 0.92
8–6.52	0.26, 0.34, 0.43, 0.52, 0.60, 0.69, 0.77, 0.86, 0.95
10–2.17	0.28, 0.37, 0.46, 0.56, 0.65, 0.74, 0.84, 0.93
10–4.35	0.26, 0.35, 0.44, 0.52, 0.61, 0.70, 0.79, 0.87, 0.96
10–6.52	0.32, 0.43, 0.54, 0.64, 0.75, 0.86, 0.97

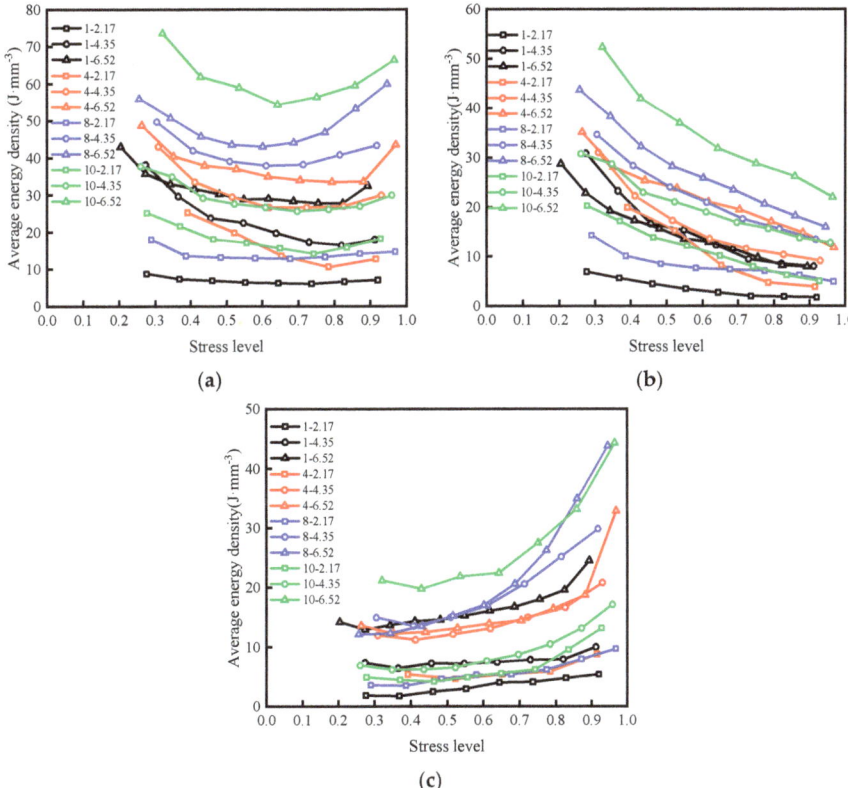

Figure 7. Relationship between energy density and stress level of samples under different working conditions: (a) u; (b) u_e; (c) u_d.

At the initial stage of fractional loading, the three energies all decreased. This phenomenon is due to the existence of many pores and cracks in the sample itself. In the process of first-stage cyclic loading and unloading, the closure and friction between these pores and cracks need to absorb more energy, and the absorbed energy is mainly stored in the sample in the form of elastic energy. Thus, the energy of each part of the sample is high in the first stage, and after entering the next stage of the cycle, most of the pores inside the sample have closed, and the degree of compaction is significantly improved, resulting in a decrease in the energy of all parts. With the further increase in stress, the primary cracks in the sample begin to expand, and new cracks are constantly initiated, leading to the decline of the ability of the sample to accumulate elastic energy. The energy consumed by internal friction and plastic failure increases gradually, which implies that the elastic energy density decreases continuously, while the dissipated energy density increases gradually.

4.4. Energy Distribution in Samples

In the closed test system, the energy input to rock samples by the testing machine is mainly transformed into elastic energy and dissipated energy, which will affect the deformation and failure of rock samples. The relationship between the proportion division of types of energy in samples and stress level is shown in Figure 8. With the constant increase in stress, the proportion of elastic energy and the proportion of dissipated energy in the sample show a non-linear trend, and the elastic energy presents a downward trend as a whole, while the dissipated energy shows the opposite trend, and the rates of change of both gradually increase.

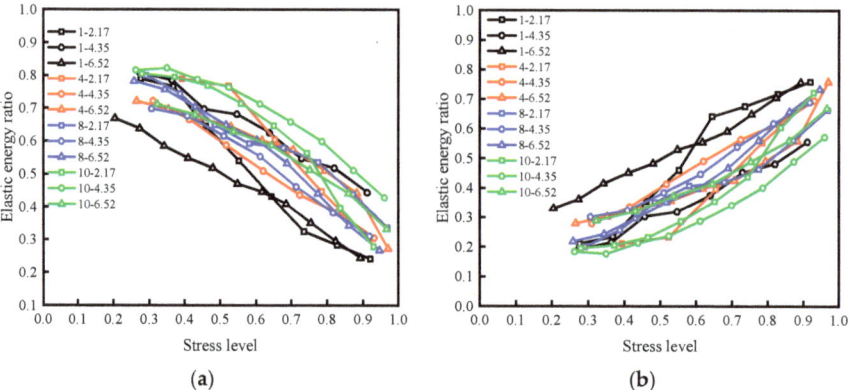

Figure 8. Relationship between energy ratio and stress level under different working conditions: (a) u_e; (b) u_d.

At the initial stage of fractional loading, the internal pores of the sample are compressed and compacted, and the proportion of elastic energy is much higher than that of dissipated energy, indicating that most of the energy input from the outside is converted into elastic energy and stored in the sample. The energy consumed by internal crack closure and friction slip is small. With the increase in stress, the elastic energy ratio decreases step-by-step at a low rate, while the dissipated energy ratio changes in the opposite way. At this stage, although most of the pores inside the sample are closed, the stress concentration leads to the expansion and initiation of micro-cracks, increasing the dissipated energy ratio, but most of the energy from the external input is still accumulated in the sample. When the ratio of dissipated energy and elastic energy approach each other, the connection of micro-cracks in the sample and the formation and unstable expansion of macro-cracks lead to the dissipation of most of the external input energy. The elastic energy accumulated in the sample begins to release, showing that the proportion of dissipated energy decreases rapidly.

4.5. Energy Criterion for Rock Discontinuity Failure

For an ideal material, any form of energy applied to it can all be converted into releasable elastic strain energy within the material, and during the deformation process, its internal structure does not suffer damage, and the absorption and storage of elastic energy will not dissipate. However, for rock materials, the friction of pores or micro-cracks inside the specimen in the compaction stage, and the expansion of micro-cracks and connection in the elastic and plastic stages are all accompanied by energy dissipation. The occurrence and accumulation of irreversible deformation are the direct causes of specimen failure [31], and the dissipative energy can indirectly reflect the irreversible deformation generated in the specimen. The accumulation of dissipative energy will facilitate the sample in its gradual transformation from the initial stable state to an unstable state, and then to another stable state (the strength therein being the residual strength after the main fractures have inter-connected and split into multiple rock blocks) through the reorganization of the internal structure of the sample. The change from a stable state to an unstable state entails the process of internal damage accumulation to unstable failure, as well as the process of internal energy transformation. Therefore, the energy dissipation ratio K (the ratio of the dissipated energy density to the elastic energy density of the loaded rock sample) is used to characterize the deterioration of the sample, and also indirectly reflect the state of the sample. K is given by

$$K = \frac{u_d}{u_e} \quad (4)$$

When $K < 1$, it can be considered that the internal structure of the sample is in a relatively stable state, and the damage to it is small; when $K = 1$, it can be considered that the loaded rock sample reaches its energy storage limit, is in a critical state, and is about to enter the unstable development stage; when $K > 1$, it can be considered that the rock sample is in an unstable state.

Figure 9 shows the stress level–energy dissipation ratio deformation of samples under different working conditions and Figure 10 demonstrates the energy consumption ratio of each sample that varies with the stress level.

According to the sample energy dissipation ratio and shear deformation seen in Figure 9, the failure process of the sample can be roughly divided into three stages. In the early stage of fractional loading, the sample is in the extrusion compaction stage (the first stage), and the internal pores and cracks of the sample are closed in this stage. The hardening of the sample is significantly improved, resulting in the irreversible deformation of the rock discontinuity which decreases significantly, and the energy dissipation ratio remains quasi-constant. Then the sample enters the stage of steady crack propagation (the second stage), and the internal damage intensifies, and the dissipated energy increases continuously due to the increase of internal friction and plastic deformation. The sample gradually reaches its energy storage limit ($K = 1$), which shows that the deformation of the structural plane and energy dissipation ratio increase linearly with the increasing stress. With the further increase of the stress level, the sample enters the unstable crack development stage (the third stage). The connection and penetration of cracks results in the worsening of the plastic failure of the sample, the rapid increase of dissipative energy, and the energy accumulated in the sample begins to release. When the failure is near, both the deformation and energy consumption ratio of the rock discontinuity change dramatically, indicating that a large amount of plastic failure occurs in the sample. The stored energy in the sample is released instantly, leading to the rapid loss of bearing capacity and instability failure.

Figure 10 illustrates that the energy dissipation ratio of all samples has roughly the same variation with stress, showing slow growth in the early stage, and a significant increase in the growth rate after reaching the energy storage limit ($K = 1$). However, there are significant differences in the stress on all samples when reaching the energy storage limit.

Instability and failure of specimens occurred after reaching the energy storage limit and are the root cause of a rapid release of elastic energy. Energy storage limits of the samples at the corresponding stress level can describe the samples' accumulated elastic energy capacity: the higher the stress level, the more the sample storage limits the strength or stiffness, and instability and failure are less likely.

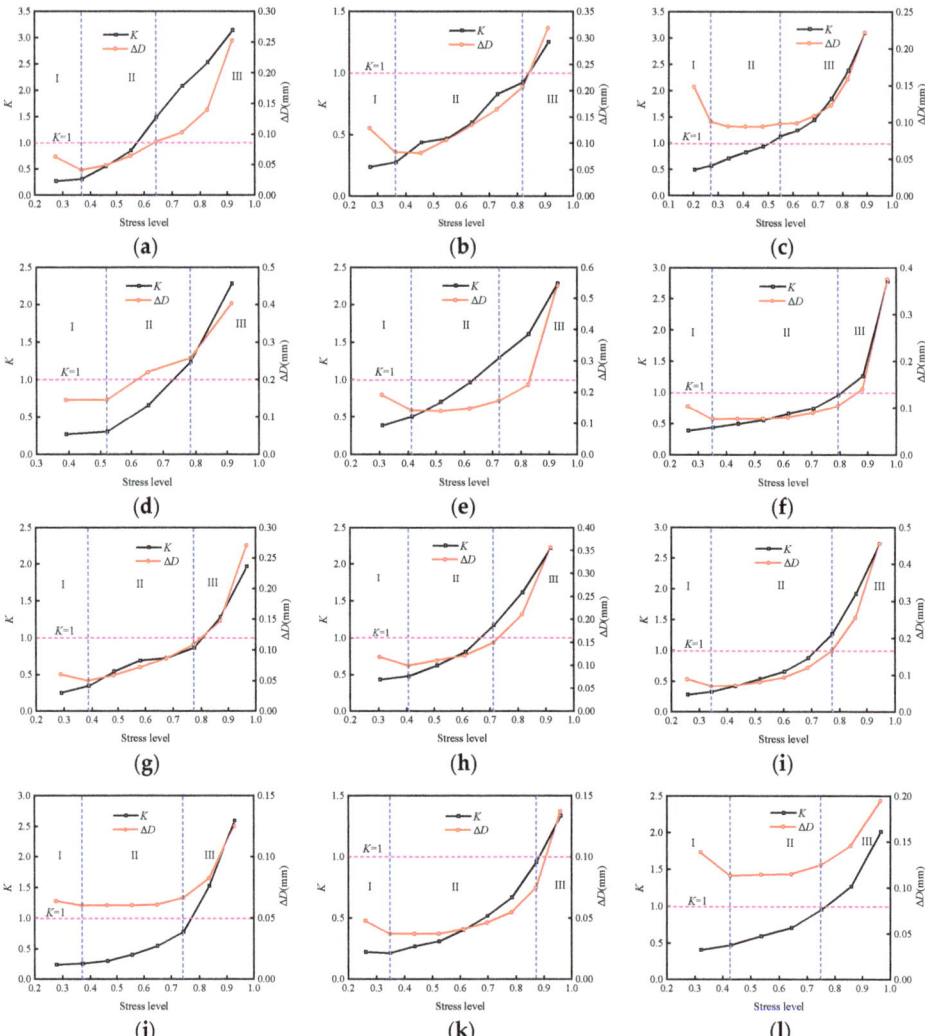

Figure 9. Stress level–energy dissipation ratio deformation of samples under different working conditions (ΔD is the total deformation of 10 cycles under different stress levels): (**a**) 1–2.17; (**b**) 1–4.35; (**c**) 1–6.52; (**d**) 4–2.17; (**e**) 4–4.35; (**f**) 4–6.52; (**g**) 8–2.17; (**h**) 8–4.35; (**i**) 8–6.52; (**j**) 10–2.17; (**k**) 10–4.35; (**l**) 10–6.52.

To explore the influence of roughness on the energy storage capacity of a specimen, the stress corresponding to each sample at $K = 1$ in Figure 10 was taken as the ultimate stress associated with its energy storage, and the results under the action of three different normal stresses under the same roughness were averaged to determine the variations in the ultimate stress of the energy storage of the sample with the roughness (Figure 11). The energy storage limiting stress for the specimen shows a positive correlation with the increase of JRC; that is, the rougher the rock discontinuity, the greater its energy storage limit and stiffness. According to the linear fitting results in the figure, the energy storage capacity of the rock discontinuity sample increases by about 0.75% when the value of JRC increases by 1.

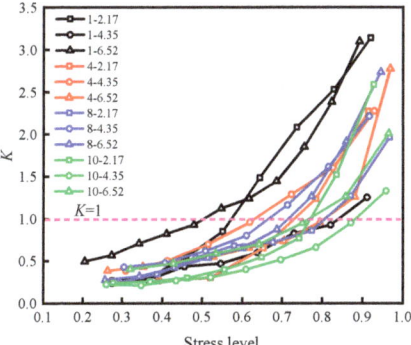

Figure 10. The energy consumption ratio of each sample varies with the stress level.

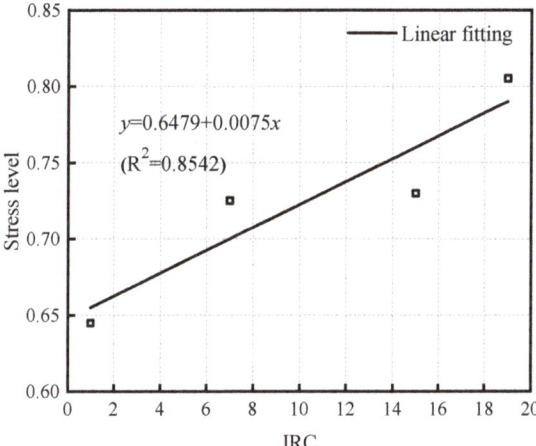

Figure 11. Variations in stress at the energy storage limit of samples with JRC.

5. Conclusions

The following conclusions can be drawn following hierarchical cyclic loading and unloading tests and energy calculation and analysis of specimens containing discontinuities:

(1) Compared with the stress–strain curve, the energy density can clearly reflect the internal deterioration of the rock discontinuity, so as to predict the failure of the rock discontinuity more accurately.

(2) Under the same cyclic stress, the specimen gradually hardens with the increase in the number of cycles. With the increase of cyclic stress, the dissipated energy density of each stage gradually exceeds the elastic energy density and occupies a dominant position and increases rapidly as failure becomes imminent.

(3) With the increase of stress level, the elastic energy proportion of the sample presents a downward trend, with a slow rate in the early stage, but decreases significantly as the sample approaches failure; the variation in the proportion of energy dissipated shows the opposite trend.

(4) The energy dissipation ratio can be used to characterize internal damage to the sample under load. In the initial stage of loading, the sample is in the extrusion

and compaction stage, and the energy dissipation ratio remains unchanged. Then, the fracture develops steadily, the damage in the sample intensifies, and the energy dissipation ratio increases linearly (albeit at a low rate). Before the specimen is about to fail, the change rate is accelerated, and then a sudden change occurs, indicating that the rapid release of energy is the fundamental reason for the failure of the rock discontinuity.

(5) The increase of the roughness of rock discontinuity samples will improve their energy storage capacity to a certain extent: the higher the JRC of the rock discontinuity, the greater the energy storage limit and stiffness of the specimen.

Author Contributions: Writing—original draft preparation, conceptualization, W.Z.; writing—review and editing, methodology, L.G., Z.W.; data curation, supervision, J.M., H.L., H.Z. All authors have read and agreed to the published version of the manuscript.

Funding: This work was financially supported by the National Natural Science Foundation of China (Grant No. 42002266), the Fundamental Research Funds for the Central Universities (Grant No.30922010918) and the Chinese Postdoctoral Science Foundation (2020M673654).

Institutional Review Board Statement: Not applicable.

Informed Consent Statement: Informed consent was obtained from all subjects involved in the study.

Data Availability Statement: The data used to support the findings of this study are included within the article.

Conflicts of Interest: The authors declare no conflict of interest.

References

1. Sagong, M.; Bobet, A. Coalescence of multiple flaws in a rock-model material in uniaxial compression. *Int. J. Rock Mech. Min. Sci.* **2002**, *39*, 229–241. [CrossRef]
2. Meng, Q.B.; Zhang, M.W.; Han, L.J.; Pu, H.; Nie, T.Y. Effects of Acoustic Emission and Energy Evolution of Rock Specimens Under the Uniaxial Cyclic Loading and Unloading Compression. *Rock Mech. Rock Eng.* **2016**, *49*, 3873–3886. [CrossRef]
3. Yang, S.Q.; Yin, P.F.; Zhang, Y.C.; Chen, M.; Zhou, X.P.; Jing, H.W.; Zhang, Q.Y. Failure behavior and crack evolution mechanism of a non-persistent jointed rock mass containing a circular hole. *Int. J. Rock Mech. Min. Sci.* **2019**, *114*, 101–121. [CrossRef]
4. He, M.C.; Sousa, L.R.E.; Miranda, T.; Zhu, G.L. Rockburst laboratory tests database—Application of data mining techniques. *Eng. Geol.* **2015**, *185*, 116–130. [CrossRef]
5. Liu, Y.; Dai, F.; Fan, P.; Dong, L. Experimental Investigation of the Influence of Joint Geometric Configurations on the Mechanical Properties of Intermittent Jointed Rock Models under Cyclic Uniaxial Compression. *Rock Mech. Rock Eng.* **2017**, *50*, 1453–1471. [CrossRef]
6. Xie, H.P.; Li, L.Y.; Yang, J.; Peng, R.D.; Yang, Y.M. Energy analysis for damage and catastrophic failure of rocks. *Sci. China Technol. Sci.* **2011**, *54*, 199–209. [CrossRef]
7. Sujatha, V.; Kishen, J. Energy Release Rate due to Friction at Bimaterial Interface in Dams. *J. Eng. Mech.* **2003**, *129*, 793–800. [CrossRef]
8. Huang, D.; Li, Y.R. Conversion of strain energy in triaxial unloading tests on marble. *Int. J. Rock Mech. Min. Sci.* **2014**, *66*, 160–168. [CrossRef]
9. Xie, H.P.; Peng, R.D.; Yang, J.U.; Zhou, H.W. On energy analysis of rock failure. *Chin. J. Rock Mech. Eng.* **2005**, *15*, 2603–2608. (In Chinese)
10. Xie, H.P.; Ju, Y.; Li, L.Y.; Peng, R.D. Energy mechanism of deformation and failure of rock masses. *Chin. J. Rock Mech. Eng.* **2008**, *27*, 1729–1740. (In Chinese)
11. Tu, Y.L.; Liu, X.R.; Zhong, Z.L.; Li, Y.Y. New criteria for defining slope failure using the strength reduction method. *Eng. Geol.* **2016**, *212*, 63–71. [CrossRef]
12. Peng, K.; Wang, Y.Q.; Zou, Q.L.; Liu, Z.P.; Mou, J.H. Effect of crack angles on energy characteristics of sandstones under a complex stress path. *Eng. Fract. Mech.* **2019**, *218*, 106577. [CrossRef]
13. Wang, J.A.; Park, H.D. Comprehensive prediction of rockburst based on analysis of strain energy in rocks. *Tunn. Undergr. Space Technol.* **2001**, *16*, 49–57. [CrossRef]
14. Wu, Z.H.; Song, Z.Y.; Tan, J.; Zhang, Y.Z.; Qi, Z.J. The evolution law of rock energy under different graded cyclic loading and unloading modes. *J. Min. Saf. Eng.* **2020**, *37*, 836–844+851. (In Chinese)
15. Wang, H.; Yang, T.H.; Liu, H.L.; Zhao, Y.C.; Deng, W.X.; Hou, X.G. Mechanical properties and energy evolution of dry and saturated sandstones under cyclic loading. *Rock Soil Mech.* **2017**, *38*, 1600–1608. (In Chinese)

16. Jia, Z.Q.; Li, C.B.; Zhang, R.; Wang, M.; Gao, M.Z.; Zhang, Z.T.; Zhang, Z.P.; Ren, L.; Xie, J. Energy Evolution of Coal at Different Depths Under Unloading Conditions. *Rock Mech. Rock Eng.* **2019**, *52*, 4637–4649. [CrossRef]
17. Deng, Y.; Chen, M.; Jin, Y.; Zou, D.W. Theoretical analysis and experimental research on the energy dissipation of rock crushing based on fractal theory. *J. Nat. Gas Sci. Eng.* **2016**, *33*, 231–239. [CrossRef]
18. Zhang, Z.Z.; Gao, F. Confining pressure effect on rock energy. *Chin. J. Rock Mech. Eng.* **2015**, *34*, 1–11. (In Chinese)
19. Munoz, H.; Taheri, A.; Chanda, E.K. Rock Drilling Performance Evaluation by an Energy Dissipation Based Rock Brittleness Index. *Rock Mech. Rock Eng.* **2016**, *49*, 3343–3355. [CrossRef]
20. Munoz, H.; Taheri, A.; Chanda, E.K. Fracture Energy-Based Brittleness Index Development and Brittleness Quantification by Pre-peak Strength Parameters in Rock Uniaxial Compression. *Rock Mech. Rock Eng.* **2016**, *49*, 4587–4606. [CrossRef]
21. Song, D.Z.; Wang, E.Y.; Liu, J. Relationship between EMR and dissipated energy of coal rock mass during cyclic loading process. *Saf. Sci.* **2012**, *50*, 751–760. [CrossRef]
22. Patton, F.D. Multiple modes of shear failure in rock. In Proceedings of the 1st ISRM Congress, Lisbon, Portugal, 25 September–1 October 1966; International Society for Rock Mechanics and Rock Engineering: Lisbon, Portugal, 1966.
23. Wang, Z.; Gu, L.L.; Shen, M.R.; Zhang, F.; Zhang, G.K.; Wang, X. Shear stress relaxation behavior of rock discontinuities with different joint roughness coefficient and stress histories. *J. Struct. Geol.* **2019**, *126*, 272–285. [CrossRef]
24. Barton, N. Suggested methods for the quantitative description of discontinuities in rock masses: International Society for Rock Mechanics. *Int. J. Rock Mech. Min. Sci. Geomech. Abstr.* **1978**, *15*, 319–368.
25. Liu, X.S.; Ning, J.G.; Tan, Y.L.; Gu, Q.H. Damage constitutive model based on energy dissipation for intact rock subjected to cyclic loading. *Int. J. Rock Mech. Min. Sci.* **2016**, *85*, 27–32. [CrossRef]
26. Meng, Q.B.; Zhang, M.W.; Zhang, Z.Z.; Han, L.J.; Pu, H. Research on non-linear characteristics of rock energy evolution under uniaxial cyclic loading and unloading conditions. *Environ. Earth Sci.* **2019**, *78*, 650. [CrossRef]
27. Li, P.; Cai, M.F. Energy evolution mechanism and failure criteria of jointed surrounding rock under uniaxial compression. *J. Cent. South Univ.* **2021**, *28*, 1857–1874. [CrossRef]
28. Zhang, M.W.; Meng, Q.B.; Liu, S.D. Energy Evolution Characteristics and Distribution Laws of Rock Materials under Triaxial Cyclic Loading and Unloading Compression. *Adv. Mater. Sci. Eng.* **2017**, *2017*, 5471571. [CrossRef]
29. Li, T.T.; Pei, X.J.; Wang, D.P.; Huang, R.Q.; Tang, H. Nonlinear behavior and damage model for fractured rock under cyclic loading based on energy dissipation principle. *Eng. Fract. Mech.* **2019**, *206*, 330–341. [CrossRef]
30. Zhang, Z.Z.; Gao, F. Experimental investigation on the energy evolution of dry and water-saturated red sandstones. *Int. J. Min. Sci. Technol.* **2015**, *25*, 383–388. [CrossRef]
31. Meng, Q.B.; Zhang, M.W.; Han, L.J.; Pu, H.; Chen, Y.L. Acoustic Emission Characteristics of Red Sandstone Specimens under Uniaxial Cyclic Loading and Unloading Compression. *Rock Mech. Rock Eng.* **2018**, *51*, 969–988. [CrossRef]

Comparative Study on the Seepage Characteristics of Gas-Containing Briquette and Raw Coal in Complete Stress–Strain Process

Ke Ding [1], Lianguo Wang [1,*], Zhaolin Li [2,3,*], Jiaxing Guo [1], Bo Ren [1], Chongyang Jiang [1] and Shuai Wang [1]

1. State Key Laboratory for Geomechanics and Deep Underground Engineering, China University of Mining and Technology, Xuzhou 221116, China
2. School of Mines, China University of Mining and Technology, Xuzhou 221116, China
3. State Key Laboratory of Mining Response and Disaster Prevention and Control in Deep Coal Mines, Anhui University of Science and Technology, Huainan 232001, China
* Correspondence: cumt_lgwang@163.com (L.W.); lzhlcumt@163.com (Z.L.); Tel.: +86-131-1522-5568 (L.W.)

Abstract: In this study, triaxial compression and seepage tests were conducted on briquette and raw coal samples using a coal rock mechanics-seepage triaxial test system (TAWD-2000) to obtain the complete stress–strain curves of the two samples under certain conditions. On this basis, the different damage forms of the two coal samples and the effect of their deformation and damage on their permeability were analyzed from the perspective of fine-scale damage mechanics. Moreover, the sensitivity of permeability to external variables and the suddenness of coal and gas outbursts were discussed. The results show that the compressive strength of raw coal is 27.1 MPa and the compressive strength of briquette is 17.3 MPa, the complete stress–strain curves of the two coal samples can be divided into four stages and show a good correspondence to the permeability–axial strain curves. Since briquette and raw coal have different structural properties, they present different damage mechanisms under load, thus showing great diversity in the permeability-axial strain curve, especially in the damage stage. The deformation affects the seepage characteristics of briquette mainly in the latter two stages, while it affects raw coal throughout the test. The four stages of the complete stress–strain seepage test of raw coal can well explain the four stages of coal and gas outburst process, i.e., preparation, initiation, development, and termination. Hence, the law of coal permeability to gas variation can be utilized for the coal and gas outburst prediction and forecast. The research results are valuable for exploring the real law of gas migration in coal seams.

Keywords: coal and gas outburst; gas-containing coal; complete stress–strain process; permeability; solid-gas coupling

1. Introduction

Energy, a main driver of the economy, plays a pivotal role in national economic development [1]. Along with the increasing amount and depth of coal mining, problems such as high ground stress and high gas pressure emerge one after another. In the underground mining process in coal mines, the coal body is deformed and damaged by the mining, accompanied by changes in gas permeability characteristics, which is a major cause of gas dynamic damages such as gas gushing from the working face and coal and gas outbursts [2–6]. Therefore, for an effective reduction of the gas-induced adverse effects in underground coal mining, it is of great theoretical significance and practical value to study the law of gas seepage during coal deformation and understand the coal and gas outburst mechanism.

Many scholars believe that briquette shares a similar variation law with raw coal, despite their significant difference in true relative density and apparent relative density (their porosities differ about 4 times and their pore volumes differ 4–10 times). Besides, briquette is easy and efficient to be processed into coal samples. Therefore, briquette can be used as a research object on the general law of gas-containing coal, and the yielded law is applicable to coal seams [7]. At present, researches have been conducted on both briquette and raw coal. By performing comparative tests on the permeabilities of gas-containing briquette and raw coal, Ge et al. [8] found that using briquette samples in place of raw coal samples in the laboratory can only obtain a rough variation law. Pang et al. [9] studied the deformation of coal after gas adsorption during gas pressure change using a self-developed experimental device and discussed the effect of gas on coal based on the existing studies. They concluded that the adsorbed gas would cause expansion and deformation of coal, which weakens its strength and increase its brittleness, so that it becomes more prone to sudden instability damage. Gan et al. [10] investigated the effect of gas pressure on the coal permeability to gas characteristics of coal rock materials in the complete stress–strain process. The results show that at a constant confining pressure, the complete stress–strain curve corresponds well to the seepage rate-axial strain curve, and raising the gas pressure within a certain range can enhance the permeability of coal. Sun et al. [11] conducted triaxial compression and seepage tests on briquette and raw coal using a self-developed gravitational device and found a good correspondence between their complete stress–strain curves and seepage rate-axial strain curves. Zhang et al. [12] explored the deformation characteristics and compressive strengths of briquette and raw coal under triaxial stress conditions based on the results of gas-containing triaxial tests, and concluded that although the two coal samples are similar in the above two properties, they differ significantly in the permeability characteristics. Katarzyna et al. [13] used Brazilian tests to evaluate the work of disintegration of rock resulting from the stresses produced by gas present in its porous structure. However, these research results did not provide a comprehensive comparative analysis on the permeability characteristics of briquette and raw coal. In fact, it is still necessary to further explore the differences in the permeability characteristics between the two during the complete stress–strain process and determine which one is more in line with the actual situation [14–27].

In this paper, gas-containing complete stress–strain tests were conducted on raw coal and briquette by using a coal rock mechanics-seepage triaxial test system (TAWD-2000) to study the similarity and difference in seepage characteristics of the two coal samples during the deformation process. The research results are expected to provide reference for the experimental study using briquette in place of raw coal and serve as an experimental basis for further exploration on the real law of gas migration and the mechanism of coal and gas outbursts in coal seams.

2. Test Process

2.1. Test Materials

The test coal samples were taken from the No. 8_2 coal seam of Zouzhuang Coal Mine of Huabei Mining Stock Corporation, China. The average buried depth of the coal seam is 853.7 m, as shown in Figure 1. This coal mine is bounded by the F22 fault and the Dual-Stack fault in the east, the first limestone cropline at the top of the Carboniferous Taiyuan Formation in the south, the Nanping fault in the west, and the #27 exploration line in the north. The No. 8_2 coal seam is 2.48 m thick on average. The coal body, which belongs to semi-bright coal due to its weak glassy luster, is black, pulverized-fragmented, and has developed endogenous fractures. Soft and fragile, this coal seam is typically prone to coal and gas outbursts.

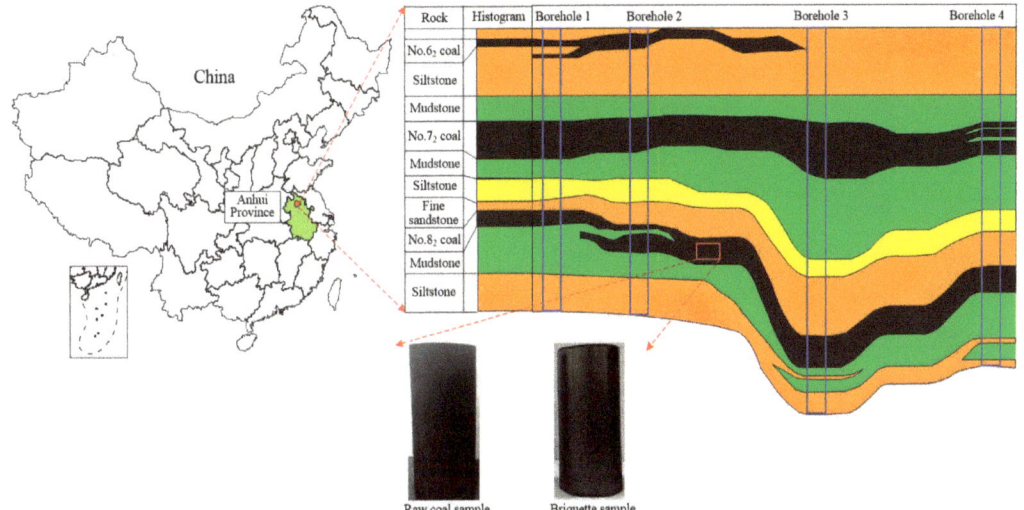

Figure 1. Diagram of the sampling site.

The preparation of raw coal samples is a difficult task: The raw coal blocks taken from the site were sealed with plastic film and transported directly to the mechanics experiment center of the China University of Mining and Technology where they were cored and processed under the condition that the stratification was perpendicular to the processing axis. According to the test platform and test specifications, the prepared raw coal samples were cylindrical with a height of 95–102 mm, a diameter of about 50 mm, and a parallelism below ±0.05 mm between the upper and lower end surfaces, and they were sealed and stored before the test.

In contrast, the preparation of briquette samples is easier: The raw coal blocks were polished into 40–60 mesh pulverized coal and mixed well with a little water and binder. Then, a certain amount of the mixture was weighed with a balance by experience and put into the mold. After being shaped by a pressure of 200 kN for 20 min, they were made into cylinders with a height of 100 mm, a diameter of 50 mm, and a parallelism of 0.02 mm between the upper and lower end surfaces.

Comparing the two samples, it is clear that the briquette sample had a standard size, a smooth surface and uniform texture, while the raw coal sample had primary damages such as vertical and horizontal fractures, cracks, and holes. Therefore, the raw coal samples with similar density, cracks, and longitudinal wave velocities were selected for the test.

2.2. Test Device and Principle

The coal rock mechanics-seepage triaxial test system (TAWD-2000) of the China University of Mining and Technology was used for the test (Figure 2). The system, which mainly consists of a pressure host system, a pressure and temperature control system, and a microcomputer operating system, can determine the permeabilities of coal rock under different pressure conditions. The maximum working pressures of confining pressure, injection pressure, and axial pressure are 70 MPa, 70 MPa, and 800 MPa, respectively, with the pressures fluctuating within 0.5% in 48 h. The test was conducted at a constant temperature of 25 °C with CH_4 as the seepage medium.

Figure 2. TAWD-2000 coal rock mechanics-seepage triaxial test system.

Coal is a porous medium in which the seepage characteristics of gas depend on the number, size, and connectivity of pores and the pressure at both ends of pores in the flow direction. Hence, the permeability characteristics of the sample can be reflected by the relation curve between permeability and other related physical quantities. The principle of the coal permeability measurement test is shown in Figure 3. According to the principle, the steady-state method was adopted in the permeability test: First, different gas pressures were applied to the two ends of a coal sample with a constant pressure difference, so that a certain pressure gradient was formed in the coal sample to promote the flow of gas through the fractures. Meanwhile, the gas flow was measured. When the flow within the coal sample developed into a steady-state flow, the amount of gas flowing through the sample over a while was recorded. The recorded amount can be substituted into the governing Equation (1) to calculate the permeability of the coal sample [19].

$$K = \frac{2p_0 Q L_{coal} \mu_{CH_4}}{A(p_1^2 - p_2^2)}, \qquad (1)$$

where K is the permeability, 10^{-15} m^2; p_0 is the atmospheric pressure, 0.1 MPa; Q is the gas flow through the coal sample, cm^3/s; L_{coal} is the standard length of the coal sample, mm; μ_{CH_4} is the gas dynamic viscosity coefficient, MPa·s; A is the cross sectional area of the coal sample, mm^2; p_1 is the inlet pressure, MPa; and p_2 is the outlet pressure, MPa.

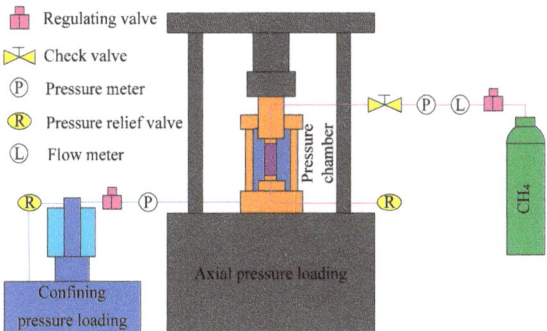

Figure 3. Schematic diagram of the principle of the permeability determination test.

2.3. Test Procedure

In this test, to investigate the changes in gas seepage velocities of briquette and raw coal samples in the complete stress–strain process, a comparative test was carried out according to the geological situation of the No. 8_2 coal seam in Zouzhuang Coal Mine at a gas pressure of 1.2 MPa and a confining pressure of 4 MPa. The specific procedure and precautions are as follows:

(1) The sample, whose cling film was unwrapped before the test, was fixed with the upper and lower ventilative plates by a thin heat shrinkable film. Then, the fixed sample was wrapped with insulating tape against the hydraulic oil in the pressure chamber during the test. Finally, the sample was wrapped with a thick heat shrinkable film to ensure its air tightness (Figure 4);

(2) Test instruments including the gas pipe, the flow meter, etc., were connected. Axial pressure, confining pressure, and gas pressure were applied to the coal sample in turn, where the amounts of three pressures followed the order: axial pressure of 4 MPa = confining pressure of 4 MPa > gas pressure of 1.2 MPa. Afterwards, the air tightness of the equipment was checked again. After each equipment in the system operated normally, the sample was allowed to fully adsorb gas for 24 h;

(3) The outlet valve was opened to release gas for 30 min until the gas flow stabilized, and then the test started. The loading, controlled by the displacement, proceeded at a rate of 0.002 mm/s until the coal sample finally failed.

Figure 4. Sealed samples.

3. Test Results and Analysis

3.1. Comparative Analysis on the Complete Stress–Strain Curves

In the triaxial compression test, the complete stress–strain curves of briquette and raw coal show similar variation trends (Figure 5). Both of them can be divided into four development stages: initial compaction stage, elastic deformation stage, plastic deformation stage, and instability damage stage. In the initial compaction stage, the elastic modulus increases with the increase of axial stress and strain, and the stress–strain curve shows a slight upsweep, which results from the compaction of pores and fractures inside the coal sample. In the elastic deformation stage, the stress and strain are linearly correlated with each other, and the elastic modulus becomes constant, following the Hoek–Brown criterion. In the plastic deformation stage, when the axial stress reaches the yield strength, internal damage occurs inside the coal sample, leading to a reduction in the sample's load-carrying capacity. At this time, the elastic modulus decreases, and the stress–strain curve is no longer linear and curves downward, which is caused by the continuously developing internal damage and new fractures in the sample. In the instability damage stage, after reaching the strength limit, the axial stress begins to decrease with the increase of the strain, which is attributed to the macroscopic cracks penetrating the sample.

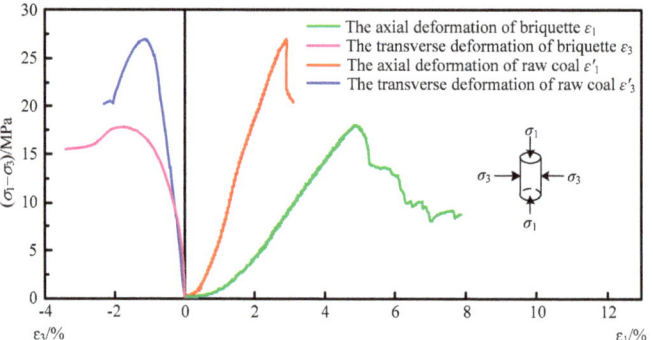

Figure 5. Complete stress–strain curves of briquette and raw coal.

The comparative analysis suggests many differences between the test results of the two samples. First, briquette experiences more severe transverse deformation and axial deformation than raw coal, its transverse deformation being twice that of raw coal and its axial deformation being three times that of raw coal. Second, briquette has a lower compressive strength than raw coal, as the compressive strength of raw coal is 27.1 MPa and the compressive strength of briquette is 17.3 MPa, the former being 63.7% of the latter. The elastic modulus of briquette obtained from the test is 0.517 GPa, the elastic modulus of raw coal is 2.306 GPa, the Poisson's ratio of briquette is 0.22, and the Poisson's ratio of raw coal is 0.16. Third, the two coal samples exhibit obvious differences in the post-peak deformation and damage stage. Specifically, after the peak, briquette presents a strain softening phenomenon and a gentle decrease in axial pressure, while raw coal shows a stress drop and significant and sharp change in axial pressure, just like the instantaneously and rapidly occurring coal and gas outburst on-site.

The factor that most directly determines coal permeability to gas in coal seams is the development degree of pores and fractures. In the laboratory, both briquette and raw coal have certain primary micropores and microfractures, which are referred to as primary damages [20]. The lower initial permeability of raw coal indicates its much slighter primary damages than briquette. However, as the applied load continues to grow, these primary damages will further develop, extend, and finally connect with each other, leading to a change in permeability of the coal sample.

Based on the changes in strains and permeabilities of briquette and raw coal during the loading stage, the complete stress–strain curves and permeability-strain curves of the two samples were obtained (Figure 6). A correspondence can be found between the complete stress–strain curves and the permeability-strain curves of briquette and raw coal, but the two coal samples have different gas permeability variation laws due to their different damage forms:

(1) Initial compaction stage (OA section): With the rise of axial pressure, the stiffnesses of the two coal samples are enhanced gradually, and the primary fractures are gradually compacted and closed, leading to the shrink of seepage channels. As a result, the permeabilities of both briquette and raw coal decline to some extent;

(2) Elastic deformation stage (AB section): The stress–strain curves of the two coal samples show approximately linear variations. Raw coal is barely damaged internally, so all its primary damages only deform elastically. As its primary micropores and microfractures further close, the coal permeability to gas continues to decline, but such a decline is insignificant owing to its low initial permeability. In contrast, under the action of the external load, the cohesive force of briquette is reduced by the extrusion and dislocation of its particles. Resultantly, the primary fractures between the particles are filled, leading to a rapid decline in its permeability. Besides, its permeability is the most sensitive to stress in this stage;

(3) Plastic deformation stage (BC section): The permeabilities of the two coal samples begin to grow. With the rise of axial pressure, the continuous distributed damages that occurred inside the raw coal create a condition for the stable extension of more and more microfractures, causing plastic deformation. At this time, the permeability of raw coal grows rapidly due to the further development of primary fractures and the formation of new fractures, and the permeability is the most sensitive to stress in the plastic deformation stage. For briquette, the shear movement of its particles facilitates the stable extension of fractures. However, the newly generated fractures are blocked by the detached particles (as they squeeze and displace each other), and thus the permeability grows slowly;

(4) Instability damage stage (CD section): As raw coal experiences a stress drop, where its damage develops from continuous damage to local damage, its fractures with elastic deformation undergo elastic unloading deformation. Consequently, the inelastic strain borne by primary fractures gradually focuses on few fractures generated by the local damage. These large instability-induced fractures enable the gas to pass through smoothly and promote the permeability of raw coal rapidly. However, as briquette only develops based on shear damage, its bearing capacity begins to decline, and its internal structure disenables a sudden stress drop, so its permeability grows only gently.

3.2. Relation between Permeability and Axial Stress

Based on the data of strains and permeabilities of briquette and raw coal in the loading stage, the permeability-axial stress curves of the two samples were obtained (Figure 7). When the axial stress is below the yield stress, the permeabilities of raw coal and briquette show essentially the same variation trend, that is, they decline with the increase of axial pressure and reach the minimum at the yield stress point. The difference is that the permeability of raw coal diminishes more gently than that of briquette. This shows that briquette is loose and soft, with a large number of voids and a large compressible space; although raw coal has primary fractures, its initial permeability is low and cannot be enhanced obviously by compression.

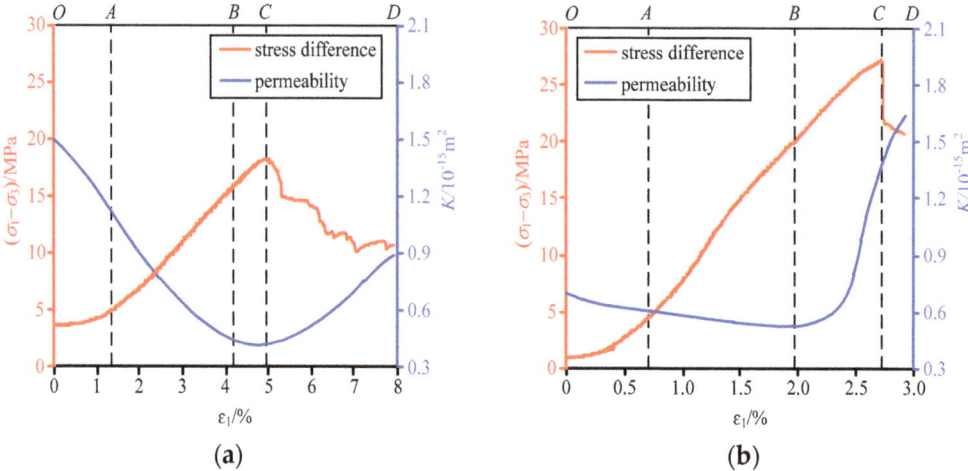

Figure 6. Axial pressure-permeability-strain curves of briquette and raw coal. (**a**) Briquette; (**b**) raw coal.

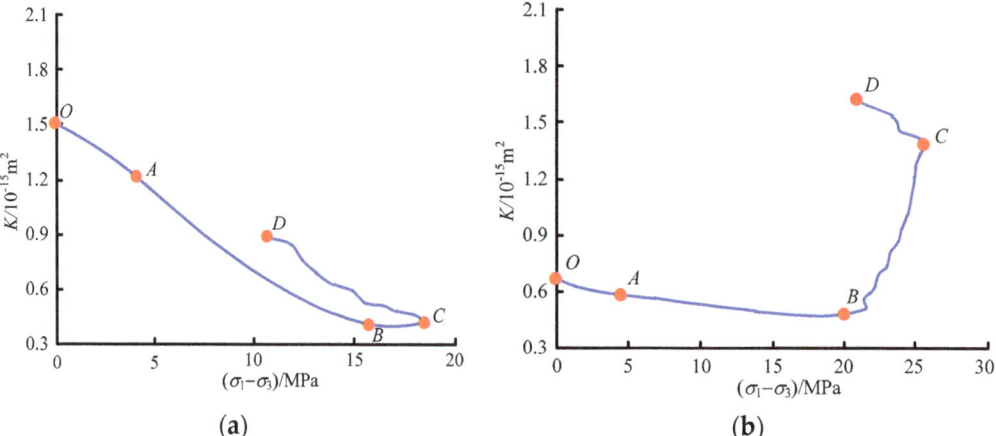

Figure 7. Axial pressure-permeability curves of briquette and raw coal. (**a**) Briquette; (**b**) raw coal.

When the axial pressure exceeds the yield stress, the permeabilities of both coal samples begin to grow, but in obviously different variation trends. In the plastic deformation stage (BC section), briquette exhibits only a smooth rise in permeability, while raw coal shows a steep rise due to the fracture seepage caused by the development of primary fractures and the generation of new fractures.

The peak stress point indicates that the sample reaches its maximum bearing capacity. At this time, the fractures accumulated before the peak reach a critical number, and the sample is on the verge of complete damage, which is a turning point for permeability. The instability damage stage (CD section) reflects the post-peak permeability variation trend. From Figure 7a, it is observed that after the peak stress, raw coal exhibits a rapid stress drop and a steep permeability rise, and this phenomenon indicates that its main fracture emerges and extends suddenly, so the damage is sudden. After raw coal is damaged, its gas pressure gradient soars, thus raising the risk of coal and gas outbursts. From Figure 7b, it can be seen that, unlike raw coal, briquette does not feature suddenness in its parameters, that is, the two samples differ essentially in terms of damage modes and seepage characteristics.

4. Discussion

4.1. Sensitivity Analysis on Gas Permeabilities of Briquette and Raw Coal

In this test, the change in axial pressure affects the permeability of the coal sample at all stages. The entire variation of axial pressure was normalized to analyze the sensitivity of coal permeability to axial pressure in each stage. The axial pressure and permeability exhibit different trends throughout the process and the absolute values of their variations in all stages were summed as 1 for normalization. The axial pressure and permeability gradient curves of the two coal samples are given in Figure 8.

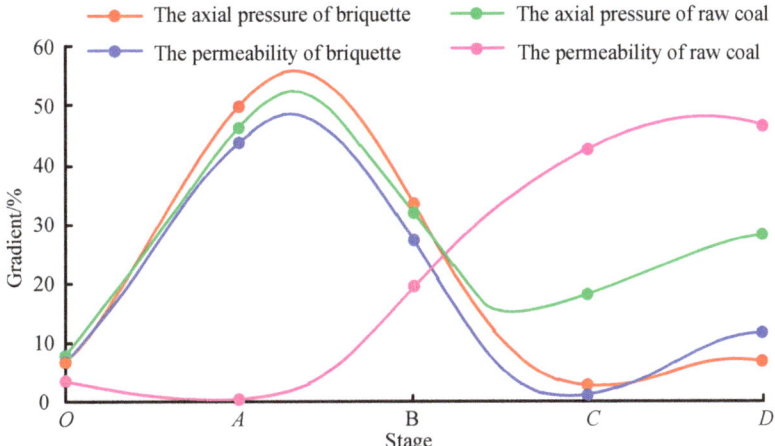

Figure 8. Axial pressure-permeability gradient curves of briquette and raw coal.

For briquette, its permeability and axial pressure gradients have the same variation trend, both being the largest (55% and 63% respectively) in the elastic deformation stage (AB section) and relatively small in the other stages. This shows that, under triaxial compression, the permeability of briquette is significantly affected by the axial stress and is the most sensitive to axial pressure in the elastic deformation stage.

In contrast, for raw coal, the variation laws of permeability and axial pressure are relatively inconsistent. Nevertheless, in the plastic deformation stage (BC section) and instability damage stage (CD section), the variation laws of the two parameters are roughly consistent, which indicates a relatively high sensitivity of its permeability to axial pressure in these two stages.

4.2. Analysis of the Suddenness of Coal and Gas Outbursts

Generally, the occurrence of coal and gas outbursts can be divided into four stages: preparation, initiation, development, and termination. However, some scholars believe that initiation and termination are only two mutation points, while only preparation and development are continuous processes [21]. The seepage tests on briquette and raw coal under triaxial compression can provide new insights into the mechanism of coal and gas outbursts.

Based on the complete stress–strain test and seepage test on raw coal, it is found that the initial compaction stage and elastic deformation stage of the stress–strain-seepage velocity variation belong to the preparation stage of coal and gas outbursts. In this stage, with the rise of axial pressure, the increasing elastic energy from the elastic deformation blocks the gas flow channel, and thus the gas internal energy mounts up obviously. These changes create conditions for the initiation of an outburst. In the plastic deformation stage, plastic deformation occurs because of stable extension of continuously distributed microfractures. At this time, the stress of the sample reaches a limited equilibrium state.

In the instability damage stage, the stress drop is a mutation point of outburst initiation. Under the action of axial pressure, the limited equilibrium state of stress is damaged into an instability state. At this time, the sudden release of elastic energy and gas internal energy accumulated in the solid medium of the coal body leads to sudden coal damage, which enables a continuous outburst until the coal body completely loses its bearing capacity. When the subsequent elastic energy and gas internal energy are below the surrounding constraining force, the outburst terminates.

Coal and gas outburst is a kind of dynamic disaster that can be intensely completed within a short time. The difficulty in predicting it lies in its suddenness. As can be concluded from the test, after reaching the peak, the stress of raw coal drops sharply within a short time, accompanied by an increase in gas pressure gradient and a surge of seepage velocity, which is quite close to the occurrence of coal and gas outbursts on site, while the changes in briquette are quite different from the real outburst. Such a result shows that the suddenness of coal and gas outburst is determined by the sudden coal damage under the joint action of in situ stress (axial pressure and confining pressure) and gas pressure, which cannot be accurately reflected by briquette.

5. Conclusions

(1) The complete stress–strain curves of briquette and raw coal have similar trends and can be divided into four developing stages: initial compaction stage, elastic deformation stage, plastic deformation stage, and instability damage stage. However, due to the difference between the two samples in structural property, their deformation and damage mechanisms are different. Briquette has a lower compressive strength and experiences a much more severe deformation than raw coal. Therefore, the applicability of briquette in place of raw coal for the simulation study on mechanical behavior needs to be investigated further;

(2) Permeability changes with the deformation and damage of coal, and the variation trend corresponds to the deformation and damage developing stage under load. As the axial pressure rises in the loading process, the permeability of raw coal declines or remains almost unchanged in the initial compaction stage and the elastic deformation stage, and then surges in the plastic deformation stage and the instability damage stage. In contrast, that of briquette plunges in the initial compaction stage and the elastic deformation stage, remains almost unchanged in the plastic deformation stage, and grows steadily in the instability damage stage, but its final permeability is lower than the initial value;

(3) Under triaxial compression, the permeability of briquette is the most sensitive to axial pressure in the elastic deformation stage, while that of raw coal is the most sensitive in the plastic deformation stage. In the test process, compared with the stable damage of briquette, the sudden damage of raw coal is closer to the suddenness of on-site coal and gas outbursts;

(4) The variation law of coal permeability is related to the law of coal deformation, and the four stages of the complete stress–strain-seepage test of raw coal can well explain the four stages of the coal and gas outburst process, i.e., preparation, initiation, development, and termination. Therefore, coal and gas outbursts can be predicted by utilizing the variation laws of coal deformation and damage and permeability in the field.

Author Contributions: Conceptualization, K.D.; Formal analysis, Z.L. and C.J.; Funding acquisition, Z.L. and J.G.; Investigation, K.D. and Z.L.; Methodology, L.W.; Software, B.R.; Validation, S.W.; Visualization, J.G.; Writing – original draft, K.D.; Writing – review & editing, Z.L. and J.G. All authors have read and agreed to the published version of the manuscript.

Funding: This study was supported by Natural Science Foundation of Jiangsu Province, China (Grant No. BK20200628).

Informed Consent Statement: Not applicable.

Data Availability Statement: The data used to support the findings of this study are included within the article.

Acknowledgments: This study was supported by Natural Science Foundation of Jiangsu Province, China (Grant No. BK20200628) and the support is gratefully acknowledged.

Conflicts of Interest: The authors declare that they have no conflict of interest regarding the publication of this work.

References

1. Meng, H.; Yang, Y.; Wu, L. Strength, Deformation, and Acoustic Emission Characteristics of Raw Coal and Briquette Coal Samples under a Triaxial Compression Experiment. *ACS Omega* **2022**, *6*, 31485–31498. [CrossRef]
2. Meng, H.; Wu, L.; Yang, Y.; Wang, F.; Peng, L.; Li, L. Evolution of Mechanical Properties and Acoustic Emission Characteristics in Uniaxial Compression: Raw Coal Simulation Using Briquette Coal Samples with Different Binders. *ACS Omega* **2021**, *6*, 5518–5531. [CrossRef] [PubMed]
3. Wang, T.; Zhao, H.; Ge, L.; Zhang, H.; Liu, R. Study on Deformation Evolution Law of Cox Under Asymmetric Loading by Digital Image Correlation. *Appl. Opt.* **2020**, *59*, 10959–10966. [CrossRef] [PubMed]
4. Buravchuk, N.I.; Guryanova, O.V. Briquetting of Fine Coal Raw Materials with Binders. *Solid Fuel Chem.* **2021**, *55*, 148–153. [CrossRef]
5. Zhao, H.; Wang, T.; Li, J.; Liu, Y.; Su, B. Regional Characteristics of Deformation and Failure of Coal and Rock under Local Loading. *J. Test. Eval.* **2021**, *49*, 3701–3715. [CrossRef]
6. Zhou, A.; Wang, K.; Hu, J.; Fan, X. Experimental Research on the Law of the Deformation and Damage Characteristics of Raw Coal/briquette Adsorption-instantaneous Pressure Relief. *Fuel* **2021**, *308*, 122062. [CrossRef]
7. Braga, L.T.P.; Kudasik, M. Permeability Measurements of Raw and Briquette Coal of Various Porosities at Different Temperatures. *Mater. Res. Express* **2019**, *6*, 105609. [CrossRef]
8. Ge, L.; Yi, F.; Du, C.; Zhou, J.; Cui, Z.; Wang, T. Deformation Localization and Damage Constitutive Model of Raw Coal and Briquette Coal under Uniaxial Compression. *Geofluids* **2022**, *2022*, 4922287. [CrossRef]
9. Pang, L.; Yang, Y.; Wu, L.; Wang, F.; Meng, H. Effect of Particle Sizes on the Physical and Mechanical Properties of Briquettes. *Energies* **2019**, *12*, 3618. [CrossRef]
10. Gan, Q.; Xu, J.; Peng, S.; Yan, F.; Wang, R.; Cai, G. Effect of Molecular Carbon Structures on the Evolution of the Pores and Strength of Lignite Briquette Coal with Different Heating Rates. *Fuel* **2021**, *307*, 121917. [CrossRef]
11. Sun, X.; Chen, G.; Li, J.; Xu, X.; Fu, S.; Xie, J.; Liang, L. Propagation Characteristics of Ultrasonic P-wave Velocity in Artificial Jointed Coal Briquettes. *J. Geophys. Eng.* **2020**, *17*, 827–837. [CrossRef]
12. Zhang, H.; Zhao, H.; Li, W.; Yang, X.; Wang, T. Influence of Local Frequent Dynamic Disturbance on Micro-structure Evolution of Coal-Rock and Localization Effect. *Nat. Resour. Res.* **2020**, *29*, 3917–3942. [CrossRef]
13. Kozieł, K.; Nowakowski, A.; Sitek, L.; Skoczylas, N. Rock and Gas Outbursts in Copper Mines: Use of Brazilian Tests to Evaluate the Work of Disintegration of Rock Resulting from Stresses Produced by Gas Present in Its Porous Structure. *Rock Mech. Rock Eng.* **2022**. [CrossRef]
14. Li, M.; Liang, W.; Yue, G.; Zheng, X.; Liu, H. Dynamic Mechanical Properties of Structural Anisotropic Coal Under Low and Medium Strain Rates. *PLoS ONE* **2020**, *15*, e0236802. [CrossRef] [PubMed]
15. Ma, H.; Wang, L.; Niu, X.G.; Yao, F.; Zhang, K.; Chang, J.; Li, Y.; Zhang, X.; Li, C.; Hu, Z. Mechanical Characteristics of Coal and Rock in Mining under Thermal-Hydraulic-Mechanical Coupling and Dynamic Disaster Control. *Math. Probl. Eng.* **2021**, *2021*, 9991425. [CrossRef]
16. Zhao, Y.; Sun, X.; Meng, W. Research on the Axial Velocity of the Raw Coal Particles in Vertical Screw Conveyor by Using the Discrete Element Method. *J. Mech. Sci. Technol.* **2021**, *35*, 2551–2560.
17. Wang, D.; Lv, R.; Wei, J.; Zhang, P.; Yu, C.; Yao, B. An Experimental Study of the Anisotropic Permeability Rule of Coal Containing Gas. *J. Nat. Gas Sci. Eng.* **2018**, *53*, 67–73. [CrossRef]
18. Gan, Q.; Xu, J.; Peng, S.; Yan, F.; Wang, R.; Cai, G. Effects of Heating Temperature on Pore Structure Evolution of Briquette Coals. *Fuel* **2021**, *296*, 120651. [CrossRef]
19. Liu, N.; Li, C.; Feng, R.; Xia, X.; Gao, X. Experimental Study on the Influence of Moisture Content on the Mechanical Properties of Coal. *Geofluids* **2021**, *2021*, 6838092. [CrossRef]
20. Hou, W.; Wang, H.; Wang, W.; Liu, Z.; Li, Q. A Uniaxial Compression Experiment with CO_2-Bearing Coal Using a Visualized and Constant-Volume Gas-Solid Coupling Test System. *JOVE-J. Vis. Exp.* **2019**, *148*, e59405.
21. Meng, H.; Wu, L.; Yang, Y.; Wang, F.; Peng, L.; Li, L. Experimental Study on Briquette Coal Sample Mechanics and Acoustic Emission Characteristics under Different Binder Ratios. *ACS Omega* **2021**, *6*, 8919–8932. [CrossRef] [PubMed]
22. Demirel, C.; Gürdil, G.A.K.; Kabutey, A.; Herak, D. Effects of Forces, Particle Sizes, and Moisture Contents on Mechanical Behaviour of Densified Briquettes from Ground Sunflower Stalks and Hazelnut Husks. *Energies* **2020**, *13*, 2542. [CrossRef]
23. Ma, H.; Yao, F.; Niu, X.; Guo, J.; Li, Y.; Yin, Z.; Li, C. Experimental Study on Seepage Characteristics of Fractured Rock Mass under Different Stress Conditions. *Geofluids* **2021**, *2021*, 6381549.

24. Liu, W.; Chen, J.; Luo, Y.; Chen, L.; Shi, Z.; Wu, Y. Deformation Behaviors and Mechanical Mechanisms of Double Primary Linings for Large-Span Tunnels in Squeezing Rock: A Case Study. *Rock Mech. Rock Eng.* **2021**, *54*, 2291–2310. [CrossRef]
25. Xie, B.; Yan, Z.; Du, Y.; Zhao, Z.; Zhang, X. Determination of Holmquist-Johnson-Cook Constitutive Parameters of Coal: Laboratory Study and Numerical Simulation. *Processes* **2019**, *7*, 386. [CrossRef]
26. Gan, Q.; Xu, J.; Peng, S.; Yan, F.; Wang, R.; Cai, G. Effect of Heating on the Molecular Carbon Structure and the Evolution of Mechanical Properties of Briquette Coal. *Energy* **2021**, *237*, 121548. [CrossRef]
27. Lv, H.; Tang, Y.; Zhang, L.; Cheng, Z.; Zhang, Y. Analysis for Mechanical Characteristics and Failure Models of Coal Specimens with Non-penetrating Single Crack. *Geomech. Eng.* **2019**, *17*, 355–365.

Article

Mechanical Properties and Failure Mechanism of Anchored Bedding Rock Material under Impact Loading

Yunhao Wu [1], Xuesheng Liu [1,2,*], Yunliang Tan [1,2,*], Qing Ma [1], Deyuan Fan [1], Mingjie Yang [3], Xin Wang [1] and Guoqing Li [1]

1. College of Energy and Mining Engineering, Shandong University of Science and Technology, Qingdao 266590, China
2. State Key Laboratory of Mining Disaster Prevention and Control Co-Founded by Shandong Province and the Ministry of Science and Technology, Qingdao 266590, China
3. School of Resources and Environmental Engineering, East China University of Technology, Shanghai 200237, China
* Correspondence: xuesheng1134@163.com (X.L.); yunliangtan@163.com (Y.T.)

Abstract: In view of the problem that anchored bedding rock material is prone to instability and failure under impact loading in the process of deep coal mining, and taking the lower roadway of a deep 2424 coal working face in the Suncun coal mine as the engineering background, a mechanical model of anchored bedding rock material was established, and the instability criterion of compression and shear failure of anchored bedding rock material was obtained. Then, the separated Hopkinson pressure bar was used to carry out an impact-loading test on the anchored bedding rock material, and the dynamic mechanical properties of the rock with different anchoring modes and bolt bedding angles were studied; the evolution law of the strain field of the anchored bedding rock material was also obtained. The results show the following: (1) The bolt support could effectively improve the dynamic load strength and dynamic elastic modulus of the rock material with anchorage bedding, the degree of improvement increased with the increase in the angle of the bolt bedding, and the full anchorage effect was much higher than the end anchorage effect was. (2) The bolt bedding angle and anchorage mode greatly influenced crack development and displacement characteristics. After an impact, the bedding rock material had obvious shear displacement along the bedding direction, and obvious macroscopic cracks were produced in the bedding plane. The research results offer theoretical guidance to and have reference significance for deep roadway anchorage support engineering.

Keywords: anchored bedding rock material; impact loading; instability mechanism; complex stress environment; shear failure

Citation: Wu, Y.; Liu, X.; Tan, Y.; Ma, Q.; Fan, D.; Yang, M.; Wang, X.; Li, G. Mechanical Properties and Failure Mechanism of Anchored Bedding Rock Material under Impact Loading. *Materials* 2022, *15*, 6560. https://doi.org/10.3390/ma15196560

Academic Editor: Tomasz Sadowski

Received: 29 August 2022
Accepted: 19 September 2022
Published: 21 September 2022

Publisher's Note: MDPI stays neutral with regard to jurisdictional claims in published maps and institutional affiliations.

Copyright: © 2022 by the authors. Licensee MDPI, Basel, Switzerland. This article is an open access article distributed under the terms and conditions of the Creative Commons Attribution (CC BY) license (https://creativecommons.org/licenses/by/4.0/).

1. Introduction

Deep coal resources are an important guarantee as China's main energy sources [1–3]. Statistics show that coal resources with a depth of more than 600 m account for about 73% of the total coal, while coal resources with a depth of more than 1000 m account for about 53%. Deep mining is becoming the new normal in coal-resource development [4,5].

In the process of deep coal mining, the surrounding rocks of deep roadways bear large initial ground stress, and the superposition of supporting pressure leads to stress concentration, which places the surrounding rock in a state of high strength compression for a long time [6–8]. In addition, the mining process is often accompanied by the periodic fracturing and the lateral collapse of the overlying strata, fault slips, blasting, and other activities [9,10]. The impact loading generated by these activities causes strong disturbance to the surrounding rock of the roadway [11–14]. When the static load and impact loading of the surrounding rock in the mining space exceed the critical load of the impact failure of coal and rock material, the surrounding rock becomes unstable, resulting in roof collapses, hydraulic support compression frames, partial sidewalls, and other rock burst accidents [15–18]. The complex geological environment of the surrounding rock is also an important factor that threatens the stability of the surrounding rock of roadways [19,20]. As layered composite rock material is widely distributed in underground roadways, and the bedding plane of the rock stratum is its natural weak surface, the existence of bedding weakens the antideformation ability of the rock material and reduces the stability of the surrounding rock [21–25]. Therefore, the layered surrounding rock of deep coal mining roadways is prone to instability and failure under impact loading, and supporting it is very difficult [26,27], which has become one of the main bottlenecks restricting the safety of deep coal mining.

Aiming at the failure problem of anchorage structures under impact loading, Wu Yongzheng [28,29] analyzed the dynamic load response characteristics of surrounding anchorage rock under impact loading, and studied the dynamic response law of anchorage rock under lateral impact load, and the influence of impact energy and bolt mechanical properties on rock material mechanical behavior. Mu Zonglong [30] analyzed the failure conditions of roadway surrounding rock under static load, and dynamic and static loads, and experimentally studied the failure conditions of dynamic load on surrounding roadway rock. Wang Zhengyi [31] analyzed the dynamic effect of longitudinal waves, established the criterion of rock burst of anchoring roadway support structures, and formulated the corresponding preventive measures. Wang Aiwen et al. [32] found that there is an obvious time difference effect between the vibration of the bolt and the surrounding rock under impact loads, which leads to the asynchronous vibration of the bolt and the surrounding rock, resulting in the dynamic shear of the anchorage agent. Qiu Pengqi [33] believed that improving the antisliding characteristics and coordinated deformation ability of a rock, the anchoring agent and bolt can effectively prolong the anti-impact aging time of anchoring rock and reduce the impact of dynamic load on the supporting structure of surrounding anchoring rock. Skrzypkowski K [34,35] believed that the places of particular exposure to shear stresses are faults and layered roof layers; between them, there are surfaces of reduced cohesion. Jiao Jiankang [36] analyzed the dynamic load response characteristics and impact failure evolution process of roadway anchorage bearing structures under dynamic load disturbances, and put forward the impact-failure criterion, and the criterion of anchorage bearing structures under dynamic load disturbance and the control technology of surrounding roadway anchorage rock under dynamic load disturbance.

These studies revealed the failure mechanism of anchorage structures under impact loading to a certain extent, and put forward the corresponding control technology of anchorage rocks of rock burst roadways, but failed to fully reveal the overall instability mechanism of anchorage bodies. There is little research on dynamic instability of anchorage bodies under complex stratum conditions, which fails to solve the problem of the instability and failure of anchored bedding rock material that widely exists in engineering geology.

Therefore, the lower roadway of 2424 coal working face in Suncun Coal Mine was taken as the engineering background, the mechanical model of the anchored bedding rock material is established, the compression and shear instability criterion of the anchored bedding rock material is put forward, and the split Hopkinson pressure bar test device was used to carry out an impact loading test on the anchored bedding rock material and reveal the influence law and instability mechanism of the angle between different bolts and bedding and the anchoring mode on the mechanical properties of the bedding rock material.

2. Stress Analysis and Instability Mechanism of Anchored Bedding Rock Material

2.1. Engineering Background

Taking the lower roadway of 2424 coal working face in the Suncun coal mine as an example, the coal seam of the −800 m level 2424 coal working face was four coals, and the rock column diagram is shown in Figure 1.

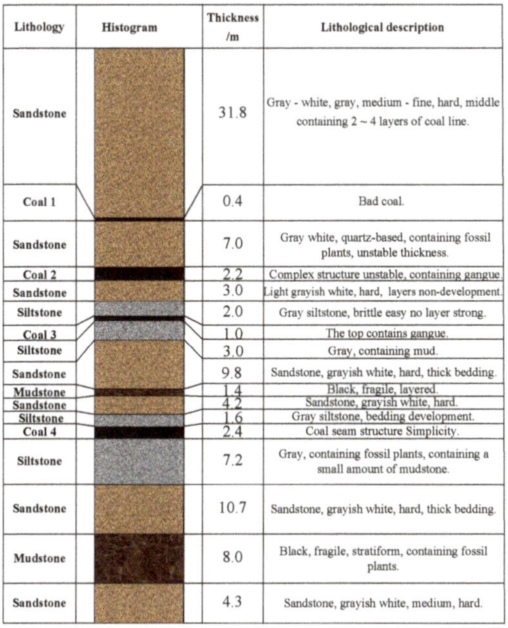

Lithology
Histogram
Thickness /m
Lithological description
Sandstone
31.8
Gray - white, gray, medium - fine, hard, middle containing 2 ~ 4 layers of coal line.
Coal 1
0.4
Bad coal.
Sandstone
7.0
Gray white, quartz-based, containing fossil plants, unstable thickness.
Coal 2
2.2
Complex structure unstable, containing gangue.
Sandstone
3.0
Light grayish white, hard, layers non-development.
Siltstone
2.0
Gray siltstone, brittle easy no layer strong.
Coal 3
1.0
The top contains gangue.
Siltstone
3.0
Gray, containing mud.
Sandstone
9.8
Sandstone, grayish white, hard, thick bedding.
Mudstone
1.4
Black, fragile, layered.
Sandstone
4.2
Sandstone, grayish white, hard.
Siltstone
1.6
Gray siltstone, bedding development.
Coal 4
2.4
Coal seam structure Simplicity.
Siltstone
7.2
Gray, containing fossil plants, containing a small amount of mudstone.
Sandstone
10.7
Sandstone, grayish white, hard, thick bedding.
Mudstone
8.0
Black, fragile, stratiform, containing fossil plants.
Sandstone
4.3
Sandstone, grayish white, medium, hard.

Figure 1. 2424 Coal working face rock histogram.

The average thickness of the coal seam was 2.4 m, and the dip angle was 26.5°. The direct top was siltstone and sandstone, thickness was about 5.8 m, bedding development was relatively hard, and compressive strength was 21.2 MPa. The basic roof was mainly sandstone with about 11.2 m thick and hard bedding. The buried depth of the lower roadway in the 2424 coal working face was 1254 m, and the total length was 1132 m. The section was trapezoidal, and the net width was 4 m. The height of the left side was 3 m, and the height of the right side was 4 m. The roof was an arc section, as shown in Figure 2.

20°
210
950
950
950
950
20°
250
30°
500
1000
1000
1000
500
30°
Exposed Bolt
100 mm
500
30°
1000
Bolt
(2.2 m)
1000
500
30°
4000
4200

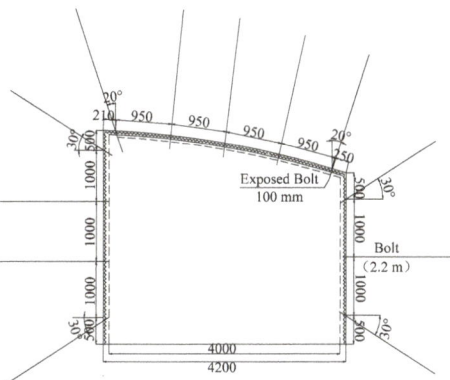

Figure 2. Schematic diagram of roadway support.

The mining process of the coal working face uses a Φ22 × 2400 mm HRB 600 resin anchored anchor bolt + diamond mesh + W steel belt + tray support. The anchor bolt anchorage length was 1.2 m. The three middle rows of anchor bolts were installed on the vertical roof (at an angle of 14° with the vertical line), and the outermost anchor bolts were installed at an angle of 20° with the vertical line.

This was adopted by two gangs, namely, Φ22-2200 mm MSGLD-600(x) equal strength threaded steel bolt + diamond mesh + W steel strip + high-strength tray support. The lowest row of bolts on the two sides of the roadway was 500 mm from the roadway floor and inclined downward at an angle of 30° with the horizontal line. The lowest row of bolts on the two sides of the roadway was 500 mm from the roadway roof and inclined upward at an angle of 30° with the horizontal line.

A half-length anchor was adopted for the two sides and the roof. The anchoring length was 1.4 m. Fast-setting resin was selected as the anchoring agent, composed of two kinds of materials, namely, resin cement (cement containing resin, dolomite powder, silica, and accelerator) and a curing agent (containing benzoyl peroxide, light calcium, and water).

2.2. Stress Analysis of the Anchored Rock Material

The dynamic action in the mining process mainly comes from mining activity and the stress response of surrounding rock to the mining activity. The mining process of a coal seam is always accompanied by the movement of the overlying strata. The high stress waves generated by the roof breaking in the movement of the overlying strata form mutual interference and superposition in the coal and rock, and have a strong disturbance effect on the surrounding rock of a roadway.

Taking the lower drift of coal working face 2424 of the Suncun coal mine as an example, when the layered composite surrounding rock roof and two sides of the deep mining roadway were supported by bolts, the anchored surrounding rock was subject to high ground stress and impact loading, which led to the surrounding anchored layered rock very easily losing stability and breaking the ring, as shown in Figure 3a.

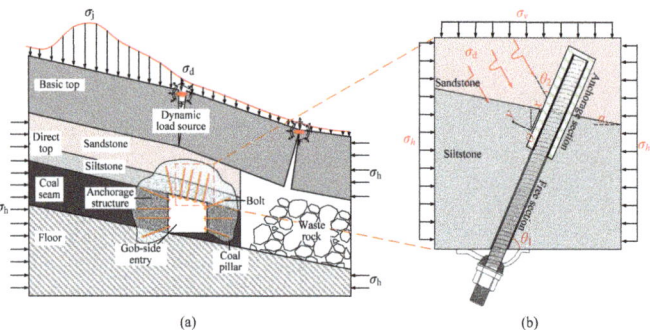

Figure 3. Stress environment of the surrounding rock and mechanical model of anchored bedding rock material. (**a**) Anchorage structure and stress environment of layered composite surrounding rock in Inclined Mining Roadway; (**b**) Mechanical model of anchorage bedding rock mass.

In order to study the stress condition and instability mechanism of the anchored surrounding rock at the bedding development under impact loading, the mechanical model of the anchor under impact loading was established with the roof anchor as the research object, as shown in Figure 3b. The inclination angle of the anchor rod was θ_1, and bedding dip angle was α. The anchor body was subjected to vertical stress of surrounding rock σ_v, and horizontal stress σ_h was also affected by the dynamic load disturbance caused by roof fracture instability. The dynamic load was σ_d, which included angle with anchor rod θ_2.

A dynamic load generated by roof fracture instability mainly propagates in the form of stress waves in the rock structure. As the surrounding rock in the actual project is heterogeneous and anisotropic material, it is difficult to accurately analyze it. Therefore, the following were assumed: ① The surrounding rock of the roadway is an ideal homogeneous elastic–plastic body; ② at a certain distance from the source, the stress waves propagating in the rock can be regarded as plane waves [37]. The shear wave effect of the stress waves was not considered, and the dynamic load (σ_p) generated by longitudinal waves (P) P can be expressed as follows:

$$\sigma_p = \rho C_p v_p \tag{1}$$

where C_p is the propagation velocity of P waves, m/s; v_p is the peak vibration velocity of the particle at the boundary of the dynamic load source, m/s; and ρ is the density of rock mass, kg/m^3.

The research shows that the propagation of stress waves in a coal rock medium presents a power function attenuation law [38], and the attenuation law of particle vibration velocity can be expressed as follows:

$$v = v_p L^{-\eta} \tag{2}$$

where v is the peak vibration velocity of the particle at the propagation of the stress wave, m/s; L is the distance from the boundary of a moving source, m; η is the peak velocity attenuation coefficient related to the propagation medium; attenuation coefficients η of the intact rock mass, intact coal body, and fractured coal body are [38] 0.5755, 0.867, 1.0655, respectively.

Combining Formulas (1) and (2), the dynamic load caused by stress-wave propagation to the surface of the anchorage structure is:

$$\sigma_d = \rho C_p v_p L^{-\eta} \tag{3}$$

At this time, stress at any point O on the anchorage body can be characterized along the inclined direction (x axis) and the vertical direction (y axis) of the bolt as follows:

$$\begin{cases} \sigma_1 = \sigma_x = \sigma_v \cdot \sin\theta_1 + \rho C_p v_p L^{-\eta} \cdot \cos\theta_2 \\ \sigma_3 = \sigma_y = \sigma_v \cdot \cos\theta_1 + \rho C_p v_p L^{-\eta} \cdot \sin\theta_2 \end{cases} \tag{4}$$

According to the Mohr–Coulomb criterion, because of $\sigma_3 > 0$, when $\sigma_1 = \sigma_v \cdot \sin\theta_1 + \rho C_p v_p L^{-\eta} \cdot \cos\theta_2 \geq \sigma_c$ (σ_c is uniaxial compressive strength of anchorage), compression fracture occurs in the surrounding rock of an anchor solid.

2.3. Shear Failure Mechanism of Anchorage Rock Material

Since the dynamic load propagation process is transferred from sandstone to siltstone, and the propagation velocity of stress waves in different strata is different, shear dislocation occurs on the joint surface, and shear failure occurs on the anchorage body. In order to analyze this problem, the instantaneous shear model of anchor solid bedding surface under dynamic load was established, as shown in Figure 4. Assuming that the transverse displacement of the structural plane was U_1, the longitudinal displacement was U_2, and $U_2 = U_1 \tan\psi$, ψ is the shear angle of the structural plane.

Then, the lateral deformation of the bolt is u_1, and the axial deformation is u_2. According to the geometric relationship:

$$\begin{aligned} u_1 &= U_1 \sin(\alpha + \theta_1)/\cos\alpha \\ u_2 &= U_2/\sin(\alpha + \theta_1) \end{aligned} \tag{5}$$

The shear strength of the anchored bedding rock material is generally composed of four parts: the shear strength of the bedding plane itself τ_j, the shear strength contributed by the 'pin' action of the bolt body τ_{bd}, and the shear strength contributed by the normal and tangential components of the axial load of the bolt along the joint surface τ_{bd}, τ_{bs} [39]. Therefore, shear strength τ_{bd} of the anchored bedding rock material can be expressed as follows:

$$\tau_{bj} = \tau_j + \tau_{bd} + \tau_{bt} + \tau_{bs} \tag{6}$$

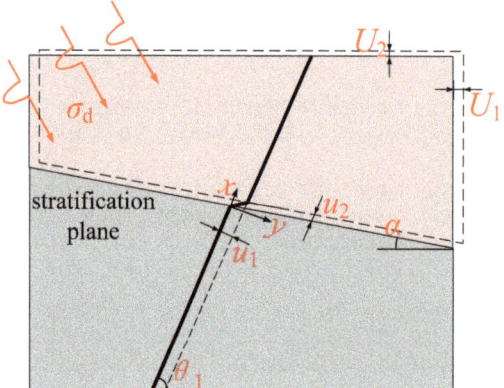

Figure 4. Instantaneous shear model of anchor solid bedding plane.

According to the Coulomb–Mohr criterion, the shear strength τ_j of the bedding plane itself is:

$$\tau_j = \sigma_j \tan(\psi + \varphi') + c_j \tag{7}$$

where σ_j is the normal stress of the structural plane, c_j is the structural surface adhesion, and φ' is the structural surface friction angle.

τ_{bd}, τ_{bd}, and τ_{bd} can be calculated with the following formula:

$$\begin{cases} \tau_{bd} = \tau_b \cdot \eta [\sin(\alpha + \theta_1)/\cos \alpha] \\ \tau_{bt} = \sigma_b \cdot \sin(\alpha + \theta_1) \cdot \eta \cdot \tan \varphi' \\ \tau_{bs} = \sigma_b \cdot \cos(\alpha + \theta_1) \cdot \eta \end{cases} \tag{8}$$

where σ_b is axial stress of the bolt, τ_b is the average shear stress on the cross-section of the bolt, and η is the ratio of the cross-sectional area of the bolt to the area of the bedding plane containing a single bolt.

In order to obtain the axial stress of the bolt, a bolt element was taken for stress analysis, as shown in Figure 5.

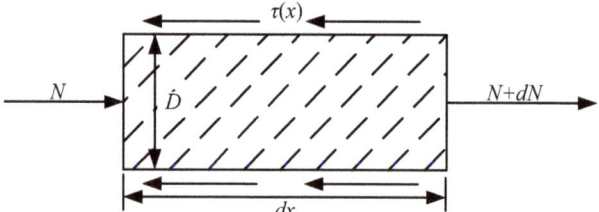

Figure 5. Mechanical model of a bolt element.

According to the unit force balance, the differential equation is established:

$$N(x) - \pi D \tau(x) dx = N(x) + dN(x) \tag{9}$$

where $N(x)$ is the axial force of the bolt at position x from the bedding plane, D is the bolt diameter, and $\tau(x)$ is the interfacial shear stress at x position from the bedding plane.

Through the linear relationship between load and displacement, and the stress equilibrium state of bolt, the equation of interfacial shear stress with displacement is obtained:

$$\tau(x) = Ku(x) \tag{10}$$

where $u(x)$ is the axial normal strain of the bolt; and K is the comprehensive shear stiffness, $K = \frac{K_1 K_2}{K_1 + K_2}$ (K_1 and K_2 are shear stiffness of slurry and shear stiffness of rock material, respectively).

Combining Formulas (9) and (10), we can obtain the following:

$$\frac{dN(x)}{dx} = \pi DKu(x) \tag{11}$$

Then, the axial strain $\varepsilon(x)$ of the bolt at a distance from the bedding plane x can be expressed as follows:

$$\varepsilon(x) = \frac{du(x)}{dx} = \frac{\sigma_x}{E} = \frac{N(x)}{EA} \tag{12}$$

where E is the elastic modulus of the anchor, and A is the cross-sectional area of the anchor.

From the above analysis:

$$\frac{d^2 N(x)}{dx^2} - \frac{K \cdot N(x)}{EA} = 0 \tag{13}$$

When the bedding plane is damaged, it is assumed that the axial force of the bolt at the bedding plane is pretightening force N_0, and the axial force at the distance from the joint plane u_2 (axial deformation of the bolt) is 0. Therefore, the boundary conditions are $N(x = 0) = N_0$, $N(x = l) = 0$, and the solution is:

$$N(x) = N_0 \frac{\sinh\left[\sqrt{\frac{K}{EA}}(u_2 - x)\right]}{\sin\left(u_2 \sqrt{\frac{K}{EA}}\right)} \tag{14}$$

Then the axial stress σ_b of bolt is:

$$\sigma_b = \frac{N_0 \sinh\left[\sqrt{\frac{4K}{E\pi D^2}}(u_2 - x)\right]}{\frac{\pi D^2}{4} \sin\left(u_2 \sqrt{\frac{4K}{E\pi D^2}}\right)} \tag{15}$$

Before the failure of the bedding anchor solid, the anchor and rock near the bedding plane are in an elastic state. According to the principle of elastic mechanics, the lateral deformation of the anchor satisfies:

$$\frac{d^4 u_1}{dy^4} + \frac{Dk}{EI} u_1 = 0 \tag{16}$$

where $k = 300\, \sigma_c/D$, the rock reaction coefficient; and EI is the anchor bending stiffness.

Analysis shows that, when $y = 0$, the bending moment is 0; when $y \to \infty$, the bending moment is 0, and the shear force is 0:

$$\begin{cases} M_{y=0} = -EI \frac{d^2 u_1}{dy^2} = 0 \\ M_{y \to \infty} = -EI \frac{d^2 u_1}{dy^2} = 0 \\ \tau_{y \to \infty} = -EI \frac{d^3 u_1}{dy^3} = 0 \end{cases} \tag{17}$$

From these boundary conditions, the relationship between the lateral displacement of the bolt and shear stress τ_b in the shear-slip process of the bedding plane can be obtained:

$$\tau_b = \frac{8 EI u_1(y)}{pD^2 \left(\frac{4EI}{kD}\right)^{\frac{3}{4}}} \tag{18}$$

The reaction force provided by the rock in the elastic state is $p = ku_1$. When the compressive stress is too large, the rock produces plastic failure and reaches the ultimate reaction, $p = n\sigma_c$. At this time, the lateral elastic displacement limit of the bolt end is $u_1 = nD/300$. n is a coefficient dependent on the internal friction angle of rock, and its value is 2–5 according to the degree of rock hardness.

Comprehensive Formulas (6)–(8), (15) and (18) can be substituted into numerical values to obtain the shear strength at any point (x, y) of the rock material with anchor bedding:

$$\tau_{bj}(x,y) = \sigma_j \tan(\psi + \varphi') + \frac{N_0 \sinh\left[\sqrt{\frac{4K}{E\pi D^2}}(u_2-x)\right]}{\frac{\pi D^2}{4}\sin\left(u_2\sqrt{\frac{4K}{E\pi D^2}}\right)} \cdot \eta[\sin(\alpha + \theta_1)/\cos\alpha] + \frac{8EIu_1(y)\eta}{pD^2\left(\frac{4EI}{kD}\right)^{\frac{3}{4}}} \cdot [\sin(\alpha + \theta_1) \cdot \tan\varphi' + \cos(\alpha + \theta_1)] \quad (19)$$

When $\sigma_d \sin(\theta_1 + \alpha) + \sigma_v \sin\alpha = \rho C_p v_p L^{-\eta} \sin(\theta_1 + \alpha) + \sigma_v \sin\alpha \geq \tau_{bjmax}$, the bedding plane shear failure occurs in the anchored bedding rock material under dynamic load disturbance, resulting in anchoring failure.

On the basis of the stress analysis of anchored bedding rock material, the instability criterion of anchored bedding rock material in a complex stress environment can be obtained:

$$\begin{cases} \sigma_v \cdot \sin\theta_1 + \rho C_p v_p L^{-\eta} \cdot \cos\theta_2 \geq \sigma_c, \text{ Compression Failure} \\ \rho C_p v_p L^{-\eta} \sin(\theta_1 + \alpha) + \sigma_v \sin\alpha \geq \tau_{bjmax}, \text{ Shear Failure} \end{cases} \quad (20)$$

where σ_c is the uniaxial compressive strength of rock mass.

According to this formula, the parameters of the bedding anchor solid can be obtained with field measurements, and physical and mechanical tests, and the instability of the anchored bedding rock material can be judged:

① When $\sigma_v \sin\theta_1 + \rho C_p v_p L^{-\eta} \cdot \cos\theta_2 \geq \sigma_c$, $\rho C_p v_p L^{-\eta} \sin(\theta_1 + \alpha) + \sigma_v \sin\alpha \geq \tau_{bjmax}$, shear compression failure occurred in the anchored bedding rock material.

② When $\sigma_v \sin\theta_1 + \rho C_p v_p L^{-\eta} \cdot \cos\theta_2 < \sigma_c$, $\rho C_p v_p L^{-\eta} \sin(\theta_1 + \alpha) + \sigma_v \sin\alpha \geq \tau_{bjmax}$, shear failure occurred in the anchored bedding rock material.

③ When $\sigma_v \sin\theta_1 + \rho C_p v_p L^{-\eta} \cdot \cos\theta_2 \geq \sigma_c$, $\rho C_p v_p L^{-\eta} \sin(\theta_1 + \alpha) + \sigma_v \sin\alpha < \tau_{bjmax}$, compression failure of the anchored bedding rock material.

④ When $\sigma_v \sin\theta_1 + \rho C_p v_p L^{-\eta} \cdot \cos\theta_2 < \sigma_c$, $\rho C_p v_p L^{-\eta} \sin(\theta_1 + \alpha) + \sigma_v \sin\alpha < \tau_{bjmax}$, the anchored bedding rock material is not damaged.

3. Anchored Bedding Rock Material Impact Loading Test

3.1. Sample Preparation

According to the instability criterion of anchored bedding rock material, under the condition of a constant surrounding rock stress environment, the angle between bolt and bedding, and the physical and mechanical properties of anchorage body are important factors affecting the instability of anchored bedding rock material. In order to further study the mechanical properties and instability mechanism of anchored bedding rock material, the impact loading test of anchored bedding rock material was carried out.

According to the stress conditions of the anchor solid, as shown in Figure 3b, assuming that the dynamic load of the rock fracture acted vertically on the anchor, the anchorage bedding rock specimen was produced as shown in Figure 6, where the angle between the anchor and the bedding plane was θ, $\theta = \alpha + \theta_1$.

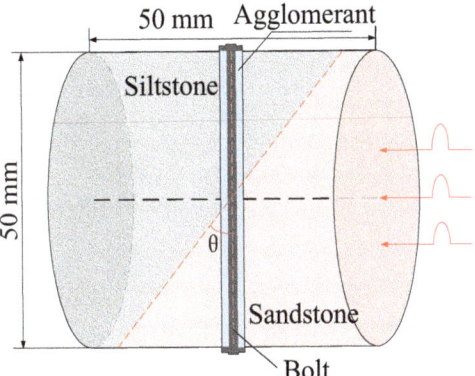

Figure 6. Schematic diagram of inclined bedding surrounding rock anchorage specimen.

Taking the surrounding rock of the lower roadway in the 2424 working face of the Suncun coal mine as the research object, the siltstone and sandstone of the roof were selected. After drilling, cutting, and grinding, columnar specimens with different inclination angles were produced. The siltstone and sandstone specimens were bonded with epoxy resin, and specimens of $\Phi 50 \times h\ 50$ mm were produced. In order to study the influence of the bolt bedding angle and anchorage method on the mechanical properties of anchorage rock material, three groups of specimens were designed. As shown in Table 1, Group A was the control group and the full-length anchorage. The bolt bedding angle was 15°. Test Group 1 was anchorage specimens with different bolt bedding angles, and the bolt bedding angles were 30° and 45°, which were all full-length anchorage specimens. Test Group 2 was the anchorage specimens with different anchorage modes, and the angle between the bolt beddings was 15°. The anchorage modes were nonanchor and end anchor.

Table 1. Anchored bedding rock material impact loading test design table.

Group	Schematic Diagram 1	Schematic Diagram 2	Schematic Diagram 3
Test 1: bedding angle.	A: 15°	B: 30°	C: 45°
Test 2: anchored form.	A: full-length anchor.	D: end anchor.	E: nonanchor.

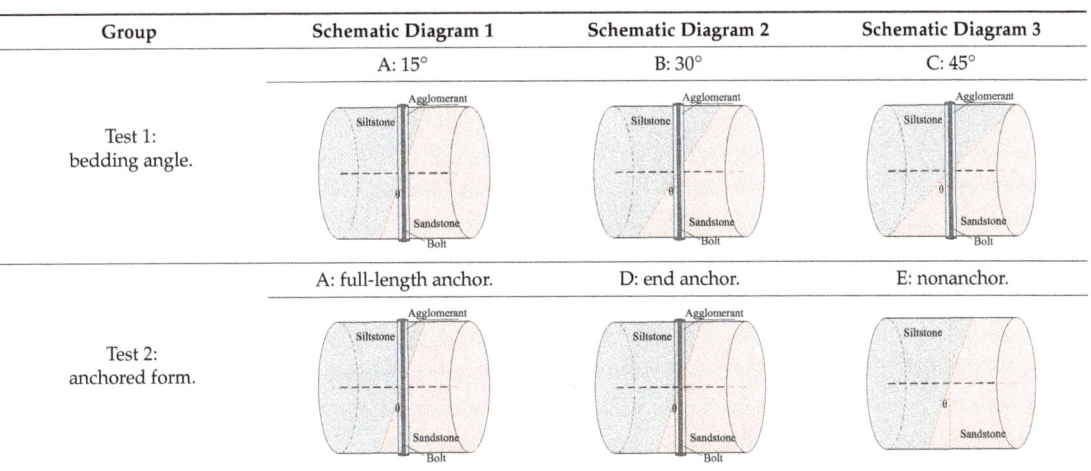

The anchoring process of the full-anchor specimen was as follows. First, a 5 mm drilling rig was used to drill axially into the middle of the side of the composite specimen. A 4 mm high-strength rebar was used as the indoor experimental anchor, and a flat cushion was used to simulate the tray. A mechanical torque wrench was used to impose preload torque on the anchor. Then, an epoxy resin adhesive was selected as the anchoring agent and was injected into the pores with a needle pipe. After a period of time, it solidified. The end-anchored specimen only used high-strength rebar as the anchor rod, and the flat pad simulated the tray. After applying the preload torque, no anchoring agent was added. The size parameters and preload torque are shown in Table 2.

Table 2. Test specimen parameters.

Number	Anchored Form	Bedding Angle/°	Tightening Torque/N·m	Diameter/mm	Height/mm	Dynamic Speed/m/s
A-1	Full-length anchor	15	19.9	49.9	49.9	9.24
A-2			20.1	50.0	50.0	9.21
B-1	Full-length anchor	30	20.0	50.0	49.9	9.27
B-2			20.0	50.0	50.0	9.22
C-1	Full-length anchor	45	19.9	50.0	50.0	9.23
C-2			20.0	50.1	50.0	9.20
D-1	End anchor	15	20.1	49.9	50.0	9.19
D-2			20.0	50.0	50.0	9.22
E-1	Nonanchor	15	0	50.0	49.9	9.25
E-2			0	49.9	50.0	9.28

After the preparation of each type of specimen had been completed, the strain gauge (SG1) was pasted perpendicular to the loading direction at the center of the specimen disk surface to monitor the strain of the anchor solid matrix during the test. The strain gauge (SG2) was pasted onto the middle surface of the bolt to monitor the tensile strain of the bolt during the test.

3.2. Testing Equipment

The test equipment comprised the Φ50 mm split Hopkinson pressure bar (SHPB) system, as shown in Figure 7. The incident rod, transmission rod, and absorption rod are produced with a Φ50 mm steel rod with an elastic modulus of 206 GPa, and the compression rod material was 48CrMoA. The length of the man-shot and transmission rods was 1500 mm, and the length of the absorption rod was 1000 mm. In order to produce sine waves with a slow loading section, we adopted a spindle-shaped bullet of the same material for the punch with a maximal diameter of 50 mm and length of 365 mm, and the strain signal transmission frequency and maximal sampling frequency of the dynamic strain gauge were 1 MHz.

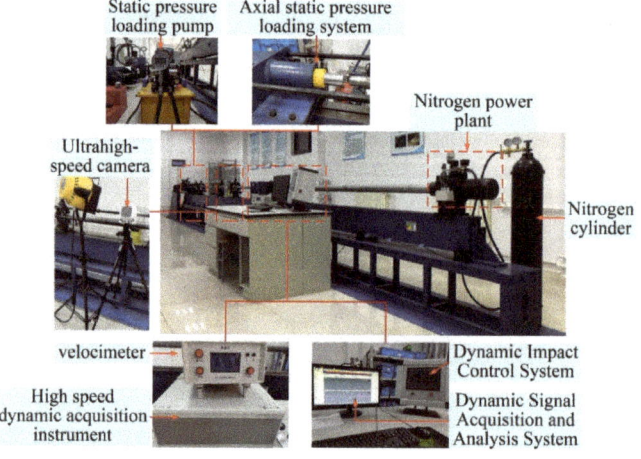

Figure 7. Separation Hopkinson pressure bar test system.

3.3. Test Scheme

Before the test, the front and rear faces of the specimen were ground, polished, and smeared with lubricant to ensure good contact between the specimen, and the input and output rods. Speckles were produced on the surface of the specimen in advance. During the test, a VisionResearch/V410L high-speed camera was used to capture the whole process of the impact failure of the specimen, and the digital image correlation method was used to obtain the crack and displacement changes during the failure of the specimen.

During the impact loading process of the rock material with anchorage bedding, the designed stress wave was incident from the vertical bolt on the side of sandstone, and the composite specimen without anchorage bedding was tested several times to determine the appropriate impact velocity and ensure the complete fracture of the specimen; the impact speed of each group of test pieces is shown in Table 2. After the preparation work had been completed, the impact loading test of the anchored bedding rock material was carried out successively. The incident and reflected waves, and transmission pulse in the transmission rod were measured and recorded with the strain gauge. According to [36], the dynamic strain $\varepsilon(t)$, strain rate $\dot{\varepsilon}$, and stress σ of the sample are indirectly obtained by using the three-wave method. The calculation formula is as follows:

$$\varepsilon(t) = c_0/l_0 \int_0^t \varepsilon_i - \varepsilon_r - \varepsilon_t d\tau \tag{21}$$

$$\dot{\varepsilon} = c_0/l_0(\varepsilon_i - \varepsilon_r - \varepsilon_t) \tag{22}$$

$$\sigma = AE/2A_0(\varepsilon_i + \varepsilon_r + \varepsilon_t) \tag{23}$$

where c_0 is the elastic wave velocity of the bar, m/s; A_0 is the original cross-sectional area, m²; ε_i, ε_r, and ε_t are the time-history strains of the incident wave, reflected wave, and transmitted wave when they propagated independently; A is the cross-sectional area of the rod, m²; and E is the elastic modulus of the rod, Gpa.

4. Test Result Analysis
4.1. Characteristics of Stress–Strain Curve

In order to obtain the influence law of bolt bedding angle and anchorage mode on the mechanical properties of the anchored bedding rock material, the stress–strain curve characteristics of the specimen were first analyzed. In the test, all specimens were subjected to compressive shear failure under impact loading, and the fracture was relatively complete. The dynamic stress–strain curves of the specimens are shown in Figure 8. The experimental results are shown in Table 3.

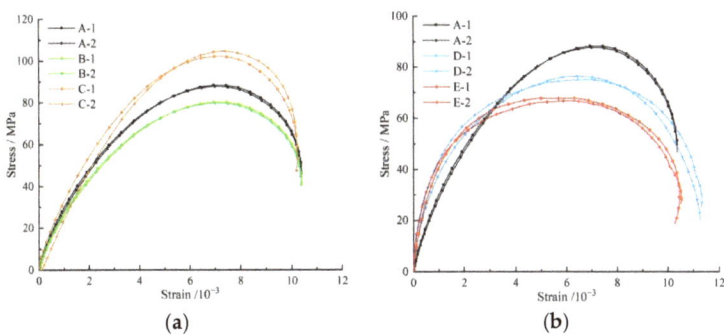

Figure 8. Stress–strain curve. (**a**) Angle between bolt and bedding; (**b**) Anchoring method.

Table 3. Dynamic strength and peak strain of anchorage rock specimens.

Groups	Number and Category		Peak Stress/MPa	Mean Peak Stress/MPa	Peak Strain/10^{-3}	Mean Peak Strain/10^{-3}
Control group	A-1 A-2	15°, full anchorage	80.28 79.75	80.02	7.06 7.08	80.28 79.75
Stratification angle	B-1 B-2	30°	88.31 87.71	88.01	7.14 7.10	7.12
	C-1 C-2	45°	102.23 104.71	103.47	7.07 7.32	7.20
Anchoring method	D-1 D-2	End anchorage	67.60 68.69	68.15	6.75 6.55	6.65
	E-1 E-2	No anchorage	61.03 60.07	60.55	6.39 6.31	6.35

Figure 8 and Table 3 show that all specimens had slight crack compaction signs at the initial loading stage and then quickly entered the elastic state; stress and strain were basically linearly increased, and then the stress increase tended to be flat. When the stress reached the peak, the rock was destroyed, and the stress dropped to the plastic state. The angle between bolt and bedding, and anchorage mode affected the dynamic stress–strain characteristics of the anchor solid.

The analysis of Experimental Group 1 shows that the average peak stress of the specimens (C-1, C-2) with an angle of 45° was 103.47 MPa, 17.57% higher than that of the specimen with 30° (B-1, B-2), and 29.31% higher than that of the specimen with 15° (A-1, A-2). The dynamic load strength of anchorage body increased with the increase in the angle between anchor and bedding, but the influence is limited. The analysis of Experimental Group 2 shows that the average peak stress of the fully anchored specimen (A-1, A-2) was 80.02 MPa, which is 17.42% higher than that of the end-anchored specimen (D-1, D-2), and 32.16% higher than that of the nonanchored specimen. The bolt support could effectively improve the dynamic load strength of the inclined bedding rock material, and the fully anchored effect was much higher than that of the end-anchored effect.

4.2. Dynamic Elastic Modulus of the Specimen

In order to investigate the influence of different bedding angles and anchorage modes on the dynamic deformation capacity of anchorage body, the dynamic elastic modulus E_d of the specimen was calculated according to Formula (24).

$$E_d = \frac{2P(t)}{\pi DL} \frac{2D[1 - 0.7854(1 - \mu)]}{d_{ox}} \tag{24}$$

where $P(t)$ is the failure load, μ is Poisson's ratio of the rock mass before peak stress, d_{ox} is the total displacement of the specimen center along the diameter direction in the vertical loading direction when the specimen is damaged, D is the specimen diameter, and L is the length of specimen.

Dynamic elastic modulus E_d under different conditions was obtained with numerical calculation, as shown in Figure 9.

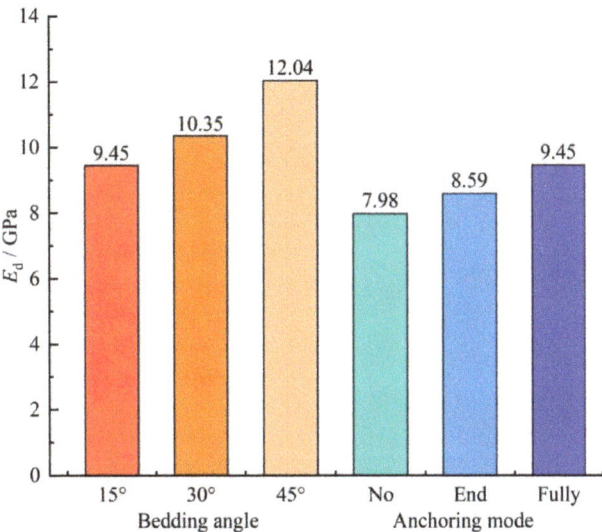

Figure 9. Dynamic elastic modulus (E_d).

Figure 9 shows that the angle between the bolt and bedding, and the anchoring mode impacted the dynamic elastic modulus of the specimen. The E_d of the specimen increased with the increase in the angle between the bolt and bedding, and the increase was obvious. When the angle was 15°, the E_d of the specimen was 9.45 GPa. When the angle was 30° and 45°, the E_d of the anchorage body increased by 9.52% and 27.41%, reaching 10.35 and 12.04 GPa, respectively, and the tensile elastic modulus was effectively improved. Analysis shows that, with the increase in the angle between the bolt and the bedding, when it was subjected to dynamic load impact, macroscopic crack propagation is difficult, and it is more difficult to damage the rock material, so that the dynamic elastic modulus of the anchorage bedding specimen is improved.

In the nonanchor, the dynamic elastic modulus E_d was 7.98 GPa. After the end anchorage, the E_d of the specimen was improved to a certain extent, about 7.64%, reaching 8.59 GPa. When the full anchor was used, the E_d of the specimen was increased by 18.42% to 9.45 GPa, and the tensile elastic modulus of the specimen was significantly improved. Analysis shows that, under dynamic load impact, the bolt of the end-anchored specimen had an axial anchoring effect that changed the tensile properties of the anchorage body. For the fully anchored specimen, the bolt also had an axial and tangential anchoring effect, thus effectively improving its dynamic elastic modulus.

4.3. Evolution Law of Strain Field

In order to explore the strain field evolution law of an anchored bedding rock mass under impact dynamic load, and to study the shear displacement process of a rock mass along the bedding direction, an ultrahigh-speed camera and digital speckle measurement technology were used during the test [33,39]. By capturing the y-direction strain cloud map of the dynamic load response of the anchored bedding rock mass from prepeak to postpeak under dynamic load, the strain field evolution law of anchored bedding rock mass was obtained. The results are shown in Figure 10.

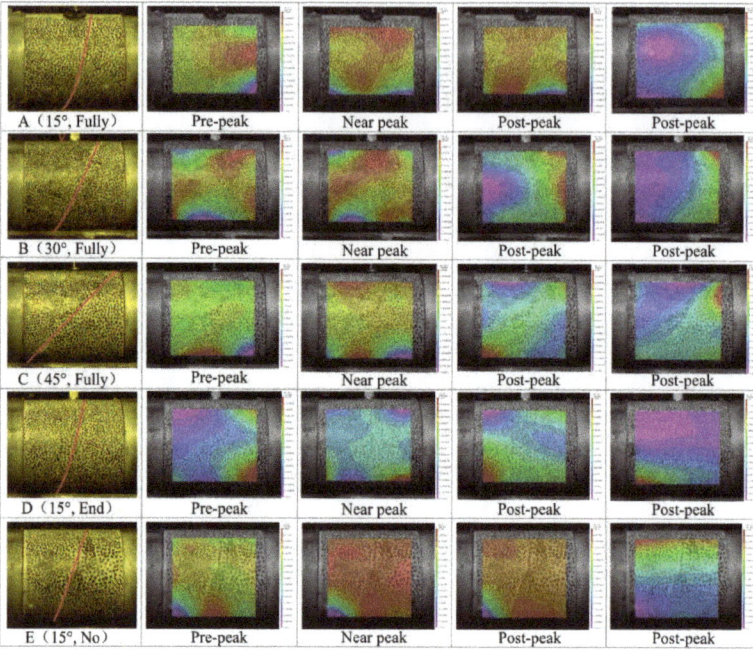

Figure 10. y-direction nephogram of the dynamic load response of an anchorage body.

Figure 10 shows that, after the full-anchor specimen with an angle of 15° between the bolt and the bedding had been subjected to impact, the central position of the impact side first formed the maximal y-direction strain concentration zone and gradually extended inward. When it extended to the vicinity of the bedding plane, the strain began to expand along the bedding face to both sides of the tip, producing macroscopic cracks and gradually extending to the end of the specimen. At that time, when the specimen was in the peak state, the failure of the specimen entered the plastic state, and the crack propagation of the bedding plane gradually stopped. The anchorage rock material formed a more obvious shear movement along the vicinity of the bedding plane, and the coarse sandstone of the impact side moved upward, while the strain of the fine sandstone side was smaller and more stable. After the full-anchor specimen with an angle of 30° had been impacted, the maximal y-direction strain concentration zone first formed near the central position of the impact side and the upper tip of the bedding plane, and gradually extended inward; then, the maximal strain zone began to expand along the bedding plane towards the central area, resulting in macroscopic cracks. The specimen broke into the postpeak plastic stage, and the crack propagation of the bedding plane gradually stopped. The anchorage rock material formed a more obvious shear movement near the bedding plane, and the coarse sandstone moved upward on the impact side, while the strain of the fine sandstone side was small and stable. After the full-anchor specimen with an angle of 45° had been impacted, the maximal y-direction strain concentration zone was first formed near the bottom bedding plane of the specimen, and the maximal strain zone began to expand along the bedding plane to the upper region. The strain near the bedding plane was much larger than that on both sides. The macroscopic cracks of the through specimen were generated at the bedding plane. Subsequently, the specimen broke into the postpeak plastic stage, and the fracture expansion of the bedding plane gradually stopped. The anchoring rock material formed a relatively obvious shear movement near the bedding plane. The coarse sandstone on the impact side moved upward at an angle of 45°, while the strain on the fine sandstone side was small and stable.

At the same time, Figure 9 also shows that, after the end-anchored specimen had been impacted, the maximal y-direction strain area formed by the impact side gradually expanded inward, and the strain near the joint surface was slightly higher than that in other areas. After entering the postpeak stage, the upper part of the specimen was affected by the end-anchored anchorage, and its strain was small, which is in sharp contrast to the free section of the lower part. In addition, the overall strain of the specimen was higher than that of the fully anchored specimen, and the fracture characteristics were more obvious.

At the same time, Figure 9 also shows that, after the end-anchored specimen had been impacted, the maximal y-direction strain area formed by the impact side gradually expanded inward, and the strain near the joint surface was slightly higher than that in other areas. After entering the postpeak stage, the upper part of the specimen was affected by the end-anchored anchorage, and its strain was small, which is in sharp contrast to the free section of the lower part. In addition, the overall strain of the specimen was higher than that of the fully anchored specimen, and the fracture characteristics were more obvious. After the impact on the nonanchor specimen, the maximal y-direction strain region of the impact side gradually extended inward. When it extended to the vicinity of the joint surface, the right surrounding rock showed obvious upward displacement compared with the left surrounding rock. With the further evolution of the strain field, the overall displacement of the unanchored specimen appeared to be larger, the strain near the joint was larger, and an obvious shear motion was formed along the joint surface. When it entered the postpeak stage, the shear motion gradually stopped, but the overall displacement of the specimen was still much larger than that of the end- and full-anchor specimens.

In summary, under impact loading, there were obvious differences in the crack development and displacement characteristics of the rock material with different angles and anchorage modes of bolt and bedding. After the impact, the bedding rock material had obvious shear displacement along the bedding direction, obvious macroscopic cracks appeared on the bedding surface, and failure and instability occurred; with end-anchored and nonanchor support, the overall displacement of the specimen was significantly increased, and the fracture characteristics were also more obvious.

Therefore, the anchored bedding rock material was prone to shear failure due to the impact loading. In field engineering practice, it is still necessary to be vigilant about the shear failure of anchored bedding rock material and compression failure. In practice, we could determine the comprehensive shear strength and uniaxial compression strength of the anchored bedding rock material through the impact loading test. After obtaining the stress environment of the surrounding rock and analyzing the stress state of the anchorage body, the strength of the anchored bedding rock material was checked, and the corresponding measures were adopted to ensure that the failure and instability of the anchoring surrounding rock did not occur.

5. Conclusions

(1) On the basis of the Coulomb–Mohr criterion and stress propagation theory, the mechanical model of the anchored bedding rock material was established, and the instability criterion of anchored bedding rock material under impact loading was obtained:

$$\begin{cases} \sigma_v \cdot \sin\theta_1 + \rho C_p v_p L^{-\eta} \cdot \cos\theta_2 \geq \sigma_c, \text{ Compression Failure} \\ \rho C_p v_p L^{-\eta} \sin(\theta_1 + \alpha) + \sigma_v \sin\alpha \geq \tau_{bj\max}, \text{ Shear Failure} \end{cases}$$

On the basis of this criterion, the stress environment of the surrounding rock, and the mechanical parameters of the rock material obtained from field and test could be used to determine whether the anchored bedding rock material would fail and the failure type.

(2) All specimens were subjected to compression shear failure under impact loading, and the dynamic load strength of the anchored bedding rock material increased with the increase in the angle between the bolt and the bedding. When the angle increased

from 15° to 45°, the dynamic load strength of the anchoring solid increased by 29.31%. In addition, bolt support could effectively improve the dynamic load strength of the rock material with anchorage bedding, and the full-length anchor effect was much higher than the end-anchor effect.

(3) With the increase in the angle between the bolt and the bedding, when it was impacted by dynamic load, the macroscopic crack propagation was more difficult, and the rock material was more difficult to damage. The dynamic elastic modulus of the anchoring bedding specimen was improved. When the angle increased from 15° to 45°, the elastic modulus of the anchoring solid increased by 27.41%. Under impact loading, the axial anchoring effect of the end-anchored specimen changed the tensile properties of the anchorage body. For the full-length anchor specimen, the bolt also had an axial and tangential anchoring effect, thus effectively improving its dynamic elastic modulus.

(4) Under impact loading, there were obvious differences in the crack development and displacement characteristics of rock material with different anchoring methods and angles between bolt and bedding. After impact, the bedding rock material had obvious shear displacement along the bedding direction, obvious macroscopic cracks were produced on the bedding plane, and failure and instability occurred. With end-anchored and nonanchored support, the overall displacement of the specimen was significantly increased, and the fracture characteristics were also more obvious.

Author Contributions: Data curation, Y.W.; Formal analysis, Q.M.; Funding acquisition, Y.T.; Investigation, Y.W.; Methodology, Y.W. and G.L.; Project administration, G.L.; Resources, X.L. and Y.T.; Software, D.F. and M.Y.; Validation, X.W.; Visualization, M.Y.; Writing—original draft, Y.W.; Writing—review & editing, X.L. All authors have read and agreed to the published version of the manuscript.

Funding: This study was financially supported by the Climbling Project of Taishan Scholar in Shandong province (no. tspd20210313) and the National Natural Science Foundation of China (grants no. 52174122 and 51874190).

Conflicts of Interest: The authors declare no conflict of interest.

References

1. Fan, D.; Liu, X.; Tan, Y.; Li, X.; Lkhamsuren, P. Instability energy mechanism of super-large section crossing chambers in deep coal mines. *Int. J. Min. Sci. Technol.* **2022**, in press. [CrossRef]
2. Xue, G.; Yilmaz, E.; Feng, G.; Cao, S.; Sun, L. Reinforcement effect of polypropylene fiber on dynamic properties of cemented tailings backfill under SHPB impact loading. *Constr. Build. Mater.* **2021**, *279*, 122417. [CrossRef]
3. Qiu, H.; Chen, B.; Wang, F.; Liao, F.; Wang, M.; Wan, D. Investigating dynamic fracture in marble-mortar interface under impact loading. *Constr. Build. Mater.* **2022**, *336*, 127548. [CrossRef]
4. Li, Y.; Yang, R.; Fang, S.; Lin, H.; Lu, S.; Zhu, Y.; Wang, M. Failure analysis and control measures of deep roadway with composite roof: A case study. *Int. J. Coal Sci. Technol.* **2022**, *9*, 2. [CrossRef]
5. Wang, P.; Jiang, Y.; Ren, Q. Roof Hydraulic Fracturing for Preventing Floor Water Inrush under Multi Aquifers and Mining Disturbance: A Case Study. *Energies* **2022**, *15*, 1187. [CrossRef]
6. Liu, X.S.; Tan, Y.L.; Ning, J.G.; Lu, Y.W.; Gu, Q.H. Mechanical properties and damage constitutive model of coal in coal-rock combined body. *Int. J. Rock Mech. Min. Sci.* **2018**, *110*, 140–150. [CrossRef]
7. Ma, Q.; Tan, Y.; Liu, X.; Gu, Q.; Li, X. Effect of coal thicknesses on energy evolution characteristics of roof rock-coal-floor rock sandwich composite structure and its damage constitutive model. *Compos. Part B Eng.* **2020**, *198*, 108086. [CrossRef]
8. Tan, Y.; Liu, X.; Ning, J.; Tian, C. Front abutment pressure concentration forecast by monitoring cable-forces in the roof. *Int. J. Rock Mech. Min. Sci.* **2015**, *77*, 202–207. [CrossRef]
9. Zhao, P.; Li, X.; Liu, J.; Zhang, D.; Qiao, H. Monitoring and analysis of the subway tunnel wall temperature and surrounding rock/soil heat absorption ratio. *Build. Environ.* **2021**, *194*, 107657. [CrossRef]
10. Yang, R.; Xu, Y.; Chen, P.; Wang, J. Experimental study on dynamic mechanics and energy evolution of rubber concrete under cyclic impact loading and dynamic splitting tension. *Constr. Build. Mater.* **2020**, *262*, 120071. [CrossRef]
11. He, J.; Dou, L.M.; Cai, W.; Li, Z.L.; Ding, Y.L. Mechanism of dynamic and static combined load inducing rock burst in thin coal seam. *J. China Coal Soc.* **2014**, *39*, 2177–2182.
12. Wang, W.; Song, Q.; Xu, C.; Gong, H. Mechanical behaviour of fully grouted GFRP rock bolts under the joint action of pre-tension load and blast dynamic load. *Tunn. Undergr. Space Technol.* **2018**, *73*, 82–91. [CrossRef]

13. Xie, C.; Lu, H.; Cao, J.; Jia, N. Study on Dynamic Load of Surrounding Rock Failure Based on Finite Element COMSOL Numerical. *IOP Conf. Ser. Earth Environ. Sci.* **2019**, *384*, 012047. [CrossRef]
14. Zhou, Z.; Chen, Z. Numerical Analysis of Dynamic Responses of Rock Containing Parallel Cracks under Combined Dynamic and Static Loading. *Geofluids* **2020**, *7*, 1–17. [CrossRef]
15. He, M.C.; Miao, J.L.; Feng, J.L. Rock burst process of limestone and its acoustic emission characteristics under true-triaxial unloading conditions. *Int. J. Rock Mech. Min. Sci.* **2010**, *47*, 286–298. [CrossRef]
16. Dou, L.-M.; Lu, C.-P.; Mu, Z.-L.; Gao, M.-S. Prevention and forecasting of rock burst hazards in coal mines. *Min. Sci. Technol.* **2009**, *19*, 585–591. [CrossRef]
17. Tan, Y.A. Analysis of fractured face of rock burst with scanning electron microscope and its progressive failure process. *J. Chin. Electron. Microsc. Soc.* **1989**, *2*, 41–48.
18. Chen, X.; Li, W.; Yan, X. Analysis on rock burst danger when fully-mechanized caving coal face passed fault with deep mining. *Saf. Sci.* **2012**, *50*, 645–648. [CrossRef]
19. Huang, S.L.; Xu, J.S.; Ding, X.L.; Wu, A. Study of layered rock material composite model based on characteristics of structural plane and its application. *Chin. J. Rock Mech. Eng.* **2010**, *29*, 743–756.
20. She, C.; Xiong, W.; Chen, S. Cosserat Medium Analysis of Deformation of Layered Rockmaterial with Bending Effects. *Rock Soil Mech.* **1994**, *15*, 12–19.
21. Xu, D.P.; Feng, X.T.; Chen, D.F.; Zhang, C.Q.; Fan, Q.X. Constitutive representation and damage degree index for the layered rock material excavation response in underground openings. *Tunn. Undergr. Space Technol. Inc. Trenchless Technol. Res.* **2017**, *64*, 133–145. [CrossRef]
22. Huang, X.; Ruan, H.; Shi, C.; Kong, Y. Numerical Simulation of Stress Arching Effect in Horizontally Layered Jointed Rock Material. *Symmetry* **2021**, *13*, 1138. [CrossRef]
23. Ma, L.H.; Jiang, X.; Chen, J.; Zhao, Y.F.; Liu, R.; Ren, S. Analysis of Damages in Layered Surrounding Rocks Induced by Blasting During Tunnel Construction. *Int. J. Struct. Stab. Dyn.* **2021**, *21*, 2150089. [CrossRef]
24. Wang, M.; Xiao, T.; Gao, J.; Liu, J. Deformation mechanism and control technology for semi coal and rock roadway with structural plane under shearing force. *J. Min. Saf. Eng.* **2017**, *34*, 527–534.
25. Chen, Y.; Xu, Y.; Feng, Y. Interaction mechanism between surrounding rock and roadside pack for gob-side entry retaining in thin coal seam. *Electron. J. Geotech. Eng.* **2015**, *20*, 4719–4734.
26. Tan, Y.L.; Wang, Z.H.; Liu, X.S.; Wang, C.W. Estimation of dynamic energy induced by coal mining and evaluation of burst risk. *J. China Coal Soc.* **2021**, *46*, 123–131.
27. Wu, Y.Z.; Fu, Y.K.; Hao, D.Y. Study on dynamic response law of anchored rock material under lateral impact load. *J. Rock Mech. Eng.* **2020**, *39*, 2014–2024. (In Chinese)
28. Wu, Y.; Chen, J.; Jiao, J.; Zheng, Y.; He, J. Damage and failure mechanism of anchored surrounding rock under impact load. *J. China Coal Soc.* **2018**, *43*, 2389–2397. (In Chinese)
29. Mu, Z.; Dou, L.; He, H.; Fan, J. F-structure model of overlying strata for dynamic disaster prevention in coal mine. *Int. J. Min. Sci. Technol.* **2013**, *23*, 513–519. [CrossRef]
30. Wang, Z.Y.; Dou, L.M.; Wang, G.F. Study on the dynamic response law of surrounding rock structure of anchoring roadway. *J. China Univ. Min. Technol.* **2016**, *45*, 9. (In Chinese)
31. Wang, A.W.; Pan, Y.S.; Zhao, B.Y. Numerical analysis of impact failure mechanism of bolt-surrounding rock structure under impact loading. *J. Earthq. Eng.* **2017**, *39*, 417–424. (In Chinese)
32. Qiu, P.Q.; Ning, J.G.; Wang, J.; Yang, S.; Hu, S.C. Experimental study on anti-impact aging of anchored rock material under impact loading. *J. China Coal Soc.* **2021**, *46*, 3433–3444. (In Chinese)
33. Skrzypkowski, K.; Zagórski, K.; Zagórska, A.; Apel, D.B.; Wang, J.; Xu, H. Choice of the Arch Yielding Support for the Preparatory Roadway Located near the Fault. *Energies* **2022**, *15*, 3774. [CrossRef]
34. Skrzypkowski, K. An experimental investigation into the stress-strain characteristic under static and quasi-static loading for partially embedded rock bolts. *Energies* **2021**, *14*, 1483. [CrossRef]
35. Jiao, J.K.; Ju, W.J. Impact failure mechanism of roadway anchorage bearing structure under dynamic load disturbance. *J. China Coal Soc.* **2021**, *46*, 94–105. (In Chinese)
36. Zhu, C.Y.; Xu, G.S. Analysis and study on approximate treatment of explosion stress wave. *Rock Soil Mech.* **2002**, *4*, 455–458. (In Chinese)
37. Ge, X.R.; Liu, J.W. A Study on the shear resistance of anchored joint surfaces. *Chin. J. Geotech. Eng.* **1988**, *1*, 8–19. (In Chinese)
38. Gao, M.S.; Dou, L.M.; Zhang, N.; Mu, Z.L.; Wang, K.; Yang, B.S. Experimental Study on earthquake tremor for transmitting law of rock burst in geomaterials. *Chin. J. Rock Mech. Eng.* **2007**, *26*, 1365–1371. (In Chinese)
39. Li, Z.; Wang, J.; Ning, J.G.; Xing, C.C.; Shen, Z. Experimental research on influence of pre-tension on dynamic load impact resistance of anchorage bodys. *J. China Univ. Min. Technol.* **2021**, *50*, 459–468. (In Chinese)

Article

Multi-Level Support Technology and Application of Deep Roadway Surrounding Rock in the Suncun Coal Mine, China

Hengbin Chu [1,2], Guoqing Li [1,*], Zhijun Liu [2], Xuesheng Liu [1,3], Yunhao Wu [1] and Shenglong Yang [1]

1. College of Energy and Mining Engineering, Shandong University of Science and Technology, Qingdao 266590, China
2. Xinwen Mining Group Co., Ltd., Xintai 271219, China
3. State Key Laboratory of Mining Disaster Prevention and Control Co-Founded by Shandong Province and the Ministry of Science and Technology, Qingdao 266590, China
* Correspondence: lgq991121@163.com

Abstract: To solve these problems of poor supporting effect and serious deformation and failure of surrounding rock of mining roadway under deep mining stress, a FLAC-3D numerical calculation model is established with −800 m level no. 2424 upper roadway in the Suncun Coal Mine as the background to compare the stress, deformation, and failure law of surrounding rock of mining roadway under once support and multi-level support with the same support strength. It is found that the multi-level support technology has obvious advantages in the surrounding rock of the horizontal roadway on the 2424 working face. From this, the key parameters of multi-level support are determined, and the field industrial test is carried out. The results show that the overall deformation of the surrounding rock is obviously reduced after multi-level support. The displacement of the two sides is reduced by about 40%, the displacement of the roof and floor is reduced by about 30%, and the plastic zone of the roadway is reduced by about 75%. The peak value of concentrated stress decreases from 98.7 MPa to 95.8 MPa, which decreases slightly. The integrity and stability of the surrounding rock are excellent, and the support effect is satisfactory. The research can provide reference and technical support for surrounding rock control of deep high-stress mining roadways.

Keywords: deep mining; mining roadway; surrounding rock; numerical analysis; multi-level support

1. Introduction

With the continuous mining of coal resources, shallow coal reserves begin to decline sharply, gradually transfer of underground mining to deep [1–4]. According to statistics, coal resources with a buried depth of more than 1000 m account for about 53% of China's total coal reserves [5–9]. After coal mining into kilometer depth, due to large rock pressure, high temperature, complex stress environment, and other factors [10–13], the stress concentration of the roadway surrounding rock intensifies after rock excavation [14,15]. Under the action of deep high stress, the problems of surrounding rock failure, large deformation, difficult support, and obvious dynamic pressure are prominent [16–19]. The stability of surrounding rock becomes a problem restricting safe and efficient deep mining of coal resources [20–23].

Around the surrounding rock control of deep roadway, domestic and foreign experts have carried out a lot of theoretical and experimental research and put forward a variety of new support concepts and technologies. Kang et al. [24] put forward the collaborative control technology of support-modification-pressure relief for strong mining roadways in the soft rock of a kilometer deep well. Malan D.F et al. [25] considered that the support structure could better control the strain of the surrounding rock within the allowable range, and the deformation of the surrounding rock of the roadway can be reduced by adding more support structures. Wang et al. [26] put forward the high resistance yield pressure and high strength supporting technology of high strength bolt, strong anchor cable, and grouting reinforcement roadway and determined the supporting time of each supporting stage. Li et al. [27] put forward the support strategy of "bottom coal grouting reinforcement + high prestressed strong bolt anchor cable timely support", which improved the strength of surrounding rock and support resistance of the mining roadway.

The current study provides strong technical support for the stability of surrounding rock in deep mining roadways. However, due to the characteristics of high stress, frequent dynamic load disturbance, and complex geological structure, it is often difficult to ensure the safety of surrounding rock with one support. Sometimes strong support can effectively control the deformation of surrounding rock, but the cost is very high. Therefore, multi-level support has been recognized by some scholars and field engineers. Liu et al. [28] adopted stepped combined support, advanced support, once support, and secondary support of the top of the heading face cooperated with each other, spray anchor grouting, and other means to realize the control of excavation construction safety and surrounding rock stability. Li et al. [29] analyzed the stress and strength adjustment process of surrounding rock in the support process of soft rock roadway and combined with the rheological mechanics model of soft rock roadway, deduced the theoretical formula of the optimal time of secondary support, and used it in engineering practice to better control the stability of roadway surrounding rock. Through practice, Li et al. [30] found that bolt length, initial anchoring force, stiffness, and other multi-level support parameters play a vital role in controlling roadway deformation and roadway safety.

However, at present, there is no better solution for how determining the support parameters in multi-level support technology. Based on the engineering background of the upper roadway of 2424 working faces at −800 m level in the Suncun Coal Mine, this paper proposes a graded support scheme through field measurement and numerical calculation. According to the deformation law of roadway surrounding rock and the design principle of roadway support, the support design is improved to ensure the safety and stability of the mining roadway in the working face.

2. Project Profile

The Suncun Coal Mine −800 m level 2424 working face roadway has a buried depth of about 1300 m, a length of about 1200 m, a trapezoidal section, a net width of 4.5 m, a left high of 2.3 m, a right high of 3.6 m. The right side of the roadway is the entity coal side. The thickness of the coal seam in the working face is 2.1~2.4 m, the average coal thickness is 2.3 m, and the average dip angle is 28°. The coal seam roof is silty sandstone, grayish black, stratified development, containing plant debris fossils, compressive strength of 21.4 MPa, and average thickness of 2.0 m; the coal seam floor is gray sandstone, stratified development, compressive strength is about 66.7 MPa, thickness 6~10 m. The mechanical parameters of the roof and floor footwall, and hanging wall are shown in Table 1.

Table 1. Mechanical parameters of the roof and floor footwall and hanging wall.

Rock Name	Thickness/m	Volumetric Weight/(kN/m^3)	Bulk Modulus/GPa	Shear Modulus/GPa	Internal Friction Angle/°	Tensile Strength/MPa	Cohesion/MPa
Sandstone	6.0~10.0	25	11.5	7.3	28	8.4	2.6
Siltstone	0~4.0	23	2.1	1.8	32	4.3	0.8
4#coal	2.1~2.4	14	2.1	0.93	24	0.7	0.5
Sandstone	5.0~6.0	25	11.5	7.3	28	8.4	2.6

3. Design of Support Parameters of Mining Roadway

3.1. Model Establishment

Based on actual conditions of no. 2424 upper roadway, considering the hosting of coal seam and mining mode, and considering the influence of boundary effect and advanced support pressure, make the model meet the full extraction [31]. The FLAC3D calculation model is established to simulate the stress distribution and deformation of surrounding rock after primary and tertiary support. The model size is 250 m × 300 m × 180 m and is divided into 2,358,781 elements, as shown in Figure 1. Using the Mohr–Coulomb model, the floor is fixed, the sides are horizontally simply supported, and the top is a free border. Owing to the roadway depth of 1300 m, the vertical stress of 29.05 MPa is applied to the upper surface of the model, and the horizontal displacement constraint is applied to the rest of the model.

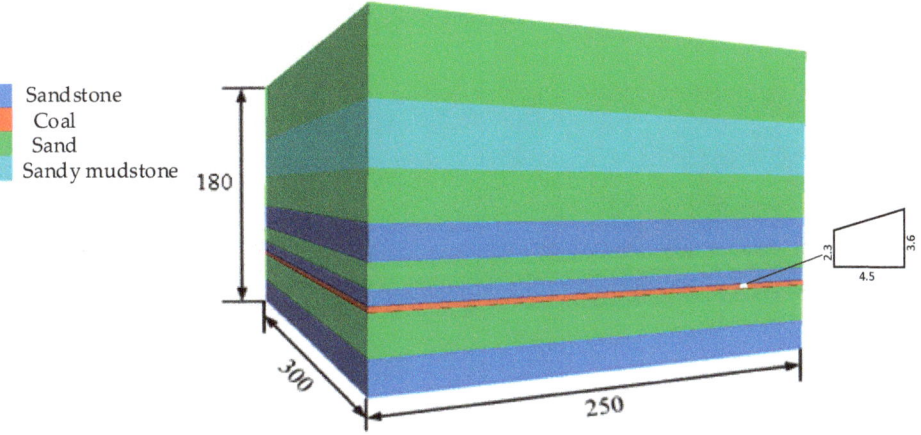

Figure 1. Diagram of the numerical model.

3.2. Numerical Calculation Scheme and Results

To compare the surrounding rock control effect of different support methods, two simulation schemes of once support and tertiary support are set up under the same support strength, and the experimental results are analyzed and compared as follows. The supporting materials used are shown in Table 2.

Table 2. Properties of supporting materials.

Support Material	Size	Yield Strength/MPa	Rod Elongation	Elastic Modulus/MPa	Poisson Ratio
mining steel strand grouting anchor cable MSGLD-600	Φ22 × 4100 mm	≥1860	≥2%	195 × 10³	0.3
equal strength thread rigid resin bolt MSGLD-335	Φ22 × 2200 mm	≥600	≥15%	206 × 10³	0.3
equal strength thread rigid resin bolt	Φ20 × 2000 mm	≥350	≥15%	200 × 10³	0.3

3.2.1. Scheme 1: Once Support

Immediately after the excavation of the roadway, no. 2424 upper roadway is supported, and bolt, anchor cable, and diamond mesh combined support are carried out in the whole roadway. Roof centered on the central axis of roadway, installation of Φ22–4100 mm mining steel strand grouting anchor on both sides, each anchor cable lengthens anchorage, anchor cable spacing 1050 × 1000 mm; MSGLD-600 equal strength thread rigid resin bolt is used in the middle of the two sides; MSGLD-335 equal strength thread rigid resin bolt is used in the top and seat angles, and the row spacing is 1000 × 1000 mm. The roadway support is shown in Figure 2, and the numerical calculation results are shown in Figure 3.

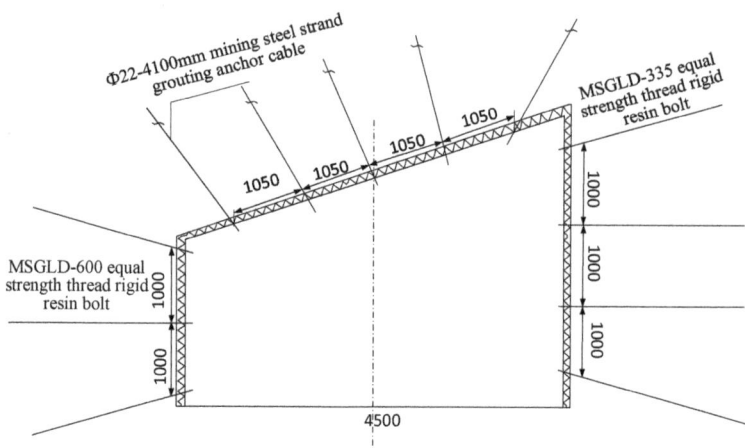

Figure 2. Roadway once support section diagram.

It can be seen from Figure 3 that the vertical stress concentration in entity coal side on the right side of the roadway and its vertical downward peak reaches 98.7 MPa. The peak stress is about 3.5 m from the right side of the roadway, which is easy to crush part of the coal body in the roadway; with once support, there is a certain degree of deformation in the roadway, and the maximum displacement of the roof is about 125 mm, the maximum deformation of the floor is about 50 mm, the maximum displacement of the two sides is 100 mm; compression-shear failure mainly occurs around the adjacent roadway of working face, but tension-compression failure occurs above the roof cutting line. The roof failure range of the plastic zone of the roadway surrounding rock is 0~20 m, and the plastic zone of the main roof of the roadway is basically bounded by the roof cutting line; the plastic yield zone is along the empty side, and the rest is the elastic zone.

Analysis shows maximum roof displacement near and above the central axis of the roadway, and the maximum displacement of surrounding rock appears in the middle of two sides. Therefore, the surrounding rock in the middle and upper part of the roof and the middle of the two sides is the key part of the support, and the bolt anchor should be used in time for support. Second, supporting should improve the support of the top angle and bottom angle on the basis of once support. The third advance support should be supported in the area with large deformation.

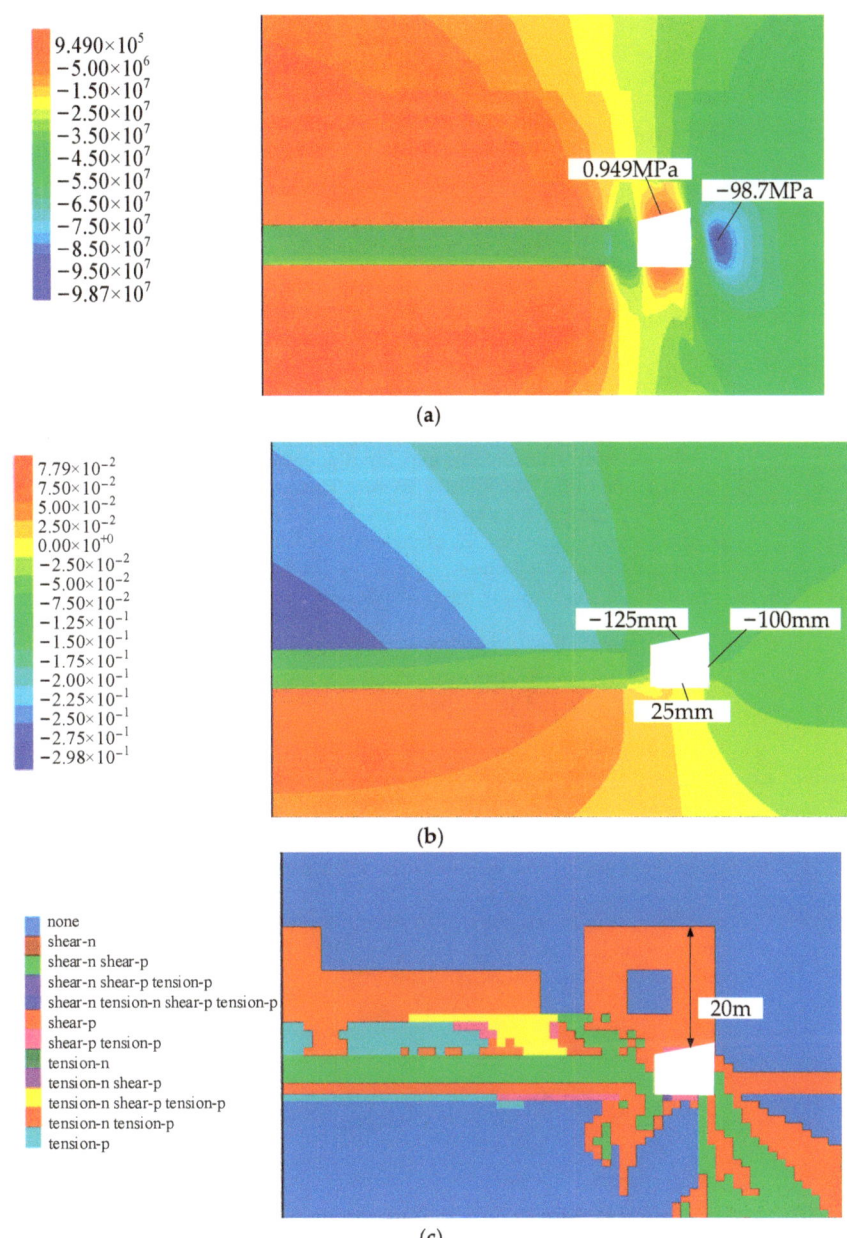

Figure 3. Simulation results of once support, (**a**) vertical stress distribution diagram, (**b**) vertical displacement distribution map, (**c**) distribution diagram of the plastic zone.

3.2.2. Scheme 2: Tertiary Support

Support no. 2424 upper roadway immediately after excavation, application of mine steel strand grouting anchor support in roof central axis and it is upper position, length of anchor cable 4.1 m, anchor cable column interval 1050 mm × 1000 mm; The middle of the two sides adopts MSGLD-600 rigid resin bolt with equal strength thread. The row spacing between the bolts is 1000 mm × 1000 mm, and the length is 2.2 m.

After once support of the roadway reaches a new balance, secondary support for no. 2424 upper roadway, installation of mine grouting anchor at top angle of roof; the top and bottom angles of the two sides are reinforced by MSGLD-335 equal strength thread rigid resin bolting, and the spacing between the equal strength thread rigid resin bolt used in the once support is 1000 mm.

When the excavation working face is 50 m, the advance support is 200 m. Install grouting anchor cable in the area with large roof deformation to complete advanced grouting reinforcement.

According to the principle of graded support, the specific multi-level support method is shown in Figure 4. The numerical calculation results are shown in Figure 5.

It can be seen from Figure 5 that the vertical stress concentration occurs on the entity coal side of the right side of the roadway. The vertical stress peak is about 3.5 m from the right side of the roadway, and the size is 95.8 MPa. In the multi-level support, there is a certain degree of deformation in the roadway surrounding rock; the roadway roof subsidence is 75~100 mm, the maximum deformation of the floor is 25 mm, the maximum deformation of the two sides is 50 mm; The adjacent roadway around the working face is mainly a compression–shear failure, but there is tension-compression failure above the cutting line. The plastic zone of the main roof of the roadway is basically bounded by the top-cut line, and the failure range of the plastic zone of the roof is between 0~5 m. The plastic yield zone is along the goaf side, and the rest is the elastic zone.

3.3. Supporting Effect Comparison

Comparison of numerical simulation results of once support and tertiary support. From Figures 3a and 5a, the peak stress concentration on the right side of the coal side is reduced from 98.7 MPa to 95.8 MPa, with a decrease of about 3%. From Figures 3b and 5b, the maximum subsidence of the roof is reduced from 125 mm to 100 mm by 20%. The displacement of the two sides is greatly reduced from the maximum displacement of the once support 100 mm to 50 mm. Although the floor is not supported, the floor deformation is almost the same, with most of the location of the maximum deformation of 25 mm; From Figures 3c and 5c, the failure range of the roof plastic zone is reduced from 0~20 m to 0~5 m with a reduction of about 75%. At the same time, the range of the plastic zone of the floor surrounding the rock of the multi-level support roadway reduces by about 20%, and the range of the plastic zone of two sides surrounding the rock is similar. The specific comparison effect is shown in Table 3.

Table 3. Comparison of relevant parameters of different schemes.

Scheme	Maximum Stress of Coal Side/MPa	Roof Subsidence/mm	Two Sides Displacement/mm	Roof Plastic Zone Range/m	Floor Deformation/m
Once support	98.7	75~125	50~100	0~20	25~50
Tertiary support	95.8	50~75	25~50	0~5	25

In summary, after multi-level support, the stress environment of the roadway surrounding rock has been greatly improved. The roof subsidence and two sides are relatively stable, but there is still a certain degree of deformation. Through the comparison of the support effect, it is determined that the three-levels support scheme is adopted in the no. 2424 upper roadway of −800 m level in the Suncun Coal Mine.

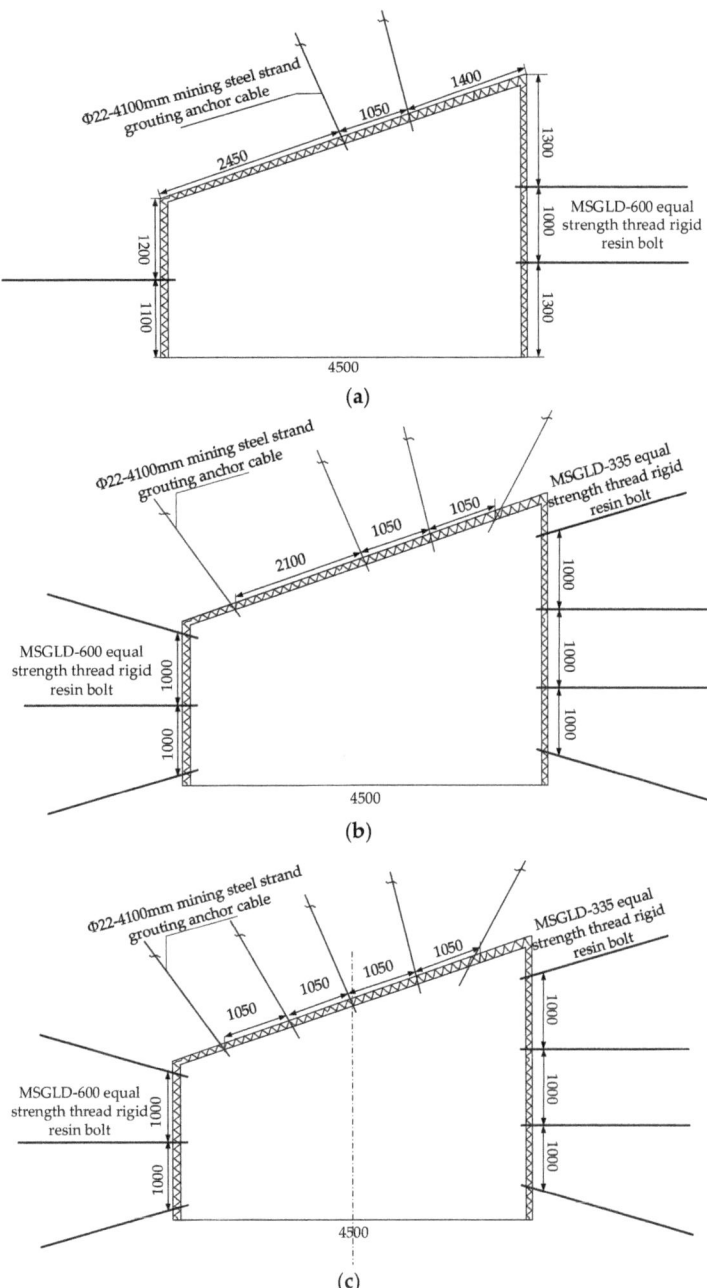

Figure 4. Tertiary support sectional drawing: (**a**) primary support, (**b**) secondary support, (**c**) tertiary support.

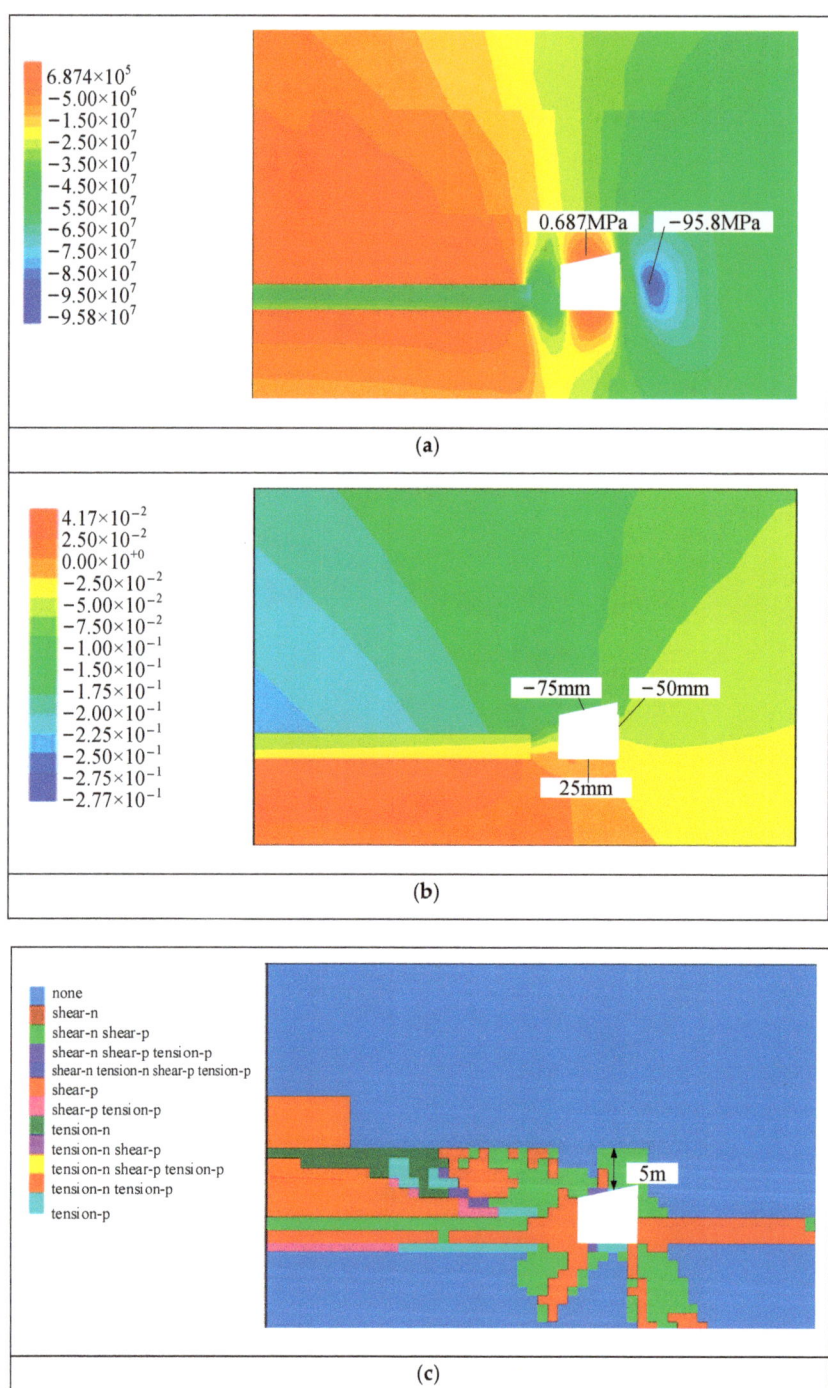

Figure 5. Tertiary support simulation results, (**a**) vertical stress distribution diagram, (**b**) vertical displacement distribution map, (**c**) distribution diagram of the plastic zone.

4. Engineering Applications

4.1. Support Scheme Design

Based on the above research, the combined support of 'grouting anchor cable + equal strength thread steel resin bolt + U-shaped steel connecting beam' was used in no. 2424 upper roadway and the tertiary support scheme were adopted, construction scheme was consistent with the tertiary support simulation scheme. Once support is the basic support, that is, the immediate support after roadway excavation, secondary support is the reinforcement support of the top and bottom angle of the roadway when once support reaches a new balance. When the secondary support was completed, the working face pushed 50 m, and the advanced support was 200 m. According to the principle of graded support, the specific multi-level support method is shown in Figure 4.

4.2. Support Effect Monitoring

To verify the effect of the control technology proposed above in field application. During the tunneling period of the working face, a surrounding rock deformation monitoring station is set up every 40 m; the station follows a head-on setting. A total of 32 deformation monitoring stations were set up at different positions of no. 2424 upper roadway roof and floor. Using a laser range finder to monitor the displacement of the roadway roof and floor and two sides to obtain the total deformation of surrounding rock at each measuring point of the upper roadway. The monitoring results are shown in Figure 6.

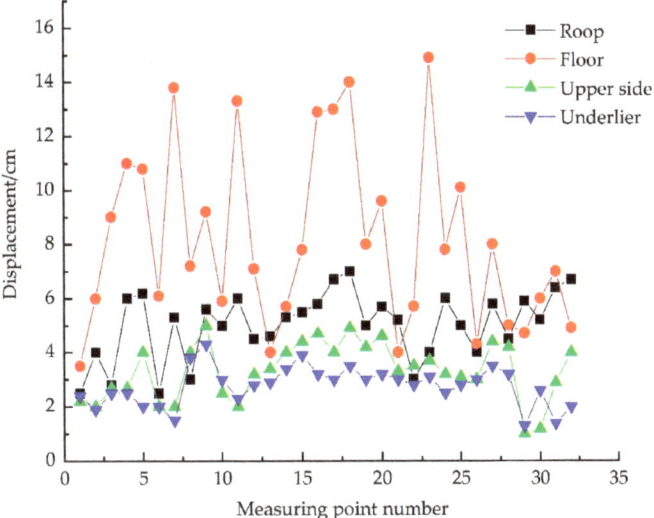

Figure 6. Total deformation of surrounding rock.

It can be seen from Figure 6, after the no. 2424 upper roadway adopts three level supports, the maximum total deformation of each measuring point on the upper side was 50 mm; maximum deformation of the underlier was 40 mm; the maximum roof subsidence was 70 mm; floor because of no support, displacement was large, many positions of more than 100 mm. After on-site detection of the roadway roof and floor and two sides of the deformation within the allowable range, it can be seen that after the three-level support, the overall deformation of the roadway surrounding rock was small, and roadway deformation was more stable.

The deformation value of the roadway surrounding rock is analyzed by monitoring, and the error of numerical simulation results is calculated based on the monitoring results. The comparison results are shown in Table 4. It is found that the error of deformation value on both sides is less than 10%, which shows that the simulation effect is good. Because the floor is not supported, it is affected by other roadways during the construction, and the error is large. The roof subsidence error is about 30%, and the deformation value is less than the numerical simulation results. In summary, the deformation of the roadway surrounding rock is well controlled when multi-stage support is adopted. Numerical simulation results can provide some theoretical support for field application.

Table 4. Comparison of numerical and analytical results.

Project	Average Roof Subsidence/mm	Average Floor Heave/mm	Average Upper Side Displacement/mm	Average Underside Displacement/mm
Numerical simulation	65	30	30	30
Analytical results	50.2	81.3	33.1	27.6
Error	29.5%	63.1%	9.4%	8.7%

2424 roadway mining process, advance 200 m surrounding rock site photos are shown in Figure 7. The roof subsidence of the roadway was small, there was no large area of coal wall falling on both sides, the floor heave was small, and the surrounding rock integrity was good. Therefore, after the three-level support, the overall deformation of the surrounding rock of the roadway was well controlled.

Figure 7. Surrounding rock control effect diagram: (a) roof, (b) upper side.

5. Conclusions

(1) The numerical calculation models of once support and three-levels support of no. 2424 upper roadway in Suncun Coal Mine are established. Simulation results show that compared with once support, the deformation of surrounding rocks is obviously reduced in three-level support, in which the convergence of two sides reduces by about 40%, and the roof subsidence reduces by about 30%. The stress concentration of surrounding rocks also reduces, in which the concentrated stress of the coal side reduces by about 3%. The maximum plastic zone height in the roof strata decreases by about 75%.

(2) The three-level support technology of 'grouting anchor cable + equal strength thread rigid resin bolt + U-shaped steel connecting beam' is determined to support no. 2424 upper roadway, which includes primary base support, secondary reinforcement support, and advanced tertiary support. Field monitoring shows that the maximum roof subsidence is about 70 mm, the side-to-side convergence is less than 50 mm, and the floor heave deformation is less than 25 mm.

(3) Compared with once support technology, the multi-level support technology on the condition with the same total support strength can reduce the roadway deformation

and stress concentration of two sides to a certain extent to improve the roadway integrity and stability.

Author Contributions: Conceptualization, H.C. and G.L.; Software, Y.W. and S.Y.; Validation, S.Y.; Formal analysis, G.L.; Investigation, Z.L. and Y.W.; Data curation, H.C., Z.L. and X.L.; Writing—original draft, H.C., G.L., Y.W. and S.Y.; Writing—review & editing, Z.L. and X.L.; Project administration, X.L. All authors have read and agreed to the published version of the manuscript.

Funding: This study was financially supported by the National Natural Science Foundation of China (nos. 52174122 and 51874190), the Excellent Youth Program of Shandong Natural Science Foundation (no. ZR2022YQ49), the Climbing Project of Taishan Scholar in Shandong province (no. tspd20210313), and the Youth Expert Project of Taishan Scholar in Shandong province.

Data Availability Statement: Not applicable.

Conflicts of Interest: The authors declare no conflict of interest.

References

1. Fan, D.; Liu, X.; Tan, Y.; Li, X.; Purev, L. Instability energy mechanism of super-large section crossing chambers in deep coal mines. *Int. J. Min. Sci. Technol.* **2022**, *32*, 1075–1086. [CrossRef]
2. Ranjith, P.G.; Zhao, J.; Ju, M.; De Silva, R.V.; Rathnaweera, T.D.; Bandara, A.K. Opportunities and Challenges in Deep Mining: A Brief Review. *Engineering* **2017**, *3*, 546–551. [CrossRef]
3. Khandelwal, M.; Ranjith, P. Correlating index properties of rocks with P-wave measurements. *J. Appl. Geophys.* **2010**, *71*, 1–5. [CrossRef]
4. Chen, S.; Qu, X.; Yin, D. Investigation lateral deformation and failure characteristics of strip coal pillar in deep mining. *Geomech. Eng.* **2018**, *14*, 421–428.
5. Wang, Q.; He, M.; Li, S. Comparative study of model tests on automatically formed roadway and gob-side entry driving in deep coal mines. *Int. J. Min. Sci. Technol.* **2021**, *31*, 591–601. [CrossRef]
6. Liu, X.; Fan, D.; Tan, Y.; Ning, J. New Detecting Method on the Connecting Fractured Zone Above the Coal Face and a Case Study. *Rock Mech. Rock Eng.* **2021**, *54*, 4379–4391. [CrossRef]
7. Rathnaweera, T.; Ranjith, P.; Perera, M. Investigation of relative flow characteristics of brine-saturated reservoir formation: A numerical study of the Hawkesbury formation. *J. Nat. Gas Sci. Eng.* **2017**, *45*, 609–624. [CrossRef]
8. Wasantha, P.; Ranjith, P.; Zhang, Q.; Xu, T. Do joint geometrical properties influence the fracturing behaviour of jointed rock? An investigation through joint orientation. *Geomech. Geophys. Geo-Energy Geo-Resour.* **2015**, *1*, 3–14. [CrossRef]
9. He, M.; Gong, W.; Wang, J.; Qi, P.; Tao, Z.; Du, S.; Peng, Y. Development of a novel energy-absorbing bolt with extraordinarily large elongation and constant resistance. *Int. J. Rock Mech. Min. Sci. Geomech. Abstr.* **2014**, *67*, 29–42. [CrossRef]
10. Zhang, S.; Lu, L.; Wang, Z. A physical model study of surrounding rock failure near a fault under the influence of footwall coal mining. *Int. J. Coal Sci. Technol.* **2021**, *8*, 626–640. [CrossRef]
11. Adrian, B.; Wang, Z. Combined support mechanism of rock bolts and anchor cables for adjacent roadways in the external staggered split-level panel layout. *Int. J. Coal Sci. Technol.* **2021**, *8*, 659–673.
12. Lu, J.; Xu, S.; Miao, C. The theory of compression-shear coupled composite wave propagation in rock. *Deep. Undergr. Sci. Eng.* **2022**, *1*, 77–86. [CrossRef]
13. Chen, S.; Jiang, T.; Wang, H.; Feng, F.; Yin, D.; Li, X. Influence of cyclic wetting drying on the mechanical strength characteristics of coal samples: A laboratory-scale study. *Energy Sci. Eng.* **2019**, *7*, 3020–3037. [CrossRef]
14. Tan, Y.; Fan, D.; Liu, X.; Song, S.; Li, X.; Wang, H. Numerical investigation on failure evolution of surrounding rock for super-large section chamber group in deep coal mine. *Energy Sci. Eng.* **2019**, *7*, 3124–3146. [CrossRef]
15. Alejano, L.R.; Muralha, J.; Ulusay, R.; Li, C.C.; Pérez-Rey, I.; Karakul, H.; Chryssanthakis, P.; Aydan, Ö. ISRM suggested method for determining the basic friction angle of planar rock surfaces by means of tilt tests. *Rock Mech. Rock Eng.* **2018**, *5*, 3853–3859. [CrossRef]
16. Fang, K.; Zhao, T.; Zhang, Y. Rock cone penetration test under lateral confining pressure. *Int. J. Rock Mech. Min. Sci.* **2019**, *11*, 149–155. [CrossRef]
17. Yang, R.; Ding, C.; Yang, L. Behavior and law of crack propagation in the dynamic-static superimposed stress field. *J. Test. Eval.* **2018**, *4*, 2540–2548. [CrossRef]
18. Tan, Y.; Liu, X.; Shen, B.; Ning, J.; Gu, Q. New approaches to testing and evaluating the impact capability of coal seam with hard roof and/or floor in coal mines. *Geomech. Eng.* **2018**, *14*, 367–376.
19. Gao, M.; Zhao, H.; Zhao, Y. Investigation on the vibration effect of shock wave in rock burst by in situ microseismic monitoring. *Shock. Vib.* **2018**, *2018*, 8517806. [CrossRef]
20. Tan, Y.; Fan, D.; Liu, X. Research progress on chain instability control of surrounding rock of super large section chambers in deep coal mine. *J. China Coal Soc.* **2022**, *47*, 180–199.

21. Ma, Q.; Tan, Y.; Liu, X.S. Experiment and numerical simulation of influence of loading rate on failure and strain energy characteristics of coal-rock combined specimens. *J. Cent. South Univ.* **2021**, *28*, 3207–3222. [CrossRef]
22. Jiang, B.; Gu, S.; Wang, L.; Zhang, G. Strainburst process of marble in tunnel-excavation-induced stress path considering in-termediate principal stress. *J. Cent. South Univ.* **2019**, *26*, 984–999. [CrossRef]
23. Feng, X.; Valter, C. Additive manufacturing technology in mining engineering research. *Deep. Undergr. Sci. Eng.* **2022**, *1*, 15–24. [CrossRef]
24. Kang, H.; Jiang, P.; Huang, B. Cooperative control technology of surrounding rock support-modification-pressure relief in kilometer deep roadway of coal mine. *J. China Coal Soc.* **2020**, *45*, 845–864.
25. Malan, D. Simulation of the tiem-dependent behavior of excavations in hardrock. *Rock Mech. Rock Eng.* **2002**, *35*, 225–254. [CrossRef]
26. Wang, W.; Peng, G.; Huang, J. Research on high strength coupling support technology of roadway in high stress and extremely soft broken rock strata. *J. China Coal Soc.* **2011**, *36*, 223–228.
27. Li, P.; Suo, Y.; Guo, M.; Lu, J. Research on deformation failure and support strategy of deep bottom coal mining roadway. *Min. Saf. Environ. Prot.* **2022**, 1–6. Available online: http://kns.cnki.net/kcms/detail/50.1062.TD.20220830.1825.004.html (accessed on 19 October 2022).
28. Liu, Q.; Kang, Y.; Bai, Y. Exploration the support method of broken weak surrounding rock in deep rock roadway of Guqiao Coal Mine. *Rock Soil Mech.* **2011**, *32*, 3097–3104.
29. Li, Y.; Zhang, H.; Meng, X. Research on secondary support time of soft rock roadway. *J. China Coal Soc.* **2015**, *40*, 47–52.
30. Li, H.; Zhang, B.; Liu, J. Research on anchorage parameters of secondary support in coal mine roadway. *J. Min. Saf. Eng.* **2017**, *34*, 962–967.
31. Liu, X.; Fan, D.; Tan, Y.; Song, S. Failure evolution and instability mechanism of surrounding rock for close-distance chambers with super-large section in deep coal mines. *Int. J. Geomech.* **2021**, *21*, 04021049. [CrossRef]

Article

The Influence of Coal Tar Pitches on Thermal Behaviour of a High-Volatile Bituminous Polish Coal

Valentina Zubkova and Andrzej Strojwas *

Institute of Chemistry, Jan Kochanowski University in Kielce, 7 Uniwersytecka Str., 25-406 Kielce, Poland
* Correspondence: andrzej.strojwas@ujk.edu.pl

Abstract: The influence of three coal tar pitches (CTPs), having softening points at 86, 94, and 103 °C, on the thermal behaviour of a defrosted high-volatile coal during co-carbonization and co-pyrolysis was studied. The following research techniques were used: X-raying of the coked charge, TG/FT-IR, ATR and UV spectroscopies, extraction, SEM, STEM, and XRD. It was determined that CTP additives change the structure of the coal plastic layer, the thickness of its zones, and the ordering degree of the structure of semi-cokes to a different extent and independently from their softening points. The softening points of CTPs do not influence the composition and yield of volatile products emitted from blends with pitch as well as the composition, structural-chemical parameters, and topological structure of material extracted from coal blends. It is suggested that such a lack of existence of any correlation between the softening points of CTPs and the degree of their influence on the thermal behaviour of coal was caused by the presence of the atoms of metals (Fe and Zn) in the CTPs. These atoms change the course of the carbonization of the CTPs themselves and their influence on organic substance of coal in blends with CTPs.

Keywords: coal tar pitch; additives; co-carbonization; co-pyrolysis; interaction

1. Introduction

The changes in the situation on the global coal market and the diversification of importers of coking coals set a task for the national industry to ensure the work continuity of Polish coking plants. The research conducted by Miroshnichenko et al. [1], in different seasons, suggests that in winter the same coal blends improve their thermoplastic and coke-making properties. Our previous research, conducted on a frozen high-volatile coal, showed that after defrosting this coal significantly improves its thermoplastic properties and, during carbonization, forms the gas-saturated zone in the plastic layer [2]. However, compared to a fresh one, a defrosted coal emits more volatile products during heating. This fact lowers the profitability of the coke-making process of such coal.

The research conducted by Nomura and Arima [3] proved that the addition of a coal-derived binder can reduce the yield of volatile products from the blend during pyrolysis at the temperature below 450 °C. Coal tar and coal tar pitch (CTP) have been known for a long time as additives that improve the fluidity of coals [4–8] and their caking properties [3,4]. The authors of the works [9,10] suggest that pitch additives can be used as additives that bring back the lost thermoplastic abilities to low-oxidized coals. These additives can be used as binders for briquettes with coals having worse caking properties [11–14]. CTPs extend the plasticity range and lower the softening temperature of coals [15,16].

Despite a large amount of research on the influence of CTPs on the course of the coking process, Nomura [17] suggests that the mechanism of their influence on coal substance has not been fully explained so far. According to the authors of the works [18,19], CTPs in blends with coals participate in the processes of "hydrogen shuttling" being either hydrogen donors or acceptors. It is believed that the decomposition products of CTP interact with coal and increase its thermoplasticity this way [3,20]. As a result of such interaction, an

increase in the coking pressure of coal blends with pitch can take place [10,21–23]. Duffy et al. [15] determined an increase in the coking pressure as a result of the addition of pitch to a coal of higher fluidity. However, in Nomura's opinion [17], CTP additives can reduce coking pressure. Previous research [16] showed that the addition of CTP to coals, having a distinct content of volatiles, changes the composition of the material extracted from them in different way. In this research, attention was attracted to the fact that the addition of CTP to coals as a material containing polycyclic aromatic hydrocarbons (PAHs) led to a decrease in content of these PAHs in the extracts of plasticized coal blends.

The aforementioned shows that, despite numerous investigations connected with the use of CTPs as additives to coked coals, the phenomenon of their interaction with coal substances has not been explained fully. In order to extend the knowledge about the interaction of CTPs with coals and to specify the effect of their influence on the mechanism of changes taking place in coked coal, the research had to be conducted with the use of three CTPs having the softening points distinct but close. It was expected that the continuity of specific changes in the thermal behaviour of coal would correspond to the changes in the softening points of CTPs. Due to this, the impact effect of CTPs and their softening points on thermochemical changes in organic substance of coal would be easily determined.

The specific aims of the research were to study the influence of CTPs on:
- the changes in thickness of plastic layer and its zones and the changes in structure and texture of the material of carbonized charge;
- the changes in volume of heated charge;
- the course of coal pyrolysis and the composition of volatile products of pyrolysis;
- the yield of material extracted from the zones of plastic layer and structural chemical parameters of this material.

2. Materials and Methods

2.1. Characteristics of Studied Samples

The subjects of the investigation were blends of a commercial sample of defrosted high-volatile bituminous coal with coal pitches. The characteristics of studied coal are presented in Table 1.

Table 1. The characteristics of studied coal.

Proximate Analysis		Ultimate Analysis, %	
W^a, %	8.5	C^a	82.2
V^{daf}, %	33.0	H^a	4.82
A^d, %	7.0	N^a	1.36
FSI	7.0	S^a	0.47
RI	77	O^{diff} *	4.15

W^a—moisture content in analytical sample (air-dry basis); V^{daf}—volatiles content in analytical sample (dry, ash-free basis); A^d—ash content (dry basis); C^a—carbon atoms content in analytical sample; H^a—hydrogen atoms content in analytical sample; N^a—nitrogen atoms content in analytical sample; S^a—sulphur atoms content in analytical sample; FSI–Free Swelling Index; RI–Roga Index. * calculated by difference $O^{diff} = 100 - C^a - H^a - N^a - S^a - A^d$.

A fresh sample of commercial coal was frozen and stored at a temperature of $-15\,°C$ for 15 months. The coal was defrosted at room temperature. After drying to the air-dry state, the coal was ground to the grain size of <3 mm. The commercial samples of hard CTPs–CTP86BL (CTP86 in the text) and CTP103BL (CTP103 in the text) along with experimental pitch CTP94 were used for the investigation. The characteristics of CTPs are presented in Table 2.

Table 2. The main characteristics of coal tar pitch.

CTP86			CTP94			CTP103		
T_s, °C	TI	QI	T_s, °C	TI	QI	T_s, °C	TI	QI
86	≥38	≤18	94	≥25	≤9	103	≥18	≤5

T_s—softening point, TI—toluene insoluble, QI—quinoline insoluble.

The softening point of coal tar pitch was determined according to the ISO 5940-2:2007 standard using the Standard Test Method for Softening Point of Pitches (Mettler Softening Point Method, ASTM D3104).

The CTP samples were ground to the grain size of <0.2 mm in diameter. The samples of 2 wt.% of CTP were prepared from the samples of coal and CTPs using the standard procedure of the manual mixing of powder ingredients previously described in [16].

2.2. Carbonization Test

The carbonization of coal and its blends with CTPs was carried out in a laboratory unit in order to study the coking process using X-raying according to the methodology described previously [24]. Figure 1 presents a scheme of the laboratory unit used to study the course of the coking process with X-raying.

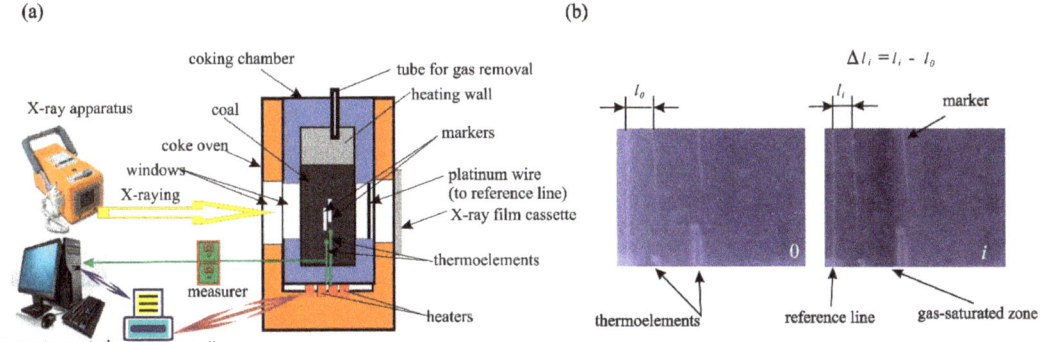

Figure 1. A scheme of the laboratory unit to study the carbonization process with X-raying (**a**) with X-ray pictures of heated charge of coals (**b**).

The unit consists of an oven with one-sided vertical heating and a coking chamber. There are some windows in the side walls of the chamber and oven through which the coked charge is X-rayed. For this purpose, an ORANGE 1040 HF X-ray apparatus was used that worked in the mode of U = 85 kV, 25 mAs. The mass of coal charge was 480 g., and the sizes of the coal grains were below 3 mm. The changes in temperature of the charge and the course of the coking process were controlled by a computer (Figure 1a). The markers, the positions of which were registered in the X-ray images taken during carbonization, were placed in the carbonized charge. The changes in position of Δl markers were calculated on the basis of elaboration of these images by a CorelDRAW version X4 software. The values of changes in position of markers were calculated by the formula: $\Delta l = l_o - l_i$. The measurements were made with respect to the vertical reference line (Figure 1b) at the designated height of charge. The marker was placed in the charge in the distance of 36 mm from the heating wall of the coking chamber. The changes in the position of the markers, depending on the temperature, reflected the changes in the volume of the charge layer between the heating wall and the markers at different temperatures of carbonization. The carbonization was carried out to reach a temperature of 950 °C on heaters with the heating rate of 4 °C min^{-1}. After this temperature was reached, the coking chamber with charge was cooled rapidly. After cooling, the charge was prepared in order to separate the zones in the composition of the plastic layer of coal. The thickness of this layer and the thickness of its zones were measured in the X-ray images taken during the penetration of the charge by a needle thermoelement according to the methodology described previously [24].

2.3. Pyrolysis in a TG/FT-IR Unit

The samples of the defrosted coal, CTPs, and blends of coal with CTPs, with a grain size of <0.2 mm, were pyrolyzed under the nitrogen atmosphere of high purity to the temperature of 750 °C according to the methodology described in the work [25]. The samples, having approximately 25 mg in weight, were heated in a platinum crucible of a Q50 thermobalance with the heating rate of 4 °C·min^{-1}. The formed volatile products were directed to the interface via a transfer line and next to a Nicolet iS10 spectrometer Madison, WI, USA). This way, the FT-IR spectra of the volatile products were registered in the wavenumber range of 4000–600 cm^{-1}. The elaboration of the FT-IR spectra was made by an OMNIC 9 software.

2.4. Extraction of Plasticized Samples

The extraction of plasticized samples of coal and its blends with CTPs was conducted using a mixture of chloroform and methanol (50:50) in an ultrasonic bath for 10 h at room temperature. After the extraction was completed, the extractant was distilled under a vacuum. The residue after distillation was kept in a vacuum dryer at a temperature of 25 °C until the constant weight was reached. The yield of extraction was calculated on the basis of mass of the obtained extract. The swollen grains, maximally swollen grains, and the material from the gas-saturated and compacted zones were treated according to the same procedure [25]. The material condensed on the surface of the swollen grains of the carbonized charge [26] was rinsed from that surface. The material extracted from the gas-saturated and compacted zones of the plastic layer was also obtained.

2.5. Spectroscopic Investigation

The samples of material that are soluble in the chloroform-methanol mixture were studied by a Nicolet iS10 spectrometer using a Smart MIRacle module (the ATR technique). In this investigation, a diamond monocrystal was used. The spectra were registered in 64 scans in the wavenumber range of 4000–600 cm^{-1}. The registered spectra were normalized with regard to the band near 1640 cm^{-1}, which was present in the spectra of all samples. The ATR spectra were elaborated by an OMNIC 9 software. The baseline was corrected in order to eliminate the non-specific background.

The same samples of extracted material, weighing 0.0001 g, were dissolved in 50 mL of acetonitrile. The UV spectra were obtained using a JASCO V630 spectrometer (Tokyo, Japan) and normalized at the wavelength of 195 nm, and next, the deconvolution option was applied using a Spectra Manager software (version 2.08.04).

2.6. Microscopic Investigation of Obtained Samples

The samples separated from the gas-saturated zone and the zone of maximally swollen grains of coal were studied using a Quanta 3D FEG scanning electron microscope (SEM, Boynton Beach, FL, USA) with an accelerating voltage of 5 keV. The samples of the material extracted from the gas-saturated zone were studied with the use of a FEI Tecnai Osiris transmission electron microscope (Lincoln, NE, USA) with an X-FEG Schottky field emitter. The accelerating voltage was 200 kV [25].

2.7. X-ray Diffraction Investigation

The X-ray phase analysis of solid residues was carried out using an internal standard. As a standard, a powdered NaF (ACS reagent, ≥99% produced by Sigma-Aldrich, St. Louis, MO, USA) was applied, the (002) line of which is near the (002) line of the studied residues by its angular position. A pyrolytic graphite powder GPR^{TM} (produced by BDH Laboratory Supplies,) Poole, UK, having a turbostratic lamellar structure, was a model substance.

The semi-coke samples from coal and its blends with CTPs heated to a temperature of 650 °C were mixed with 10% of NaF. After the obtained blends were ground in an agate mortar, some rods, of 0.58 mm in diameter and approximately 13 mm in height, were formed from them. The samples prepared this way were studied using a polycrystal diffractometer in the mode U = 25 kV and I = 40 mA. The period of counting impulses was 10 s. The diffractograms were registered in the range of angles of 2Θ 15–42. The interplanar distances d_{002} and the amount of crystalline phase C_{cryst} were calculated according to the methodology presented in work [26].

The C_{cryst} parameter was calculated according to the formula:

$$C_{cryst} = \frac{P \cdot 100}{k \cdot x \cdot (100 - P)}$$

where P is the amount of internal standard (NaF) in the test sample, x is the ratio of the integral intensity of the (002) line of the internal standard (NaF) to that of the test sample, and k is an experimentally obtained coefficient from the calibration curve. The experimentally determined coefficient k is equal to 0.88.

The interplanar distances (d_{002}) were determined according to Wulff–Bragg formula:

$$d_{002} = \frac{n \cdot \lambda}{2 \cdot \sin \theta}$$

where n is the diffraction order ($n = 1$), λ is the wavelength of radiation ($\lambda = 1.5406$ nm), and θ is the angle of scattering.

3. Results and Discussion

3.1. Influence of CTP Additives Having Different Softening Points on the Course of Co-Carbonization Process of Defrosted Coal

Figure 2 presents the results obtained during the carbonization of coal and its blends with CTPs. There is a dark band visible in the X-ray images that is a projection of the gas-saturated zone on the X-ray film (Figure 2a). The X-ray images imply that the addition of every CTP to defrosted coal causes an increase in the thickness of the gas-saturated zone in the plastic layer of blends. The greatest thickness of these zones was observed for the blend of coal with CTP103. The SEM images of material from this zone (Figure 2b) prove that the zone is filled with some foam-like plastic mass. The SEM images (taken at a magnification of M300) imply that the cells of the plastic mass in the charges of coal and its blends with CTPs have different dimensions. The thickness of the cell walls also differs. The cells in the blend with CTP 103 have the largest sizes. The thickness of the plastic layer of coal (y) and its blends with CTPs is also different (Figure 2c). The thickness of the zones of the plastic layer (swollen grains, gas-saturated, and compacted ones) also differs. The addition of CTP86 causes a decrease in the thickness of the plastic layer but the addition of CTP103 causes an increase. Figure 2d presents the changes in position of the marker in the charge of carbonized coal and its blends with CTPs.

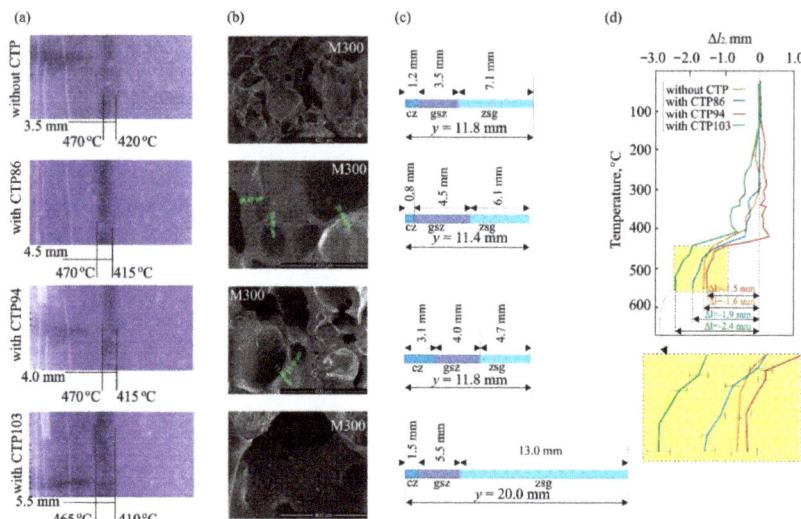

Figure 2. Results obtained during carbonization of coal and its blends with CTPs: (**a**) X-ray films of carbonized charge; (**b**) SEM images of gas-saturated zones in carbonized charge; (**c**) composition of plastic layers in carbonized charge; (**d**) changes in position of the marker in the charge of carbonized coal and its blends with CTPs.

These changes point out to changes in the volume of the layer of the heated charge between the marker and heating wall. The shape of the curve $\Delta l = f(T)$ in Figure 2d does not indicate any existence of a correlation between the changes in volume and softening points of CTPs. A greater shift in position of the marker, towards the negative values of Δl at the temperature of 550 °C, is observed in case of blends with commercial pitches, mainly with CTP86. This implies a greater volume reduction of the heated charge at the stage of its re-solidification and, hence, a greater density and compactness of the obtained semi-coke. The addition of CTP94 does not influence changes in the position of the marker towards the negative values of Δl at a temperature of 550 °C in contrast to coal without additives (Figure 2d).

3.2. Influence of CTP Additives Having Different Softening Points on the Course of Their Co-Pyrolysis with Defrosted Coal

Figure 3a presents the ATR spectra of the studied CTPs and the fragments of the normalized FT-IR spectra of their volatile products of pyrolysis emitted at a temperature of 300 °C.

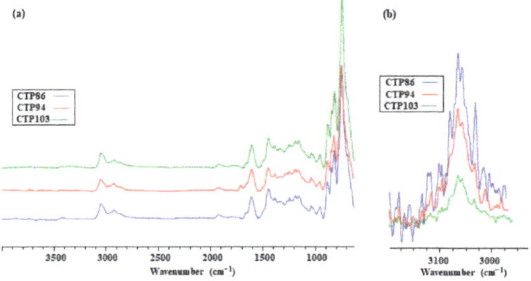

Figure 3. ATR spectra of CTPs (**a**) and fragments of FT-IR spectra of their volatile products of pyrolysis emitted at a temperature of 300 °C (**b**).

Analyzing the composition of CTPs, Diez et al. [27] stated that there are approximately 10,000 various chemical compounds having a distinct molecular mass present in them. It follows from the ATR spectra in Figure 3a that the presence of identical functional groups and groups of atoms is characteristic of the CTPs used in this research. Therefore, it was expected that these CTPs would interact with the organic substance of coal in the same way. Taking into account the differences in softening points of CTPs, it was assumed that the compounds of greater molecular mass present in CTPs having higher softening points would be included in the composition of volatiles at higher temperatures of pyrolysis. This assumption is confirmed by a fragment of the FT-IR spectra of volatile products emitted at a temperature of approximately 300 °C: more hydrocarbons are emitted from CTP86 than from CTPs having higher softening points. This implies that, under the influence of products emitted from CTP86, coal grains would soften at lower temperatures. This softening should cause an increase in the thickness of the zone of swollen grains in the plastic layer of the blend of coal with CTP86. However, it did not. It follows from Figure 2c that the zone of swollen grains in the plastic layer of the blend with CTP103 has the greatest thickness.

According to the conclusions made by the authors of works [3,20], the decomposition products of CTPs should interact with the surface of coal grains that have not yet softened. In order to prove these statements, Figure 4 presents the TGA curves of coal, CTPs, and their blends. In this figure, the course of TGA curves of blends clearly suggests that CTPs interact with coal substance in different way. This figure also gives the mass loss curves of coal blends with CTPs calculated at different temperatures based on the additivity rule according to the formula:

$$CBW(\%) = CW(\%) \cdot 0.98 + CTPsW(\%) \cdot 0.02$$

where CBW (%) is the calculated weight of blend of coal with CTPs, CW (%) is the experimentally measured weight of coal, $CTPsW$ (%) is the experimentally measured weight of coal tar pitch, and 0.98 and 0.02 are the relative proportions of coal and CTPs in the mixtures. The results of the calculations are shown in Table S1 in the Supplementary Materials.

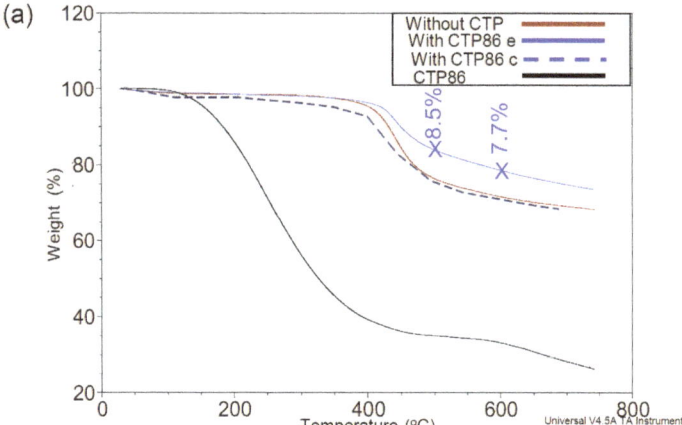

Figure 4. Cont.

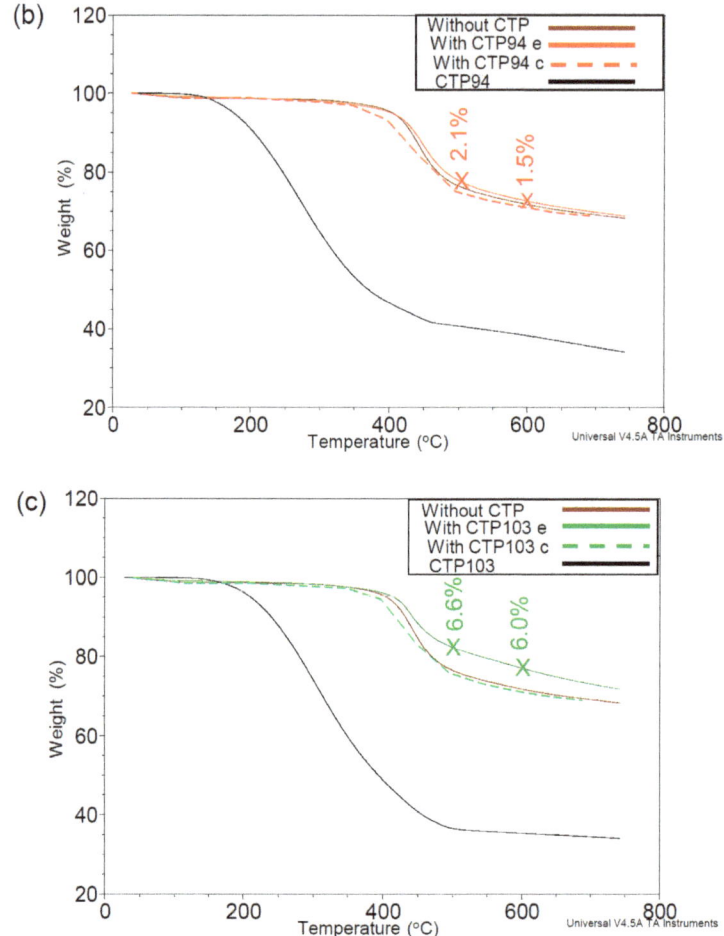

Figure 4. TGA curves of coal, its blends with CTPs and the CTPs: CTP86 (**a**), CTP94 (**b**), and CTP103 (**c**).

The values of the difference in the mass loss of experimental curves and those calculated taking into account the additivity rule are presented in the TGA curves. The values of deviations from the additivity rule were given for the temperatures of 500 and 600 °C. The experimental decrease of the mass loss for the blend of coal with CTP86 at these temperatures is 8.5 and 7.7% appropriately, for the blend of coal with CTP94–2.1 and 1.5%, and for the blend of coal with CTP103–6.6 and 6.0%.

It follows from the shape of the curves in Figure 4a,c that at temperatures above 400 °C, the CTP86 and CTP103 additives cause a decrease in the mass loss of blends compared to the mass loss of coal without additives. However, the experimental curve of blend with CTP94 almost coincides with the TGA curve of coal without additives (Figure 4b). CTP86 additive has a greater influence on the decrease in the mass loss of the blend.

Figure 5 presents the normalized FT-IR spectra of volatile products emitted from the blends of coal with CTPs at pyrolysis temperatures of 410 °C, 460 °C, and 500 °C.

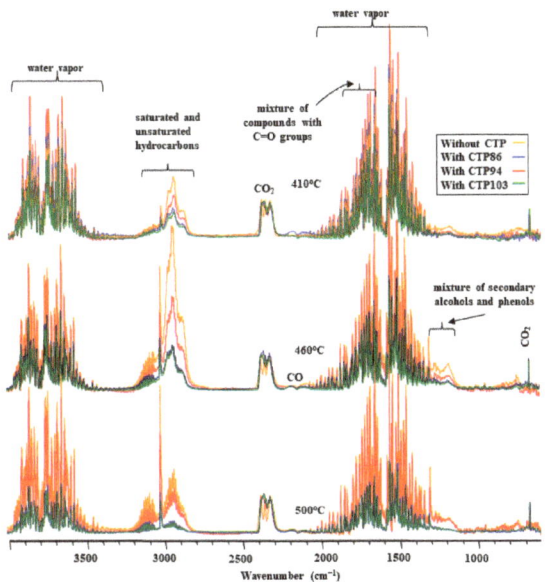

Figure 5. FT-IR spectra of volatile products emitted from coal and its blends with CTPs at pyrolysis temperatures of 410 °C, 460 °C, and 500 °C.

There are some differences in the composition of volatile products emitted from the blends at a temperature of 410 °C visible in these FT-IR spectra. This temperature corresponds to the occurrence of the zone of maximally swollen grains in the plastic layers of blends. In the FT-IR spectra of the volatile products from the blends in the wavenumber range of 3200–2800 cm^{-1}, the heights of the bands of saturated and unsaturated hydrocarbons are different. However, the heights of these bands do not show any dependence from softening points of CTPs. The greatest contribution ratio of hydrocarbons is observed in the composition of the volatiles of defrosted coal. Among the volatile products from blends of coal with CTPs, the blend of coal with CTP94 has a greater contribution ratio of hydrocarbons.

Nomura and Arima [3] stressed that the products of decomposition of coal-derived binder cause the plasticization of coal grains. Koch et al. [28] also suggested that volatile products of decomposition can migrate to the cold side of the oven and impregnate coal grains. The plasticizing influence of CTPs and the impregnation of grains should have found their reflection in the changes of relief of plasticized grains. Taking the aforementioned into account, it was reasonable to analyse the effects of this interaction with the coal grains in the zone of swollen grains of plastic layer and to present the characteristics of the material condensed on grains.

Figure 6 presents the SEM images of the fragments of the surface of swollen grains from the charge of coal and its blends with pitches. In contrast to the swollen grains of defrosted coal, there wasn't any presence of fluid drops detected on the surface of grains from the charge of blends with CTPs [2], i.e., there were no traces of leakage of thermobitumen from the swollen grains found in the charges of blends with CTPs.

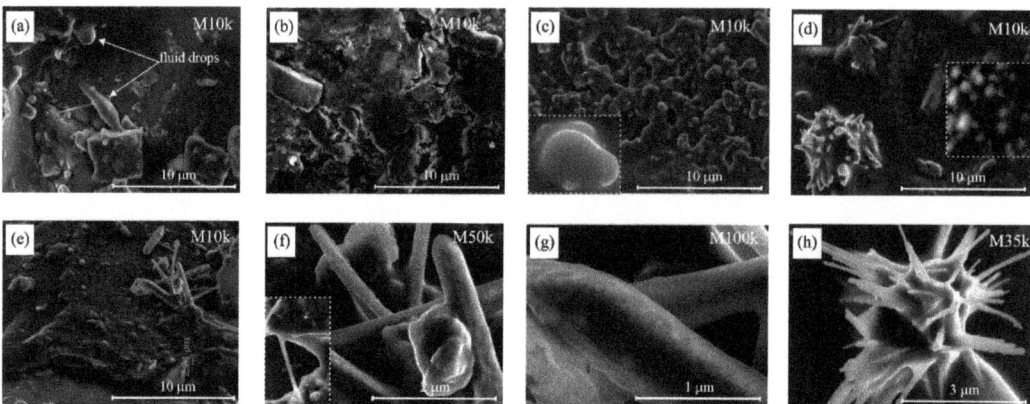

Figure 6. The SEM images of fragments of surface of swollen grains from the charge of coal and its blends with pitches. (**a**) surface of swollen grain in the charge of coal without CTPs, (**b**) surface of swollen grain in the charge of coal blend with CTP86, (**c**) surface of swollen grain in the charge of coal blend with CTP94, (**d**) surface of swollen grain in the charge of coal blend with CTP103, (**e**) crust on the surface of swollen grain in the blend with CTP103, (**f–h**) nano-objects on the surface of swollen grains in the charge of coal blend with CTP103.

It follows from Figure 6b,d that CTP additives, having distinct softening points, change the relief of the surface of swollen grains in different ways. This indicates that the volatile products formed during the carbonization of coal and blends interact with the surface of coal-swollen grains differently. This interaction results in the appearance of 'erosion' areas on the surface of the swollen grains in the charges of blends with CTP86 and CTP94 (Figure 6b,c) and the formation of a crust with nano-objects on the surface of swollen grains in the blend with CTP103 (Figure 6d–h). On the surface crust of the swollen grains from the blend with CTP 103 (Figure 6e), there are some nano-objects visible in the form of a bunch of tubes. The surface of grains from the blend with CTP94 (Figure 6c) and the tubes in Figure 6e look like clusters of connected globules approximately 100–200 nm in size (Figure 6g).

In the FT-IR spectra at a temperature of 460 °C (Figure 5), which corresponds to the occurrence of the gas-saturated zone in the plastic layer, the differences in height between the bands of hydrocarbons in the wavenumber range of 3200–2800 cm^{-1} and the bands in the range of 1300–1100 cm^{-1} become more explicit. A greater contribution ratio of saturated and unsaturated hydrocarbons is characteristic of volatile products of pyrolysis of coal without CTP additives. Wide bands that point out to the presence of hydrocarbons in the composition of volatiles from the blends of coal with CTP86 and CTP103 coincide. However, the bands of hydrocarbons, which originate from volatile products of the blend with CTP94, take the middle position. The FT-IR spectra of the volatile products imply that there is no dependence between the shape of the spectra and the softening points of the CTPs observed.

Volatile products that are formed and cumulated in the cells of the plastic mass can interact with each other and the material of the cell walls. Figure 7 presents the SEM images of the interior of cells of the material from the gas-saturated zone.

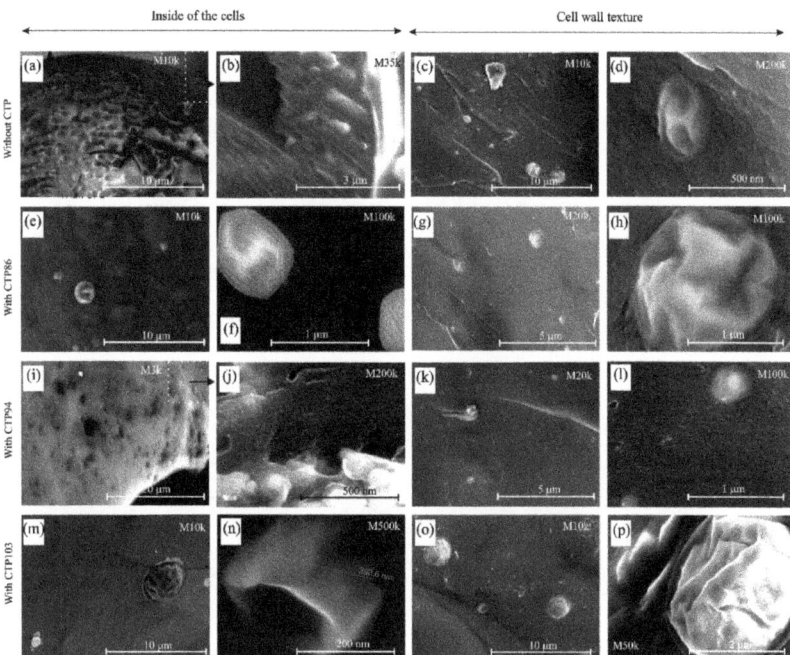

Figure 7. The SEM images of the material from the gas-saturated zone of coal and its blends with pitches. (**a,b**) insides of cells in gas-saturated zone of coal without CTPs, (**c,d**) cell wall textures in gas-saturated zone of coal without CTPs, (**e**) inside of cell in gas-saturated zone of coal with CTP86, (**f**) inside of cell in gas-saturated zone of coal with CTP86 with nano-objects, (**g,h**) cell wall textures in gas-saturated zone of coal with CTP86, (**i,j**) insides of cell in gas-saturated zone of coal with CTP94, (**k,l**) cell wall textures in gas-saturated zone of coal with CTP94, (**m**) inside of cell in gas-saturated zone of coal with CTP103, (**n**) inside of cell in gas-saturated zone of coal with CTP86 with nano-objects, (**o,p**) cell wall textures in gas-saturated zone of coal with CTP103.

At a magnification of M10k, on the surface of the interior of a cell of the plastic mass of coal without CTP (Figure 7a), there are some round concavities that were not found inside the cells of blends with CTPs. The interior of a cell of the gas-saturated zone from the blend with CTP 94 shows the signs of the gasification of the material, even at a magnification of M3k (Figure 7i). The interior of the cells in the gas-saturated zone from blends with commercial CTPs looks more compact but there are some flat nano-objects resembling yarn hanks present on the surface of a cell from the blend with CTP86. This may be the pleated balloons, the folds of which are regularly arranged. The change in direction of the arrangement of folds takes place at an angle of 120 degrees (Figure 7f). On the surface of a cell of plastic mass in the blend with CTP103, there also were some balloons that were partly deformed and partly round and empty inside (Figure 7n).

The texture of the material of the cell walls of the plastic mass of coal with CTPs (Figure 7g,k,o) has a more compact appearance than that without CTP additives. It is characteristic that in the material of the cell walls, there are some objects that resemble deformed balloons of various shapes and sizes. The fact that the balloons remained non-crushed on the surface of a broken sample points to a greater mechanical resistance of their material. It is suggested that the films making balloons were formed as a result of the condensation of the hydrocarbons from volatile products created in the gas-saturated zone of coal and its blends with CTPs. The formation of these films, as a result of polymerization of compounds present in the composition of volatile products, could have been one of the reasons for a decrease in contribution of saturated and unsaturated hydrocarbons in the composition of volatile products of pyrolysis of coal with CTPs. The permanent formation and the closing of cells in the plastic mass of coal and its blends with CTPs could have caused the presence of balloons in the cell walls of the plastic mass.

At the temperatures of occurrence of the compacted zone (T = 500 °C), the bands originating from the blends of coal with CTP86 and CTP103 coincide with the FT-IR spectra of volatiles (Figure 5), as well as the bands originating from the defrosted coal and its blend with CTP94 which almost overlap. The shape and height of the bands of normalized FT-IR spectra of the blends of coal with CTPs in the range of 3100–2800 cm^{-1} (Figure 5) do not show any continuity of changes that would imply the dependence of these changes from the softening point of CTPs.

3.3. Analysis of Condensates and Extracts

Table 3 presents the yields of material that was obtained during the treatment of plasticized samples by ultrasounds in the chloroform-methanol mixture.

Table 3. The yield of the material extracted from the zones of plastic layer.

Samples	Condensed Material		Extracted Material	
	350–380 °C	380–410 °C	430–460 °C	470–500 °C
Without CTP [2]	1.35 ± 0.10	2.69 ± 0.03	1.46 ± 0.05	0.83 ± 0.04
With CTP86	1.09 ± 0.06	0.48 ± 0.09	1.35 ± 0.11	1.44 ± 0.07
With CTP94	1.41 ± 0.13	2.79 ± 0.12	2.07 ± 0.09	1.61 ± 0.06
With CTP103	0.69 ± 0.12	1.05 ± 0.11	0.83 ± 0.06	0.98 ± 0.05

The data presented in Table 3 imply that the amounts of the material rinsed from the surface of grains and extracted from the gas-saturated and compacted zones are distinct but do not depend on the softening point of CTPs. The addition of commercial pitches (CTP86 and CTP103) causes a decrease in the amount of obtained material compared to the yield of material from defrosted coal. The only exception is the material from the compacted zone. In contrast, the addition of CTP94 causes an increase in the amount of the obtained material.

Shui et al. [29] pointed out to the existence of a dependence between the amount of extracted material and the degree of coal plasticization. It follows from Figure 2a,c that the thickness of the gas-saturated zone in the plastic layer of coal with CTPs increases, i.e., the range of appearance of the viscous-liquid state in plastic layer widens. On the other hand, the yield of the material soluble in the chloroform-methanol mixture in the presence of CTP86 and CTP103 additives decreases. It was suggested in previous research [2] that the defrosting of coal causes the mechanical destruction of the organic substance of coal and the formation of radicals. The formed radicals can easily interact with hydrocarbons with low ionization potential in the volatile products of decomposition of CTPs and initiate the reactions of dimerization and polymerization taking place with their participation [30]. In this way, compounds of greater molecular mass are formed. These compounds are not extracted by the chloroform-methanol mixture but, despite this, they play the role of plasticisers of the organic substance of coal. A similar phenomenon of the influence of pitch on a decrease in the amount of polycyclic aromatic hydrocarbons in the material extracted from plasticized coals was stated in previous research [16]. The values of the yield of the material soluble in the chloroform-methanol mixture in Table 3 cannot be considered as dependent on the softening point of CTPs.

Figure 8 presents the normalized UV spectra of material that was rinsed from the surface of swollen grains (Figure 8a,b) and extracted from the gas-saturated and compacted zones (Figure 8c,d). In the spectra of the material from the charge with CTPs that was rinsed from the layer of grains heated to the temperature of 350–380 °C (Figure 8a), there are same types of compounds with chromophore groups and unsaturated bonds that are present in the material of defrosted coal without additives. However, the shape of the normalized spectra implies that the addition of CTP103 causes the appearance of a higher absorbance in the UV spectra (the range of 225–325 nm). This points to a greater concentration of compounds with chromophore groups.

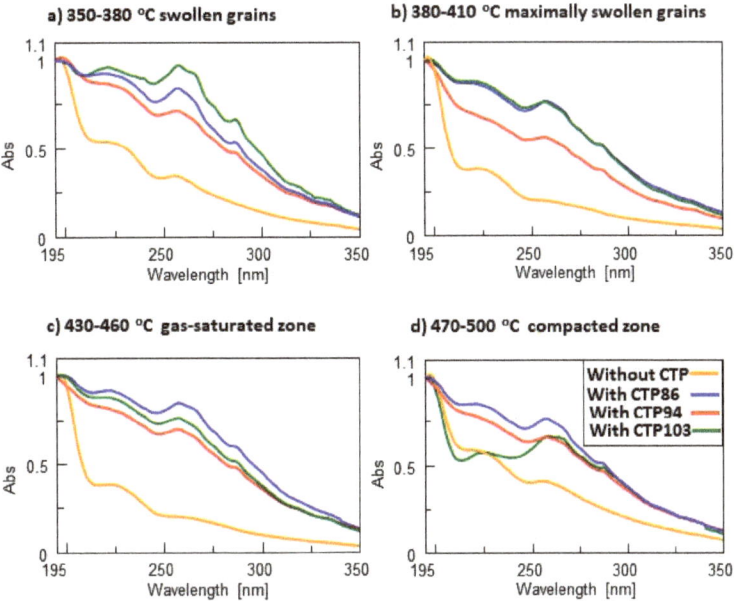

Figure 8. The normalized UV spectra of material that was rinsed from the surface of swollen grains (a,b) and extracted from the gas-saturated and compacted zones (c,d).

The material from the zone of maximally swollen grains from the charge with CTP86 and CTP103 additives (Figure 8b) has the same absorbance in the UV spectra. Among the materials from the gas-saturated and compacted zones, the material with the CTP86 additive (Figure 8c,d) has the highest absorbance in the UV spectra. In terms of absorbance, the material with the CTP94 additive takes the lowest position. However, compared with the UV spectrum of material from the charge without additives, its position is higher. This tendency does not repeat itself for the material from the compacted zone because the material from the blend with CTP 103 additive has a lower absorbance in the range of wavelength of 195–225 nm. It implies the absence of cyclic compounds and aromatic rings with chromophore groups in this material.

Figure 9 presents the ATR spectra of material that was rinsed from the surface of the grains and extracted from the gas-saturated and compacted zones. The comparison of shape of the normalized ATR spectra shows that the contribution ratio of the groups of C_{al}-H type in the range of 3000–2800 cm^{-1} decreases in the material condensed on the surface of grains from the charge with CTP additives. In the wavenumber range of 3600–3000 cm^{-1}, the contribution of H-bonds (namely self-associated −OH) increases in the material from blends with commercial pitches. In the material extracted from the blend with CTP103, the contribution ratio of bonds of C_{al}-H type (the range of 3000–2800 cm^{-1}) decreases and the contribution ratio of H-bonds (self-associated -OH and tightly bound cyclic OH-tetramers) increases [31].

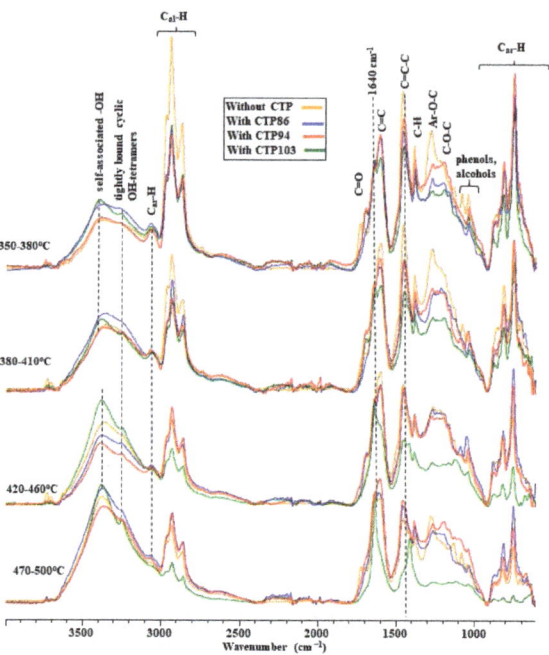

Figure 9. ATR spectra of material that was rinsed from the surface of grains and extracted from the gas-saturated and compacted zones.

In the ATR spectra of the material condensed in the wavenumber range of 1800–1600 cm^{-1}, there are some bands that point out to the presence of saturated and unsaturated esters (the exception is the material from the charge of coal with CTP86). In the material extracted from the blend with CTP103, the bands attesting to the presence of aromatic rings (C=C bonds near 1600 cm^{-1}) and the bands indicating the presence of esters disappear in this range. In the material from the blend with CTP103 in the fingerprint range of 1400–1000 cm^{-1}, the absorbance of all bands decreases gradually with the rise of temperature and the bands corresponding to the presence of deformation stretches of C$_{ar}$-H type in the range of 900–600 cm^{-1} disappear. The disappearance of bands of these stretches in the extract from the gas-saturated zone of blend with CTP103 implies that dehydrocracking processes of substituted aromatic compounds were activated. Hydrogen released during these processes can participate in the reactions of its disproportionation. The shape of ATR spectra that are presented in Figure 9 does not point to the appearance of any continuity of the changes in structural-chemical parameters for studied samples with regard to the softening points of CTP additives.

The utmost differences in shape of the bands corresponding to the appearance of hydrogen bonds are observed in the ATR spectra for the extracts from the gas-saturated zone. This implies that there are some differences in the ability of the material of the extract to aggregate and form supramolecular structures. Figure 10 presents the results of the research on material that was extracted from the gas-saturated zones of investigated charges of coal and its blends with CTPs with the use of a scanning transmission electron microscope (STEM).

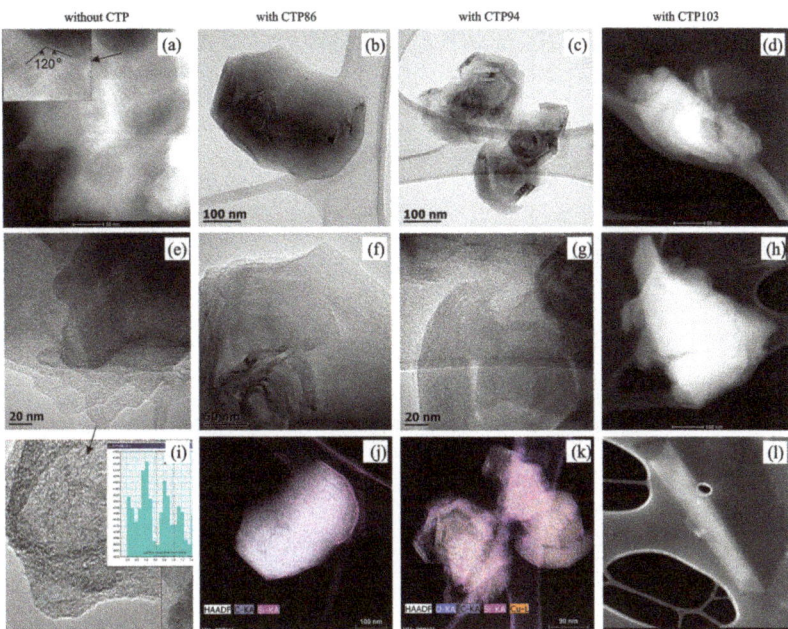

Figure 10. STEM-HAADF images of material that was extracted from the gas-saturated zones of investigated charges of coal and its blends with CTPs.

The visualization of the extracted material implies that the compounds present in it are able to form various topological structures (Figure 10a–l). It follows from the STEM-HAADF images that the material extracted from the gas-saturated zone of coal without CTP additives consists of several superimposed layers. In these layers, structural elements are arranged parallel to each other but at an angle of approximately 120° with respect to the elements of neighbouring layers (Figure 10a). Such arrangement of the elements resembles topological structures similar to the smectic mesophase.

It was determined on the basis of the measurements made in the TEM mode (Figure 10e) that the distance between the chains put together on the edges of the layers is 0.39 nm (Figure 10i). In the material extracted from the zones of the plastic layer of coal with CTP86 (Figure 10b,f) and CTP94 pitches (Figure 10c,g), there were some layers twisted similar to supramolecular structures of axialites [32]. The images in Figure 10b,f and Figure 10c,g show that the angle of twisting of layers is near to 120 degrees. The particles of the material extracted from the gas-saturated zone from the blend of coal with CTP94 have sizes smaller than the particles in the extract from the blend with CTP86. The elemental mapping proves that axialites from the blend of coal with CTP86 contain C and Si atoms (Figure 10j), whereas axialites from the blend of coal with CTP 94 (Figure 10k) contain C, Si, O, and Cu atoms. There were no topological structures and layer arrangement found in the material extracted from the blend of coal with CTP103. It is known that the interaction of polymer with low molecular weight compounds having mesophase properties leads to the occurrence of mesophase phenomenon in the polymer itself [33]. The nature of low molecular weight material present in the gas-saturated zone (the capability of formation of supramolecular structures) may give the plastic mass of the blends with CTP86 and CTP94 some unusual properties that determine the thermal behaviour of coal during the carbonization process. The material extracted from the gas-saturated zone of the blend with CTP103 does not show such properties. The disappearance of bands in the fingerprint range of the material extracted from the gas-saturated zone of the blend with CTP 103 can be a sign of the obsolescence of the mesophase properties of extracted material [34].

Taking into account the statement about the role of CTP as a plasticizing agent of the organic coal substance made by Świetlik et al. [7], an XRD investigation of the obtained semicokes was carried out. The degree of ordering of the structure of material in coal blends was evaluated on the basis of an analysis of obtained diffractograms (Figure 11).

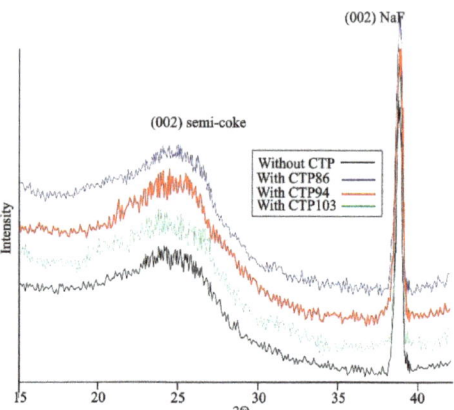

Figure 11. Diffractograms of semi-cokes from carbonized charge of coal and its blends with CTPs.

The data in Table 4 imply that all CTPs behave as plasticizing agents with regard to coal and facilitate a better ordering of the structure of semicokes.

Table 4. The structural parameters of obtained semi-cokes.

Samples	d_{002}, Å	C_{cryst}, %
Semi-coke from charge without CTPs	3.55	50.5
Semi-coke from charge with CTP86	3.55	57.4
Semi-coke from charge with CTP94	3.54	61.3
Semi-coke from charge with CTP103	3.53	59.0

More crystallites (higher value of the C_{cryst} parameter) are obtained in case of the use of CTP94, but during the formation of cokes with this additive, a greater mass loss will be observed. A better ordering of lamellas inside the crystallites (lower value of parameter d_{002}) is provided by CTP103 additive; in the semicokes from this blend the average values of interplanar distances are 3.53 Å. In the sequence: coal → blend of coal with CTP86 → blend of coal with CTP94, there is a tendency of increase of the degree of ordering into crystallites. However, the parameters of the structure of semicoke from the blend with CTP103 fall out from that tendency.

The aforementioned suggests that in the blends of coal with CTP86 and CTP94 an increase in the degree of plasticization can be caused by the mesophase properties of the material of own extracts. However, in case of the blend of coal with CTP103, an increase in degree of plasticization can be connected with the activation of the disproportionate processes of hydrogen that are facilitated by CTPs in the role of a hydrogen donor in relation to the organic substance of coal. This suggestion was based on the disappearance of the band corresponding to the presence of bonds of C_{ar}-H type in the extracts that can point to a greater contribution of groups with these bonds in the structure of crystallites having a better ordering.

3.4. Physicochemical Investigation of CTPs

Additional investigation was needed in order to explain the causes of the interaction of CTPs with the studied coal as described above. Figure 12 presents the comparison of the curves of changes in the mass loss of CTPs depending on the temperature (Figure 12a), the fragments of FT-IR spectra of CTPs at the temperatures of 440 and 560 °C (Figure 12b), and the SEM images of the relief of the pyrolyzed CTPs (Figure 12c). It follows from the data in Figure 3a that the presence of the same functional groups and the groups of atoms is characteristic of studied CTPs.

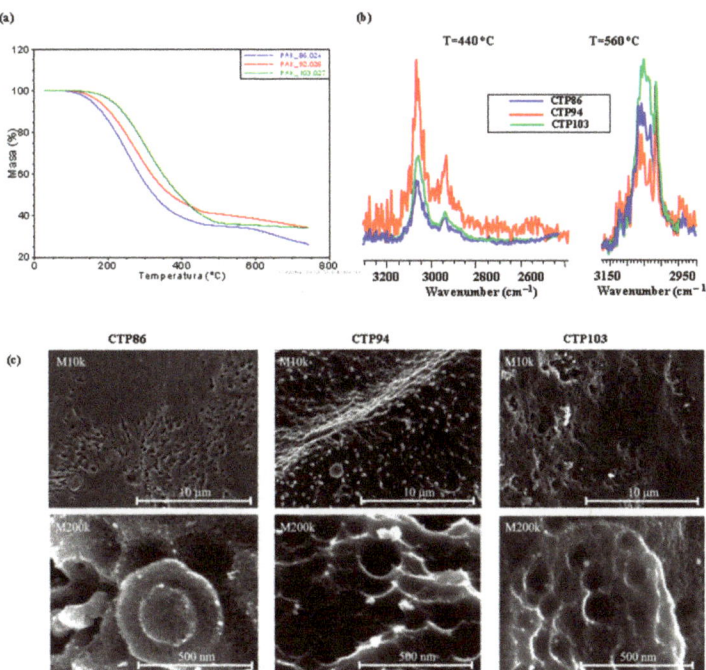

Figure 12. Results of pyrolysis of CTPs: TGA curves of CTPs: (**a**); fragments of normalised FT-IR spectra of volatile products mitted from CTPs at the pyrolysis temperatures of 440 °C and 560 °C (**b**); SEM images of the relief of pyrolyzed CTPs at a temperature of 750 °C (**c**).

This implied that during pyrolysis CTPs would behave in a similar way but the inclinations in the TGA curves for CTPs having higher softening points would be shifted towards higher temperatures. The mass losses of CTPs are consistent with their softening points to the temperature of approximately 350 °C; more volatile substances are emitted from the pitch having a lower softening point during pyrolysis (Figure 12a). Starting from a temperature of 400 °C, this dependency is disrupted. The height of the FT-IR bands of volatile products normalized with respect to the CO_2 band shows that at a temperature of 440 °C, CTP94 emits more hydrocarbons than CTP86 and CTP103, and at a temperature of 560 °C, CTP86 emits more volatile hydrocarbons than CTP94. The surface relief of the CTPs pyrolyzed at T = 750 °C differs substantially, and the changes in relief do not indicate any dependence from the softening points of CTPs. The SEM images of the surface relief of CTP86 and CTP103 have more common topological features at magnifications of both M10k and M200k.

There is an opinion that the softening point and coking yield are connected with the volatility of a CTP [35]. The data presented in Figures 4 and 12a do not allow us to support this opinion. According to another opinion, there should exist some dependency between the amount of released volatile substances in the temperature range of 400–500 °C along with the donor and acceptor abilities of pitch and the development of coke structure. However, the data in Table 3 do not prove this suggestion. It is suggested that the disproportionation processes of hydrogen may be influenced by heteroatoms that are mainly the centres of hydrogen acceptors in the structure of pitch [7].

The results of an EDX microanalysis of the pyrolyzed CTPs presented in Table 5 indicate that an increase in the softening point is accompanied by the rise of the contribution of C atoms in CTPs. The number of other atoms does not point to any dependence from their temperature. In CTP86, a high content of Zn atoms and their dispersion on the surface of pyrolyzed sample draws attention. In the pyrolyzed CTP103 sample, these atoms are observed to a much lesser degree, and in the CTP94 sample it was not identified.

Table 5. The elemental composition of pyrolyzed CTPs.

CTP	Elements, wt. %					
	C	O	S	Zn	Na	Fe
CTP86	80.5 ± 2.8	12.3 ± 2.6	1.1 ± 0.5	6.1 ± 4.9	-	0.16 *
CTP94	83.1 ± 1.3	13.5 ± 1.3	0.5 ± 0.2	-	2.87 ± 2.5	-
CTP103	86.2 ± 0.4	9.6 ± 2.5	1.0 ± 0.4	1.1 ± 0.8	1.2 ± 0.8	1.0 ± 0.9

* It occurs only in one of analyzed areas.

A great amount of the Fe atoms were determined in some areas of the pyrolyzed CTP103 sample but these atoms were also present in the CTP86 sample. Great amounts of Na atoms were identified in the CTP94 sample but they were also present in pyrolyzed CTP103 sample.

The analysis of data in Figures 3 and 4c and in Table 3 implies that the pyrolytic behaviour of CTPs can be influenced by the atoms of the metals present in them. This may also influence the interaction of CTPs with coal. It cannot be excluded that the presence of Fe atoms could have contributed to the formation of nano-structures on the surface of the swollen grains in the charge of coal with CTP103 (Figure 6e–h), and that of Zn atoms, contributing to the formation of 'yarn hanks' inside the cells of the gas-saturated zone in the blend of coal with CTP86 (Figure 7f).

4. Conclusions

The conducted research does not allow us to state the existence of a direct dependence between the softening point of the pitch and its influence on the thermal behaviour of defrosted coal. Under the influence of CTPs, the thickness of the gas-saturated zone in plastic layer increases in the sequence CTP94 → CTP86 → CTP103. Under the influence of CTP94, the thickness of the plastic layer does not change, CTP86 causes a decrease in the thickness of the plastic layer, and CTP103 causes an increase in the thickness layer.

It was determined that the additions of CTPs to the defrosted coal influenced the yield of volatile products emitted from blends with CTPs to a different degree that is independent from their softening points. CTP, having a softening point of 94 °C, does not change the yield of volatiles, but CTP additives, having a softening points of 86 and 103 °C, cause a decrease in the yield of volatile products of pyrolysis. A greater effect is observed in the case of the CTP86 additive. CTP additives decrease the contribution ratio of saturated and unsaturated hydrocarbons in the composition of volatile products of pyrolysis in the temperature range of 420–500 °C, wherein CTPs, having the softening points of 86 and 103 °C, intensify such decreases. It is suggested that the decrease in the contribution of hydrocarbons in the composition of volatiles of the blend of defrosted coal with CTP86 was caused by their condensation into nano-objects in the cells of plastic mass in the gas-saturated zone. In the blend of coal with CTP103, such decrease in the contribution of hydrocarbons can be connected with the formation of nano-objects on the surface of swollen grains. CTPs influence the composition of extracts from the gas-saturated zone of blends (change the content of compounds with chromophore groups), the topological structure of material of extracts, and their ability to form supramolecular and mesophase structures with organic substance of coal in different ways. This causes the lack of existence of any dependence between the degrees of ordering of obtained semi-cokes and the softening points of CTPs. The ordering changes in the sequence: semi-coke from blend with CTP86→ semi-coke from blend with CTP103→ semi-coke from blend with CTP94.

It is suggested that this lack of dependence between the softening points of CTPs and their influence on the thermal behaviour of coal is caused by the presence of the atoms of metals (Fe and Zn) in pitches that change the course of the carbonization of CTPs themselves and their influence on organic substance of coal in blends with CTPs.

Supplementary Materials: The following supporting information can be downloaded at: https://www.mdpi.com/article/10.3390/ma15249027/s1. Table S1. The values of calculated weight of blends of coal with CTPs.

Author Contributions: V.Z.: Conceptualization, Methodology, Writing—original draft, Text preparation, Data analysis, Funding acquisition, Project administration; A.S.: Conceptualization, Writing—original draft, Investigation, Data analysis, Project administration. All authors have read and agreed to the published version of the manuscript.

Funding: This research was funded by Jan Kochanowski University, grant number SUPB.RN .21.191 (01.01.2021-31.12.2022).

Institutional Review Board Statement: Not applicable.

Informed Consent Statement: Not applicable.

Data Availability Statement: The study did not links to publicly archived datasets analyzed or generated during the study.

Acknowledgments: The authors thank Franciszek Krok from the Institute of Physics of the Jagiellonian University in Krakow, Poland for the opportunity to use the SEM microscope.

Conflicts of Interest: The authors declare that they have no known competing financial interests or personal relationships that could have appeared to influence the work reported in this paper.

References

1. Miroshnichenko, D.V.; Desna, N.A.; Kaftan, Y.S. Oxidation of coal in industrial conditions. 2. Modification of the plastic and viscous properties in oxidation. *Coke Chem.* **2014**, *57*, 375–380. [CrossRef]
2. Zubkova, V.; Strojwas, A. The Influence of Freezing on the Course of Carbonization and Pyrolysis of a Bituminous High-Volatile Coal. *Energies* **2020**, *13*, 6476. [CrossRef]
3. Nomura, S.; Arima, T. Influence of binder (coal tar and pitch) addition on coal caking property and coke strength. *Fuel Process. Technol.* **2017**, *159*, 369–375. [CrossRef]
4. Tiwari, H.P.; Saxena, V.K. 8—Industrial Perspective of the Cokemaking Technologies. In *New Trends in Coal Conversion. Combustion, Gasification, Emissions, and Coking*; Suarez-Ruiz, I., Rubiera, F., Diez, M.A., Eds.; Woodhead Publishing: Sawston, UK, 2019; pp. 203–246. [CrossRef]
5. Fernández, A.M.; Barriocanal, C.; Díez, M.A.; Alvarez, R. Influence of additives of various origins on thermoplastic properties of coal. *Fuel* **2000**, *88*, 2365–2377. [CrossRef]
6. Fernández, A.M.; Barriocanal, M.A.; Díez, R.; Alvaerz, R. Evaluation of bituminous wastes as coal fluidity enhancers. *Fuel* **2012**, *101*, 45–52. [CrossRef]
7. Świetlik, U.; Gryglewicz, G.; Machnikowska, H.; Machnikowski, J.; Barriocanal, C.; Alvarez, R.; Diez, M.A. Modification of coking behaviour of coal blends by plasticizing additives. *J. Anal. Appl. Pyrolysis* **1999**, *52*, 15–31. [CrossRef]
8. Tramer, A.; Zubkova, V.; Prezhdo, V.; Wrobelska, K.; Kosewska, M. Influence of coal tar–water emulsion on carbonisation of caking coals. *J. Anal. Appl. Pyrolysis* **2007**, *79*, 169–182. [CrossRef]
9. Castro-Díaz, M.; Vega, M.F.; Barriocanal, C.; Snape, C.E. Utilization of Carbonaceous Materials to Restore the Coking Properties of Weathered Coals. *Energy Fuels* **2015**, *29*, 5744–5749. [CrossRef]
10. Vega, M.F.; Fernandez, A.M.; Diaz-Faes, E.; Casal, M.D.; Barriocanal, C. The effect of bituminous additives on the carbonization of oxidized coals. *Fuel Process. Technol.* **2017**, *156*, 19–26. [CrossRef]
11. Barriocanal, C.; Diaz-Faes, E.; Florentino-Madiedo, L. The effect of briquette composition on coking pressure generation. *Fuel* **2019**, *258*, 116128. [CrossRef]
12. Nomura, S. Coal briquette carbonization in a slot-type coke oven. *Fuel* **2016**, *185*, 649–655. [CrossRef]
13. Zubkova, V.; Strojwas, A. Influence of briquetting on thermal behavior of bituminous coal oxidized during storage. *J. Anal. Appl. Pyrolysis* **2020**, *152*, 104969. [CrossRef]

14. Montiano, M.G.; Diaz-Faes, E.; Barriocanal, C. Effect of briquette composition and size on the quality of the resulting coke. *Fuel Process. Technol.* **2016**, *148*, 155–162. [CrossRef]
15. Duffy, J.J.; Mahoney, M.R.; Steel, K.M. Influence of coal thermoplastic properties on coking pressure generation: Part 2—A study of binary coal blends and specific additives. *Fuel* **2010**, *89*, 1600–1615. [CrossRef]
16. Zubkova, V.; Czaplicka, M.; Puchala, A. The influence of addition of coal tar pitch (CTP) and expired pharmaceuticals (EP) on properties and composition of pyrolysis products for lower and higher rank coals. *Fuel* **2016**, *170*, 197–209. [CrossRef]
17. Nomura, S. The effect of binder (coal tar and pitch) on coking pressure. *Fuel* **2018**, *210*, 810–816. [CrossRef]
18. Mochida, I.; Marsh, H. Carbonization and liquid-crystal (mesophase) development. 11. The co-carbonization of low-rank coals with modified petroleum pitches. *Fuel* **1979**, *58*, 797–802. [CrossRef]
19. Grint, A.; Marsh, H. Carbonization and liquid-crystal (mesophase) development. 20. Co-carbonization of a high-volatile caking coal with several petroleum pitches. *Fuel* **1981**, *60*, 513–518. [CrossRef]
20. Sakurovs, R.; Lynch, L.J. Direct observations on the interaction of coals with pitches and organic compounds during co-pyrolysis. *Fuel* **1993**, *72*, 743–749. [CrossRef]
21. Nomura, S. 12—The Development of Cokemaking Technology Based on the Utilization of Semisoft Coking Coals. In *New Trends in Coal Conversion. Combustion, Gasification, Emissions, and Coking*; Suarez-Ruiz, I., Rubiera, F., Diez, M.A., Eds.; Woodhead Publishing: Sawston, UK, 2019; pp. 335–365. [CrossRef]
22. Geny, J.F.; Duchene, J.M. Effect of selected additives in the wall pressure. *Cokemak. Int.* **1992**, *4*, 21–25.
23. Marzec, A.; Alvarez, R.; Casal, D.M.; Schulten, H.R. Basic phenomena responsible for generation of coking pressure: Field ionization mass spectrometry studies. *Energy Fuels* **1995**, *9*, 834–840. [CrossRef]
24. Zubkova, V. Some peculiarities of formation mechanism of metallurgical coke from polish coals. *Fuel* **2004**, *8*, 1205–1214. [CrossRef]
25. Zubkova, V.; Strojwas, A.; Kaniewski, M.; Ziomber, S.; Indyka, P. Some aspects of influence of the composition of volatile products and extracted material on grain swelling processes and volume changes of commercial coals of different rank. *Fuel* **2019**, *243*, 554–568. [CrossRef]
26. Zubkova, V.; Prezhdo, V.; Strojwas, A. Comparative Analysis of Structural Transformations of Two Bituminous Coals with Different Maximum Fluidity during Carbonization. *Energy Fuels* **2007**, *21*, 1655–1662. [CrossRef]
27. Diez, M.A.; Garcia, R. 15—Coal Tar: A By-Product in Cokemaking and an Essential Raw Material in Carbochemistry. In *New Trends in Coal Conversion. Combustion, Gasification, Emissions, and Coking*; Suarez-Ruiz, I., Rubiera, F., Diez, M.A., Eds.; Woodhead Publishing: Sawston, UK, 2019; pp. 439–487. [CrossRef]
28. Koch, A.; Gruber, R.; Cagniant, D.; Pajak, J.; Krzton, A.; Duchene, J.M. A physicochemical study of carbonization phases. Part, I. Tars migration and coking pressure. *Fuel Process. Technol.* **1995**, *45*, 135–153. [CrossRef]
29. Shui, H.; Zheng, M.; Wang, Z.; Li, X. Effect of coal soluble constituents on caking property of coal. *Fuel* **2007**, *86*, 1396–1401. [CrossRef]
30. Kubisa, P. Ionic Polymerization. In *Chemistry of Polymers*; Florjanczyk, Z., Penczek, S., Eds.; Oficyna Wydawnicza Politechniki Warszawskiej: Warsaw, Poland, 1995; Volume 1, pp. 199–233. (In Polish)
31. Li, D.; Li, W.; Chen, H.; Li, B. The adjustment of hydrogen bonds and its effect on pyrolysis property of coal. *Fuel Process. Technol.* **2004**, *85*, 815–825. [CrossRef]
32. Rabek, J.F. *Modern Knowledge of Polymers*; Wydawnictwo Naukowe PWN SA: Warsaw, Poland, 2008. (In Polish)
33. Stanczyk, W. Liquid Crystal Polymers. In *Chemistry of Polymers*; Florjanczyk, Z., Penczek, S., Eds.; Oficyna Wydawnicza Politechniki Warszawskiej: Warsaw, Polish, 1998; Volume 3, pp. 179–192. (In Polish)
34. Fanjul, F.; Granda, M.; Santamaria, R.; Menendez, R. On the chemistry of the oxidative stabilization and carbonization of carbonaceous mesophase. *Fuel* **2002**, *81*, 2061–2070. [CrossRef]
35. Russo, C.; Ciajolo, A.; Stanzione, F.; Tregrossi, A.; Oliano, M.M.; Carpentieri, A.; Apicella, B. Investigation on chemical and structural properties of coal- and petroleum-derived pitches and implications on physico-chemical properties (solubility, softening and coking). *Fuel* **2019**, *245*, 478–487. [CrossRef]

Article

Influence of Microstructure on Dynamic Mechanical Behavior and Damage Evolution of Frozen–Thawed Sandstone Using Computed Tomography

Junce Xu [1], Hai Pu [1,2,*] and Ziheng Sha [1]

1. State Key Laboratory for Geomechanics and Deep Underground Engineering, China University of Mining and Technology, Xuzhou 221116, China
2. College of Mining Engineering and Geology, Xinjiang Institute of Engineering, Urumqi 830091, China
* Correspondence: haipu@cumt.edu.cn

Abstract: Frost-induced microstructure degradation of rocks is one of the main reasons for the changes in their dynamic mechanical behavior in cold environments. To this end, computed tomography (CT) was performed to quantify the changes in the microstructure of yellow sandstone after freeze–thaw (F–T) action. On this basis, the influence of the microscopic parameters on the dynamic mechanical behavior was studied. The results showed that the strain rate enhanced the dynamic mechanical properties, but the F–T-induced decrease in strength and elastic modulus increased with increasing strain rate. After 40 F–T cycles, the dynamic strength of the samples increased by 41% to 75.6 MPa when the strain rate was increased from 75 to 115 s^{-1}, which is 2.5 times the static strength. Moreover, the dynamic strength and elastic modulus of the sample were linearly and negatively correlated with the fractal dimension and porosity, with the largest decrease rate at 115 s^{-1}, indicating that the microscopic parameters have a crucial influence on dynamic mechanical behavior. When the fractal dimension was increased from 2.56 to 2.67, the dynamic peak strength of the samples under the three impact loads decreased by 43.7 MPa (75 s), 61.8 MPa (95 s), and 71.4 MPa (115 s), respectively. In addition, a damage evolution model under F–T and impact loading was developed considering porosity variation. It was found that the damage development in the sample was highly related to the strain rate and F–T damage. As the strain rate increases, the strain required for damage development gradually decreases with a lower increase rate. In contrast, the strain required for damage development in the sample increases with increasing F–T damage. The research results can be a reference for constructing and maintaining rock structures in cold regions.

Keywords: computed tomography; F–T action; microstructure; dynamic mechanics; damage evolution

1. Introduction

In cold regions, F–T weathering can lead to disasters such as the collapse of rock structures, landslides, and mudslides [1,2]. The microstructural remodeling caused by ice crystal pressure is the main reason for altering the mechanical behavior of rocks [3]. In addition, rock structures in cold environments are often affected by dynamic loads [4]. For example, the slopes of open-pit mines are not only affected by frost, but also susceptible to dynamic loads such as blasting and mechanical disturbances [5]. Therefore, studying microstructural changes under F–T action in response to dynamic loads is of great practical importance for the construction and maintenance of rock structures in cold regions.

The dynamic mechanical properties of rocks in cold regions have been extensively studied. The Split Hopkinson Pressure Bar (SHPB) system is one of the important devices for performing dynamic tests [6]. For instance, Xu et al. [7] investigated the dynamic mechanical properties of sandstone under F–T action using the SHPB system and found that the dynamic strength and modulus decreased exponentially with F–T damage, and developed a model to predict the decrease in strength and modulus. Li et al. [8] studied the influence of microstructure on the dynamic mechanical characteristics of the sandstone under F–T action using SHPB and nuclear magnetic resonance (NMR) systems, and pointed out that large pores are the key factor affecting the dynamic strength. Liu et al. [9] performed SHPB tests on the rock with different F–T cycles and concluded that the more severe the F–T damage, the greater the degree of rock fragmentation. In addition, Xu et al. [10] investigated the energy evolution of the sandstone during impact and suggested that F–T damage increases the dissipation energy during impact, which in turn increases the macroscopic fragmentation of the sample. Zhang et al. [11] investigated the effects of F–T action and impact loading on the brittleness of rock using the SHPB system and showed that the higher the number of F–T cycles, the lower the dynamic brittleness index, and concluded that increased porosity was the main reason. The above work mainly studied the macroscopic dynamic mechanical behavior of rocks induced by F–T action and did not quantify the effects of F–T action on microstructure. However, it is difficult to consider the damage mechanism of rocks only from the macroscopic point of view [12]. In fact, the evolution of the microstructural properties of rocks is the main cause of the changes in their macroscopic mechanical behavior [13]. Therefore, the study of the effects of F–T cycles on the dynamic behavior of rocks should begin at the microscopic level [14].

In recent years, many researchers have applied various methods, including acoustic emission (AE), scanning electron microscopy (SEM), NMR, and CT, to reveal the decay characteristics and damage mechanisms of rocks from a microscopic perspective. Fang et al. [15] used the SEM method to observe rocks subjected to different F–T cycles. They found that F–T damage is often due to pre-existing defects and that the initiation and propagation of cracks correlates closely with the original pore structure of the rock. Cheng et al. [16] reported the influence of F–T action on the microscopic pores of the rock based on NMR, and concluded that F–T damage promotes the development of various pore sizes, especially medium and large pores in the rock. Xu et al. [17] analyzed the promotion of the rock damage process using the AE method and found that the AE signals were more active in the sample with higher F–T cycles. However, the SEM method can only observe the deterioration of the sample surface, and the AE method is disturbed by environmental interference, while the NMR method cannot detect the morphological features of the pores; moreover, it is difficult to relate the changes in the parameters to the macroscopic mechanical properties [12]. The CT technique, as a non-destructive testing approach, can analyze the rock degradation process caused by F–T action from a microscopic point of view [18]. Yang et al. [19] evaluated the evolution of soft rock microstructure under F–T action using the CT technique and adopted the CT number as a variable to define F–T damage. Fan et al. [20] conducted real-time CT tests on sandstones with different freezing temperatures and noticed that the damage pattern of sandstones below $-10\ °C$ tended to be stable. In addition, Maji et al. [21] performed a 3D reconstruction of the sandstone based on the CT technique to study the microstructural deterioration of the samples with different F–T damage. Li et al. [1] combined CT and visualization techniques to analyze F–T damage and argued that inherent defects in rocks are the main factor for F–T damage. In general, the internal structural deterioration from F–T action is responsible for the variations in the macro-mechanical behavior of rocks [22]. However, these studies are limited to the F–T damage mechanisms without connecting changes in microscopic parameters to mechanical behavior, exceptionally dynamic mechanical behavior.

To this end, this study investigates the effects of microscopic parameters such as porosity, permeability, and fractal dimension on the dynamic mechanical behavior of sandstone through CT scanning and SHPB tests. Based on the CT results, a damage model of the sample under F–T and impact loading was developed. This work can serve as a reference for the construction and maintenance of slopes in open pit mines in cold areas.

2. Materials and Experimental Methods

2.1. Materials

The yellow sandstone for the work was sampled from the slope of an open pit mine in Urumqi, China. The selected yellow sandstone is a common sedimentary rock in the Xinjiang region of China. At the same time, the Xinjiang region belongs to the seasonally cold areas and the rock structures are susceptible to F–T weathering. Homogeneous blocks of the sandstone were selected to machine cylindrical samples with a diameter of 50 mm according to ISRM recommendations. Two aspect ratios were machined, namely, 1 for dynamic tests and 2 for static tests (see Figure 1). The lateral surfaces of the samples need to be vertical and smooth, with no significant structural surfaces. In addition, both ends of the samples were ground to a tolerance of 0.02 mm [8]. Then, the dry density and longitudinal wave velocity of the samples were measured to select samples with similar wave velocity and density for this study. A total of 54 samples were selected for the dynamic tests and divided into three groups for the F–T tests. Moreover, 21 samples with a length of 100 mm were selected to conduct static tests. To maximize spatial resolution, samples with a diameter of 20 mm were chosen for the CT scanning tests. Three static samples were selected for the basic physico-mechanical tests. The results are listed in Table 1. The main mineral composition of the sandstone was determined by X-ray diffraction tests (Bruker AXS, Karlsruhe, Germany, product model D8-A), and the results are presented in Figure 2.

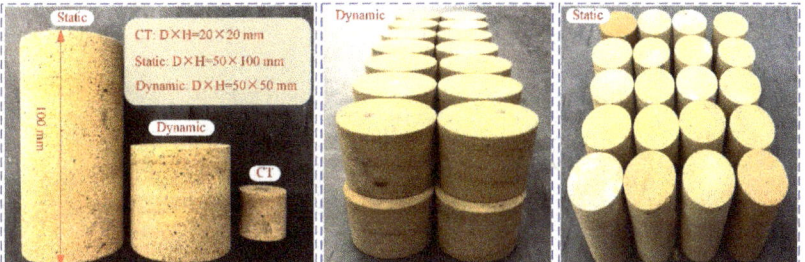

Figure 1. Diagram of static and dynamic samples.

Table 1. Mean values of physico-mechanical parameters of yellow sandstone.

V_p (km/s)	ρ_d (g/cm³)	ρ_{sat} (g/cm³)	n (%)	σ_p (MPa)
2623	2.14	2.27	11.3	58.7

Notes: V_p is the longitudinal wave velocity; ρ_d is the dry density; ρ_{sat} is the saturation density; n is the porosity; σ_p is the uniaxial compressive strength.

2.2. Experimental Methods

2.2.1. F–T Tests

To accelerate the weathering process, a rapid F–T chamber (Liaoning Fushun Instrument Co., Ltd., Fushun, China, product model QDR-50) was used in this study (see Figure 3). The chamber can operate in a temperature range of −30 to 80 °C with a control accuracy of ±0.5 °C. The cyclic F–T number in this study was set to 0, 5, 10, 20, 30, and 40 times according to the temperature variations in the cold areas. According to ASTM D5312, the F–T process was performed with a cycle of −20 °C freezing for 4 h and 20 °C thawing for 4 h. In this work, only saturated sandstone was tested for F–T cycles.

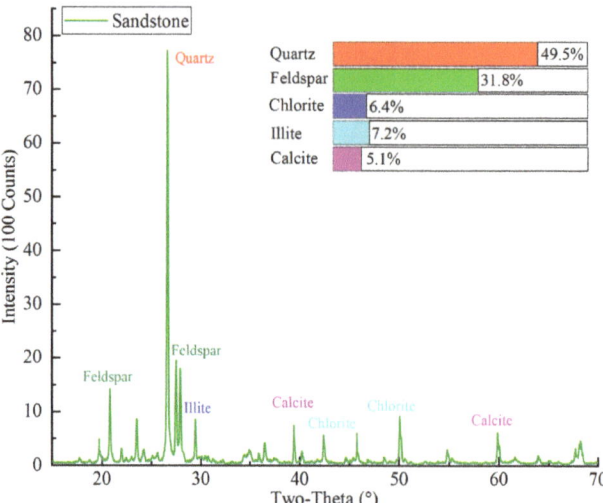

Figure 2. Results of X-ray diffraction test.

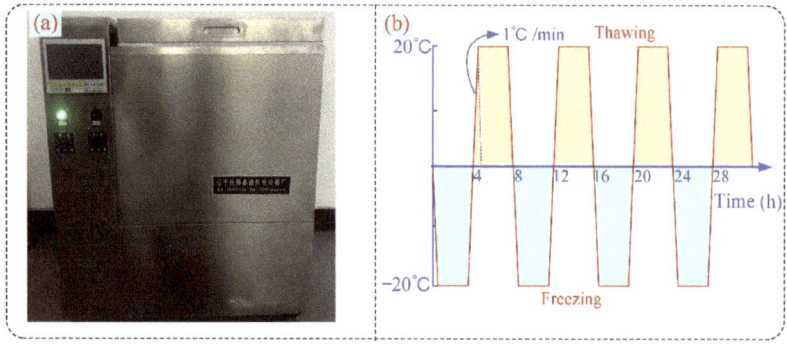

Figure 3. F–T chamber (**a**) and cyclic temperature setting (**b**).

2.2.2. CT Scanning Tests

As shown in Figure 4, a Zeiss Xradia 510 Versa high-resolution 3D X-ray microscope was used to perform CT scans of the samples with different F–T cycles. The instrument mainly consists of an X-ray source, an X-ray detector, and a multi-function base. The instrument is imaged by passing the target sample through the radiation source and measuring the attenuation energy to extract the density data of the object. In this procedure, the sample is rotated 360° to obtain thousands of vertical 2D images in all directions, which are reconstructed into approximately thousands of horizontal 2D images by computer processing [23]. When these 2D images are stacked sequentially, they can be displayed as a 3D image. In the binarization technique, pixels below the threshold are called voids, and pixels above the threshold are called rock matrix, and in turn, the voids and rock are separated [24]. That is, the pore skeleton can be extracted and used to analyze the length or connectivity of the pores. The structure of the pores can also be visualized in 3D, allowing quantification of the size, orientation, and interconnectivity of the pores in the rock. The scanning time for each sample was about 2 h. The main acquisition parameters of the system during scanning are listed in Table 2, where the current and voltage parameters can influence the number of X-rays and the intensity of penetration [25]. The CT scan of the entire rock sample yielded about 1100 horizontal images with 1024 × 1024 pixels in 16-bit grayscale. Since the CT scan is a non-destructive technique, the scan does not affect the subsequent F–T tests.

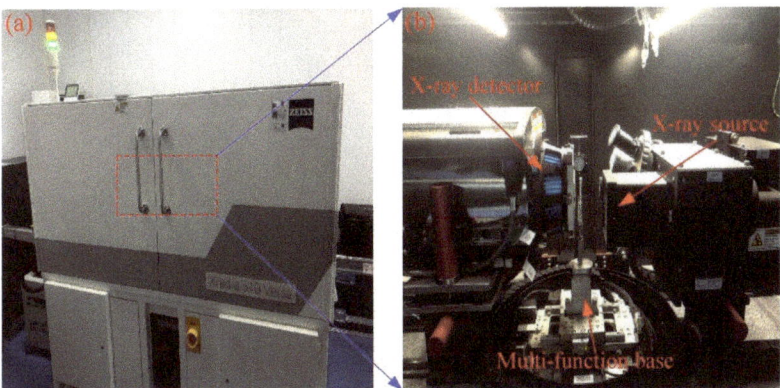

Figure 4. Zeiss Xradia 510 Versa high-resolution CT system (**a**) and interior (**b**).

Table 2. Parameters for the Zeiss Xradia 510 Versa CT.

Voltage (kV)	Current (μA)	Exposure Time (s)	Resolution (μm)	The Number of Scan Images
100	90	5	20	1100

In this study, raw data from CT tests were processed using Avizo (2019.1 version) software (Thermo Fisher Scientific Co. Shanghai, China), which provides noise reduction, threshold segmentation, binarization, and 3D reconstruction of CT images [26]. By identifying the pores and matrix of the rock sample, Avizo can calculate the porosity of the sample using Equation (1), where V and V_d denote the volume of the sample and the defect volume, respectively. Also, the permeability of the sample can be determined using the XLabSuite module integrated into Avizo. Note that the XLabSuite module uses Darcy's law (Equation (2)) to calculate permeability and can only estimate the connected pores in the 3D diagram [26].

$$P = \frac{V_d}{V} \tag{1}$$

$$k = \frac{Q\mu L}{A \Delta P} \tag{2}$$

where k is the absolute permeability, μm^2; Q is the flow rate of water, $\mu m^3/s$; μ is the viscosity of the liquid, which for water at room temperature is 0.001 Pa·s; L is the length of the seepage channel, μm; A is the outlet area of the liquid, μm^2; p is the pressure difference between import and export, Pa, which in Avizo is 10 kPa by default.

2.2.3. Static Tests

In this study, static uniaxial compression tests were performed on the freeze–thawed samples using the MTS816 system (MTS USA, Inc. Minneapolis, MN, USA), as shown in Figure 5. The MTS816 system measured the axial displacement of the samples via a linear variable displacement transducer (LVDT), recorded the axial load, and determined the stress–strain relationship of the specimens. To better control the loading process, the displacement control method was used during the test and the displacement rate was set to 0.1 mm/min to meet the requirements of quasi-static loading.

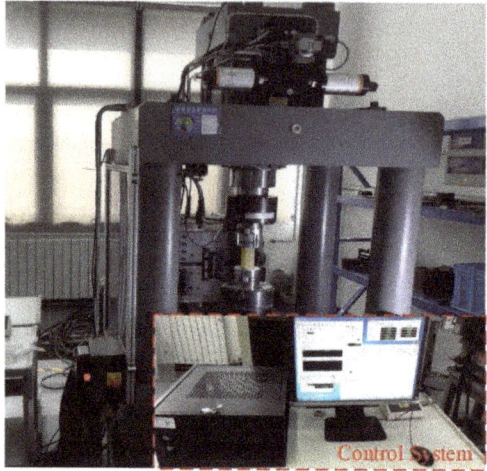

Figure 5. MTS816 experimental system.

2.2.4. Dynamic Impact Tests

As shown in Figure 6, impact tests were performed on sandstone samples using a SHPB system. The system mainly consists of four parts: an impact load system, a compression bar system, a data acquisition system, and a damping system. In the SHPB system, both the striker and the bar are 50 mm in diameter and are made of Cr40 with a P-wave velocity of 5400 m/s, and yield strength and elastic modulus of 800 MPa and 208 GPa, respectively. To obtain a suitable waveform, a circular rubber pad (D × H = 10 mm × 2 mm) was selected as the pulse shaper. Before the test, molybdenum disulfide was applied to the ends of the samples as a lubricant to reduce the friction effect [27]. When the high-pressure gas drives the striker against the incident bar, an elastic compression wave is generated. A portion of the compression wave can propagate through the sample to the transmitted bar. Based on the 1D stress wave propagation principle, the three-wave analysis method was adopted to calculate the dynamic loads P_1 and P_2, strain rate, and strain at both ends of the sample (Equation (3)) [5]. In addition, to ensure dynamic load equilibrium at both ends of the samples during the test, the stress waves at both ends need to be verified before the test [28]. After all samples had undergone the specified F–T cycles, dynamic tests with three impact loads were performed with the SHPB system to obtain the macroscopic dynamic mechanical properties of the yellow sandstone.

$$\begin{cases} P_1(t) = A_r E_0[\varepsilon_i(t) + \varepsilon_r(t)], \ P_2(t) = A_r E_0 \varepsilon_t(t) \\ \dot{\varepsilon}(t) = \frac{C_0}{L}[\varepsilon_i(t) - \varepsilon_r(t) - \varepsilon_t(t)] \\ \varepsilon(t) = \int_0^t \dot{\varepsilon}(t)dt = -2\frac{C_a}{L}\int_0^t \varepsilon_r(t)dt \end{cases} \quad (3)$$

where $\varepsilon_i(t)$, $\varepsilon_r(t)$, and $\varepsilon_t(t)$ refer to the incident, reflected, and transmitted waves, respectively; A_r is the ratio of the cross-sectional area of the bar to the sample; E_0, C_0, and L are the elastic modulus and P-wave velocity of the bar and the length of the sample, respectively.

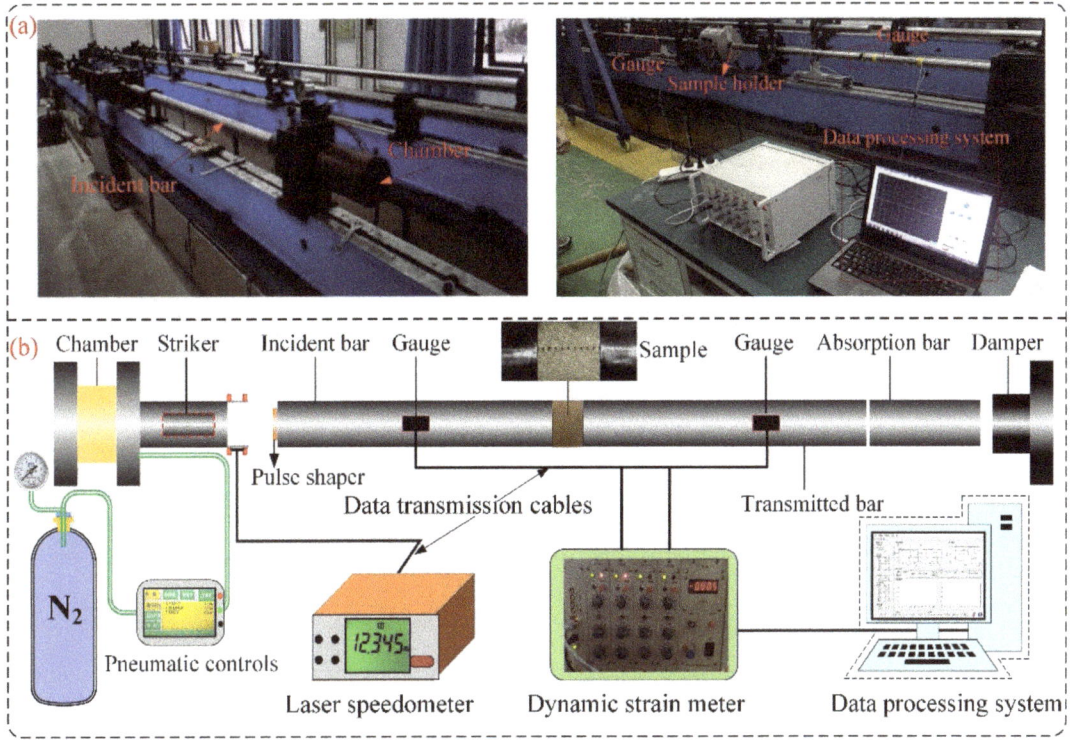

Figure 6. SHPB experimental system (**a**) and detailed schematic diagram (**b**).

3. Experimental Results and Discussion

3.1. 3D Reconstruction

As shown in Figure 7, slice data from the CT were imported into Avizo for 3D reconstruction. Compared to 2D images, 3D images can provide a visual assessment of the spatial structure and reflect the development of F–T damage in the rock. Since the artifacts at the sample edges could not be effectively avoided, and the data volume of the whole rock sample was large, the central cubic region was cropped out for 3D reconstruction and analysis with Avizo (Figure 7a). To avoid the influence of volume on the experimental results, the representative elementary volume (REV) was determined semi-automatically using the Auto-refresh tool in Avizo. This command can calculate the volume fraction of pores or matrix starting from multiple voxels with an arbitrary segmentation threshold, where the smallest volume at which the volume fraction tends to a stable value is the REV. Ultimately, a volume of $750 \times 750 \times 750$ voxels was chosen, corresponding to the actual physical size of the $15 \times 15 \times 15$ mm^3 cube. The Register Image command was used for spatial alignment to ensure the same spatial position of the rock sample under different F–T conditions [26]. It is worth mentioning that all CT results were matched in grayscale to minimize the impact of the CT scanning process on the experimental results.

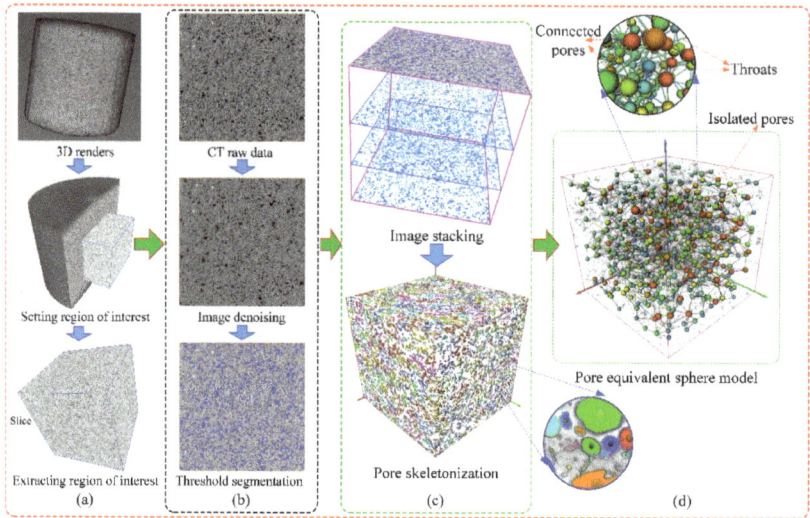

Figure 7. CT image processing and reconstruction process. (**a**) Region of interest; (**b**) 2D image processing; (**c**) 3D reconstruction; (**d**) PNM model.

In addition, the median filtering method was chosen to reduce noise points and artifacts caused by factors such as the environment and the rock itself, thus smoothing the distribution of greyscale values in the region (Figure 7b). This method can remove almost all of the noise without affecting the original image. At the same time, filtering the images can improve the image quality to speed up the processing and visualization of the experimental data from CT. In this study, the interactive segmentation algorithm in Avzio was used to perform threshold segmentation of pores and structures in CT images. To determine a suitable segmentation threshold, the actual porosity of the samples was first tested using the mercury intrusion method [12]. Then, the porosity of the 3D samples was calculated using Equation (1) based on the grayscale distribution features of the CT images, and the segmentation threshold T = 4076 between the rock matrix and the pores was determined using the backstepping method. Because greyscale matching was performed for the CT images, only the segmentation threshold T was required to complete the segmentation of all data. Finally, the 3D reconstruction model of the sandstone subjected to F–T action was developed (Figure 7c). In order to statistically capture the data of the defects in the 3D model, such as the pore equivalent diameter, throat length, etc., the PoreNetworkModeling (PNM) method in Avizo was used in this study, as shown in Figure 7d. Here, the pore is equated to a sphere using the equivalent sphere method, while the microcrack is equated to a throat [26]. The 3D images show that the sandstone pores are divided into isolated and interconnected pores, some of which are connected by throats.

3.2. Effect of F–T on Microscopic Parameters

As shown in Figure 8, based on the 3D reconstruction model, the variation of porosity and permeability of the samples was calculated using Equations (1) and (2). Porosity and permeability are closely related to the nature of the rock and are important parameters to describe the pore network in the rock. As seen in Figure 8, porosity and permeability of the samples increase with F–T cycles, and the exponential function can describe the changes of the parameters well. After 40 F–T cycles, the porosity increased to 16.25%, which is 1.3 times higher than the samples without F–T action. Similarly, the permeability of the samples subjected to 40 F–T cycles increased by 128% to 0.742 μm^2. The increased porosity and permeability of the sample provides favorable conditions for migration of water and heat during freezing or thawing, which in turn promotes damage to the rock by F–T action.

As a natural material, rocks indeed contain a large number of pores and microcracks. The pore water turns into ice at low temperatures, causing volume expansion. When the pore ice does not have enough space to expand freely, frost heaving will occur [4]. At the same time, the pores and microcracks are gradually enlarged by the frost heaving force. During thawing, pore water flows through the microcracks into other pores, and at this time, the microcracks play the role of seepage channels. For this reason, the microcracks are also called throats. In addition, the water in the rock promotes the expansion of the pores and microcracks through lubrication, erosion, hydrolysis, and F–T action [29]. Meanwhile, some isolated pores develop into interconnected pores through throats. Therefore, the porosity and permeability of the sample simulated with XLabSuite increase with the F–T action. In addition, the increase in porosity and permeability also means an increase in defects in the samples, which leads to a deterioration of the mechanical properties under load.

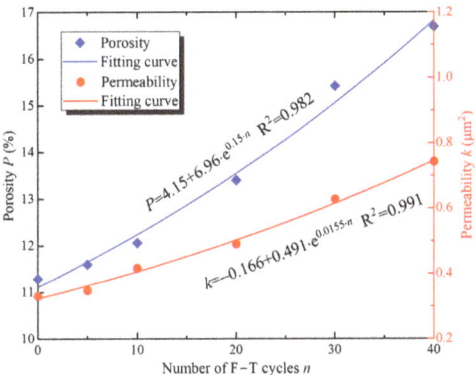

Figure 8. Changes in porosity and permeability of the sample with F–T cycles.

Additionally, the pore parameters of the sample, such as pore equivalent diameter and throat length, were determined using the PNM model. The pore diameters were divided into six ranges according to their sizes, i.e., 0–200 μm, 200–300 μm, 300–400 μm, 500–600 μm, and above 600 μm, and the results are shown in Figure 9a. After 40 F–T cycles of the sample, the total number of pores increased from 20,832 to 24,160 by about 15.98%, and the number of pores in all ranges showed an increasing trend. According to the distribution of pore diameter values, the minimum increase in the number of pores (<200 μm) was 12.9%, and the maximum (>600 μm) was 114.6%. In general, the larger the pore equivalent diameter, the more significant the percentage growth in the number of pores after F–T action. In other words, the larger the pore diameter, the more susceptible it is to F–T action, and the increase in the number of pores implies an increase in F–T weathering damage. Moreover, Figure 9b shows the maximum pore diameter and pore number with the F–T action, and the increase is consistent with the pattern of change in porosity and permeability, i.e., an exponential increase. Compared to the samples without F–T action, the maximum pore diameter of the sample increased from 605 μm to 856 μm after 40 F–T cycles, an increase of about 41.4%. The pores also showed an exponential growth pattern, increasing by 12.3% to 23,408 after 30 F–T cycles. The increase in equivalent diameter and pores reflects the increase in pore volume, expressed macroscopically as an increase in porosity in Figure 8. Overall, the increase in the number of pores and the equivalent diameter of the sample caused the expansion and development of microcracks, which in turn caused macroscopic damage to the mechanical properties.

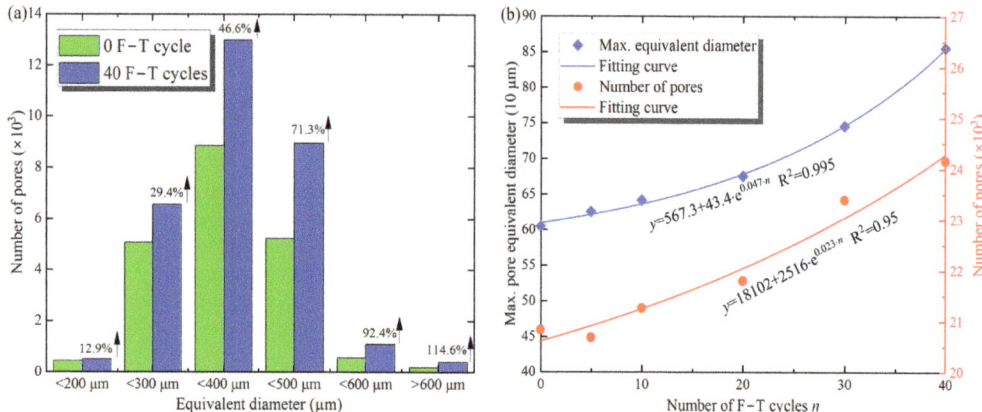

Figure 9. Distribution of the pore equivalent diameters (**a**) and the maximum and total number of pores (**b**) with F–T cycles.

Furthermore, Figure 10 shows the range of length distribution, number, and maximum length of throats with F–T cycles. From Figure 10a, the number of throat lengths shows an increasing trend in all ranges, and the greater the throat length, the more pronounced the increase. For example, the number of throats longer than 3000 μm increased by 107.2% after 40 F–T cycles, whereas the number of throats shorter than 1000 μm increased by only 19.3%, consistent with the pore growth trend. Besides, the changes in maximum length and the total number of throats with F–T action are shown in Figure 10b, and the exponential function can describe this increasing trend. After 40 F–T cycles, the maximum throat length increased from 3006 to 3285 μm, an increase of 9.3%, and the total number of throats increased by 37.4% from 4221 to 5893. Similar to the changes in pore parameters, the length and number of throats better reflect the microcrack development inside the sample. Therefore, the increase in the number and length of the throats indicates the cumulative macroscopic damage caused by the F–T action on the rock.

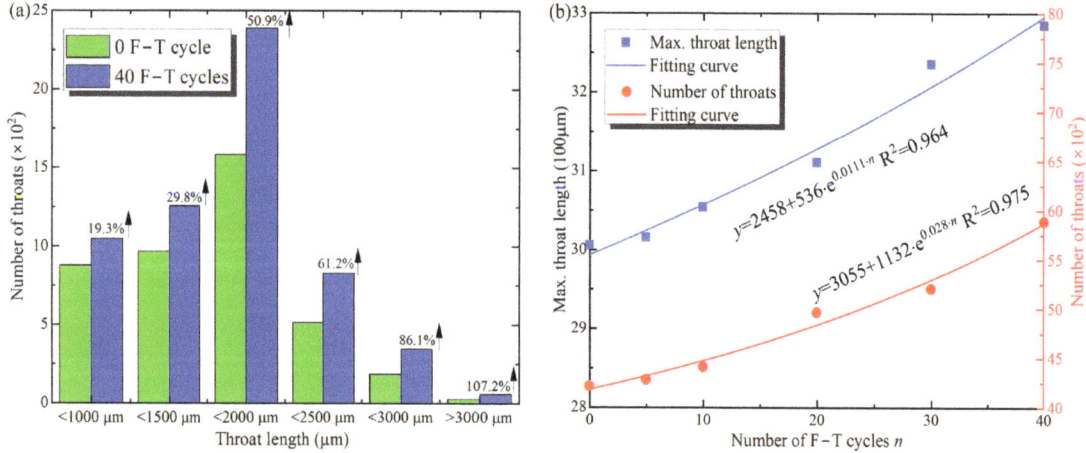

Figure 10. Distribution of the lengths (**a**) and the maximum lengths and total number (**b**) of throats with F–T cycles.

In summary, the internal damage in rocks, such as pores or microcracks, tends to increase with increasing F–T cycles, and especially larger defects increase more. The freezing of pore water in the sample at low temperature is closely related to the pore diameter, and the equation for the freezing point drop of pore water is as follows [18]:

$$T_{fre} = -273 \frac{2\sigma}{r\rho L} \tag{4}$$

where r is the equivalent diameter of the ice crystals, ρ is the ice density, σ is the pore interface tension, and L is the latent heat of the phase change. It is seen that the smaller the diameter of the ice crystals, the lower the freezing point of the pore water. Thus, as the temperature is below 0 °C, the supercooled water is in the small pores. In other words, the smaller the pore volume, the less it is affected by the cyclic F–T effect [29], which can also be seen in Figures 9 and 10, i.e., the smaller the pore volume, the lower the increase. Two main mechanisms cause F–T weathering. The first is the volume expansion mechanism, which causes a 9.05% increase in volume when the water (>91%) in pores, cracks, joints, and other voids turns to ice at below 0 °C [2]. When the pore water content exceeds 91%, it exerts a frost heaving pressure on the pore wall. At this point, in an ideal environment, the maximum stress of 207 MPa can be applied to the matrix, but in practice, a pressure of 10 to 100 MPa is applied to the matrix [14]. The second one is generated by the movement of unfrozen water in the rock. In this study, the volume expansion was considered significant due to the cyclic F–T performed on the saturated samples.

3.3. Effect of F–T on Fractal

To describe the influence of F–T action on the internal microstructural features of the sample, fractal theory, which responds to the spatial occupancy of complex shapes, was introduced as a measure of irregular, nonlinear shapes [30]. Therefore, the fractal theory is used to quantify the evolution of microstructure in rocks under F–T cycles using 3D CT technology. The fractal dimension is the most important parameter to describe the fractal. In general, the fractal dimensions mainly include the Hausdorff dimension, the assignment dimension, the box dimension, and the similarity dimension [31]. The box dimension is widely used because of its clear physical meaning [12]. In this study, the box dimension was used; that is, the 3D pore model of the sample was placed in a series of cubes with a grid side length ε. The number of grids containing pores or cracks in the cubes was $P(\varepsilon)$, and a series of $P(\varepsilon)$ was obtained by continuously adjusting the dimensions of the small grids. Then, the slope of the regression line is obtained using double logarithmic coordinates for $P(\varepsilon)$ and ε. The fractal dimension f, which corresponds to the edge length ε of the grid, is therefore given in Equation (5). Using the fractal dimension module in Avizo, the fractal dimension f of the central region was determined under F–T action [26]. The results are shown in Figure 11.

$$f = \lim_{\varepsilon \to 0} \frac{\ln P(\varepsilon)}{\ln(1/\varepsilon)} = -\lim_{\varepsilon \to 0} \frac{\ln P(\varepsilon)}{\ln(\varepsilon)} \tag{5}$$

As seen in Figure 11, the 3D fractal dimension of the samples ranged from 2.57 to 2.70, and the variation of the fractal dimension corresponded to the exponential function, which was consistent with the analytical results of porosity, permeability, and pore parameters. The greater the cyclic F–T damage, the larger the fractal dimension of the sample and close to 3. After 40 F–T cycles of the sample, the fractal dimension increases by 4.1%, from 2.574 to 2.679. During the test, the microstructure of the sample (pores or microcracks) gradually develops due to the F–T action, and the spatial distribution becomes more complex, suggesting that the pores are interconnected by throats to increase the spatial occupancy capability. The larger the fractal dimension, the greater the F–T damage to the internal microstructures of the sample. Thus, it is reasonable to characterize the damage to the sandstone microstructure by the fractal dimension. In addition, Figure 12 shows the relationship between the fractal dimension f and the porosity P, and permeability k of the samples under different cyclic F–T action.

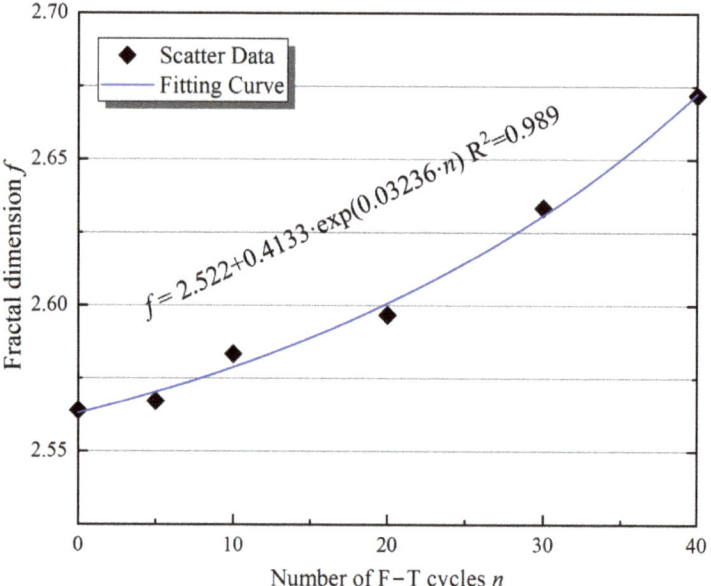

Figure 11. Variation of 3D fractal dimension of samples with different F–T cycles.

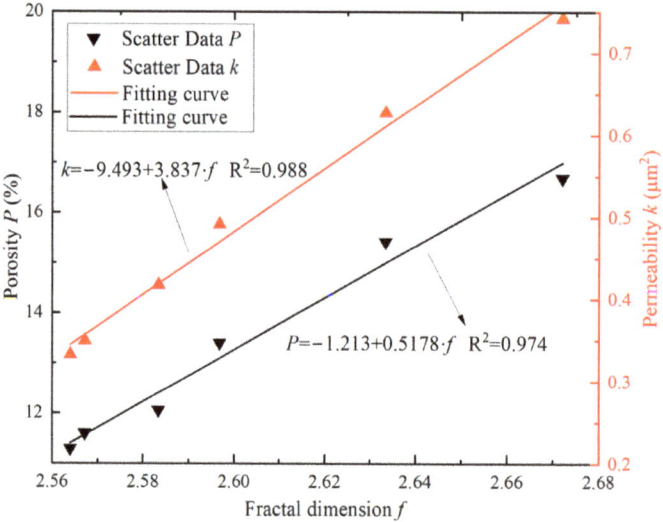

Figure 12. Changes in porosity and permeability with fractal dimension of the sample under the F–T action.

From Figure 12, the porosity and permeability of the samples are positively correlated with the fractal dimension. This is consistent with the variation of porosity, permeability, and pore parameters with the F–T cycles. Natural rock is a fractal with obvious fractal features such as porosity and permeability [32]. Hence, the larger the fractal dimension, the greater the variability of the microstructure, and the greater the porosity and permeability of the sample [12]. In other words, the larger the fractal dimension of the sample, the faster pores and microcracks (throats) develop, resulting in more significant damage to the internal structure of the sample and a more pronounced permeability phenomenon. At the same time, the increase in porosity and permeability provides convenient conditions for water and heat migration during freezing and thawing [33]. As the number of F–T cycles increases, the expansion effect of the ice gradually increases, resulting in a more rapid development of the internal structure (pores and throats) of the sample and a more complex spatial distribution, while the pores are interconnected by throats and have an increased ability to occupy space, so that the internal microstructure of the sample is increasingly damaged by the F–T action. Therefore, the fractal dimension can quantitatively describe the dynamic evolution process of the sample microstructure and effectively characterize the development of F–T damage. Since porosity and fractal dimension are positively correlated, porosity is also an effective parameter to describe F–T damage.

3.4. Macroscopic Dynamic Mechanical Properties

To investigate the effect of changes in microscopic parameters on the mechanical behavior of the samples, a series of static and dynamic tests were performed using an MTS816 and an SHPB test system. The stress–strain curves of the samples with different cyclic F–T action are shown in Figure 13. In the static tests, the peak strength of the samples decreased with the F–T damage. However, the ultimate deformation capacity (peak strain) exhibited an increasing trend, implying that the cyclic F–T damage resulted in a decrease in the bearing capacity and deformation resistance of the rock. For example, after 10 F–T cycles, the peak strength decreased by 14% to 46.2 MPa, while the peak strain increased by 15% to 1.125%. Compared with the static test, there was a strengthening effect of strain rate, i.e., the peak strength and peak strain increased with loading strain rate. When the strain rate was increased from about $75\ s^{-1}$ to $115\ s^{-1}$, the dynamic strength of the sample with 20 F–T cycles increased from 65 to 105 MPa, corresponding to 2.1 and 2.7 times the static peak strength, respectively. The strength and tangential elastic modulus under different loading conditions are presented in Figure 14. Here, the slope of the straight line segment in the stress–strain curve is used as the elastic modulus [27].

As shown in Figure 14, the peak strength (σ_P) and elastic modulus of the sample gradually decrease with the increase of F–T cycles, which is opposite to the variation of the microscopic parameters. Moreover, the linear function can well describe the changes in the strength and elastic modulus of the sandstone, and the decrease rate is positively related to the strain rate level. The peak strengths of the samples after 40 F–T cycles decreased by 52.44, 62.28, and 73.05 MPa for the three impact loads, corresponding to 1.94, 2.31, and 2.72 times the static decrease, respectively, while the changes in the elastic modulus and peak strength of the samples were similar. Instead of static loading, the dynamic mechanical behavior of sandstone is therefore more sensitive to cyclic F–T damage.

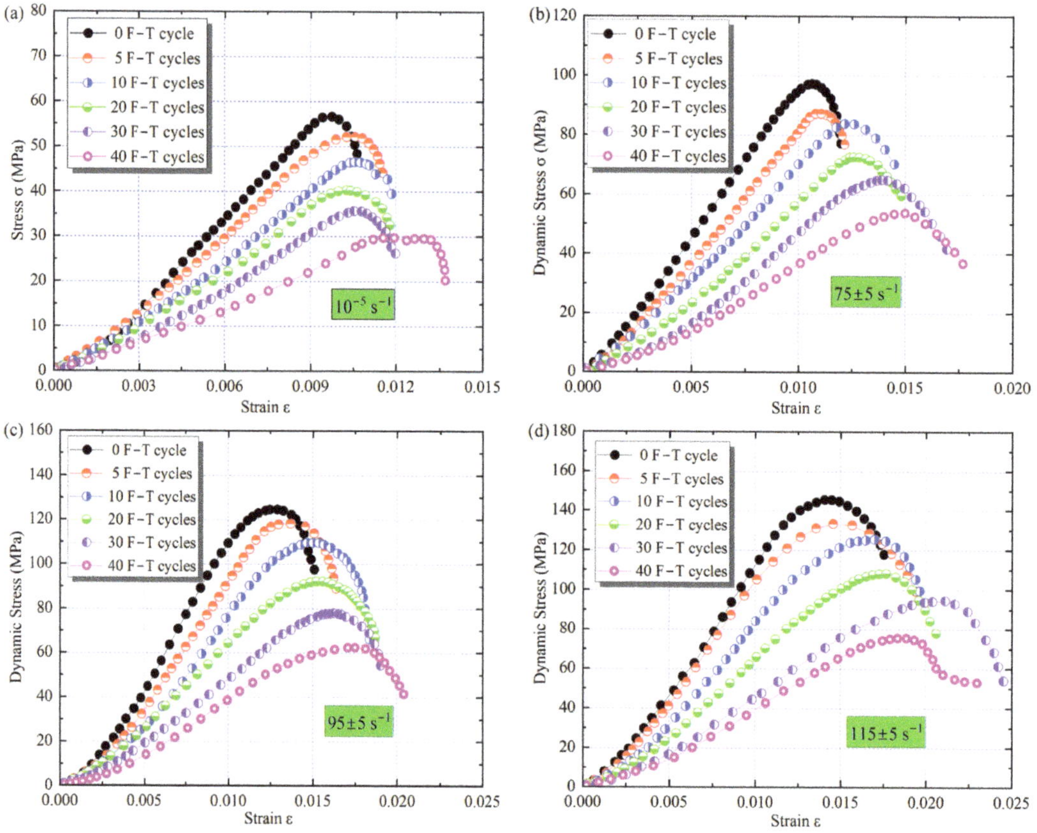

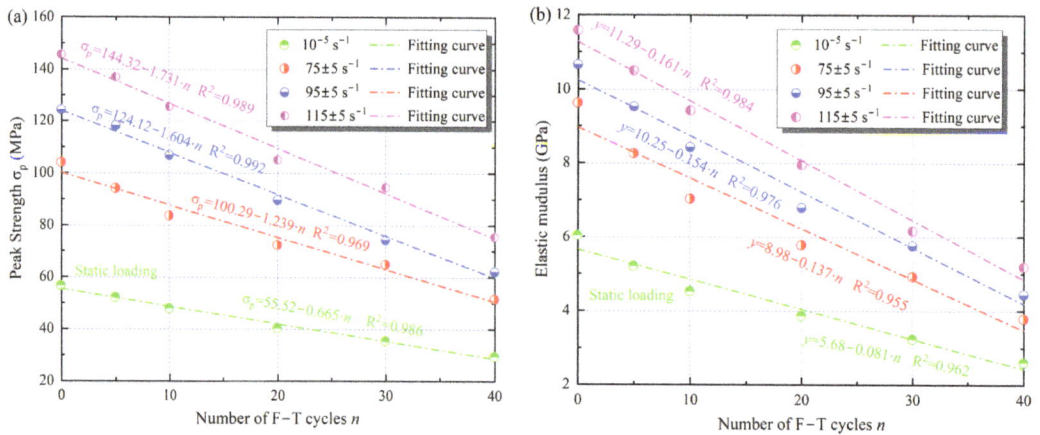

Figure 13. Stress–strain behavior under F–T action: (**a**) Static; (**b**) $75 \pm 5\ s^{-1}$; (**c**) $95 \pm 5\ s^{-1}$; (**d**) $115 \pm 5\ s^{-1}$.

Figure 14. Variations of peak strength (**a**) and elastic modulus (**b**) of samples with cyclic F–T action.

In general, F–T damage results from the interaction of capillary mechanisms, crystallization pressure mechanisms, hydrostatic pressure mechanisms, and volume expansion mechanisms [29]. Each of these mechanisms is closely associated with the properties of pores and microcracks. Due to the different chemical potentials, the water in the pores nucleates first and then freezes, while the water in the throats is supercooled and plays a role in recharging the pores [34]. As a result, the continuous increase of frost heaving force causes the pores to be connected by the throats, leading to F–T damage such as wedge cracking [14]. Under static loading, damage in the rock, such as pores and throats, reduces the bearing area, resulting in changes in mechanical behavior, with pores playing an important role due to their large volume [35]. However, under impact loading, the stress waves propagating in the rock tend to activate the throats near the pore tips, resulting in much more damage development in the rock than under static loading [6]. Moreover, the hysteresis of rock deformation with respect to impact stress waves during dynamic loading leads to strain rate strengthening effects. The dynamic mechanical behaviors of the sandstone under cyclic F–T action are more sensitive to the changes in microscopic parameters. Therefore, it is necessary to connect microscopic parameters and dynamic mechanical properties.

3.5. Macro-Micro Properties Connection

In this study, the relationship between microscopic parameters such as fractal dimension, and porosity and macroscopic mechanical parameters is shown in Figure 15. From that, the dynamic peak strength and elastic modulus decrease linearly with the fractal dimension and porosity of the samples under different loading conditions. Meanwhile, the rate of decrease correlates positively with the strain rate, which is consistent with the changes in mechanical properties during the F–T process. For example, when the strain rate increased from 75 s^{-1} to 115 s^{-1}, the decrease rate in the dynamic elastic modulus of the samples increased from 92.3 to 110.4 with increasing porosity, which is 2.02 times the decrease rate under static loading. Similarly, the decrease in dynamic strength with decreasing porosity was 1.9, 2.4, and 2.6 times higher, respectively, than the static strength for the three loading rates. Moreover, when the fractal dimension f was increased from 2.56 to 2.67, the peak strength decreased with increasing strain rate by 52.4 MPa, 62.3 MPa, and 69.4 MPa, respectively, compared with a decrease of 26.9 MPa at static loading. Zhou et al. [36] observed the effects of water saturation times on rock microstructure using SEM and found that the mechanical properties of rock were closely related to the fractal dimension; Li et al. [1] investigated the effects of porosity on the mechanical behavior of rock during F–T cycles using the CT technique and reached conclusions similar to those in this paper. The cyclic F–T action alters the internal structure of the rock by increasing the number, and size of pores and throats, and removing some of the cement [37]. As a result, the cement strength and the cohesion between the mineral particles and the skeleton are gradually weakened, increasing internal defects and damage. At the same time, the increase in porosity and fractal dimension means an increase in the complexity of the internal structure of the rock [31]. As seen in Figure 15, the complexity and inhomogeneity of the internal structure determine the degradation of the mechanical properties, so the microstructure of the sample is the key to influence the dynamic mechanical behavior. Therefore, the variation of microstructural parameters needs to be considered in the damage evolution of the rock under impact loading.

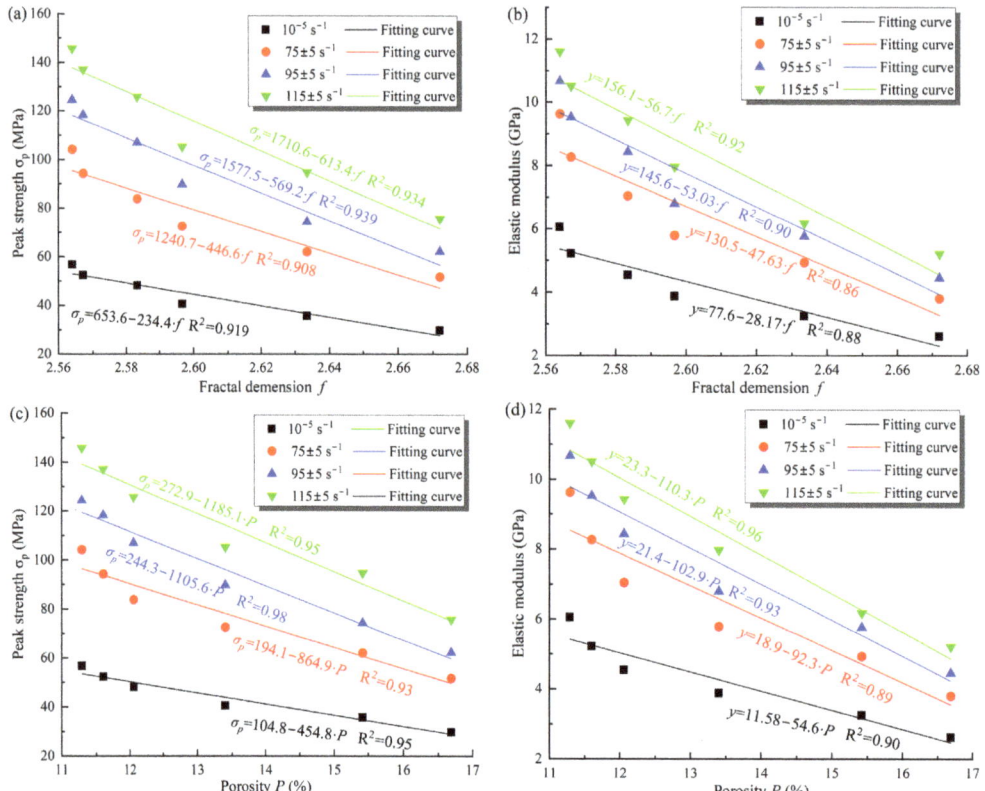

Figure 15. Connection between mechanical properties and microscopic parameters of samples. (**a**) Peak strength vs. fractal dimension; (**b**) elastic modulus vs. fractal dimension; (**c**) peak strength vs. porosity; (**d**) elastic modulus vs. porosity.

4. Damage Evolution under F–T and Impact Loading

4.1. F–T Damage D_n

From a macroscopic viewpoint, the response of the dynamic mechanical behavior can represent the rock damage [12]. Based on this, the elastic modulus was selected to describe the variable D_n for the F–T damage.

$$D_n = 1 - \alpha \cdot \frac{E_n}{E_0} \qquad (6)$$

where D_n represents the damage index of the sample caused by F–T action; E_0 refers to the initial elastic modulus; E_n indicates the dynamic elastic modulus of the sample after n F–T cycles.

Considering that the elastic modulus of the sample is obtained only from the linear section of the curves, it is not sufficient to fully reflect the F–T damage. Moreover, the random distribution of defects, the pore volume, and the degree of crack penetration in the sample significantly affect the damage process [1]. Therefore, the change in the effective bearing volume of the rock is introduced as an improvement factor α in the damage definition.

$$\alpha = \frac{V_e}{V} = \frac{V_e}{V_e + V_d - V_{di}} \qquad (7)$$

where V_e denotes the effective bearing volume (matrix volume) of the rock; V is the apparent bearing volume, which is composed of two main components: one is the matrix volume, and the other is the defect volume (pores and throats), also known as initial defects considering the various defects in the rock itself [2]. To remove the initial defects from the new damage induced by F–T cycles, the initial volume of defects V_{di} in the rock is subtracted from V. In the initial state of the sample, when V_{di} is equal to V_d, the value of D_n is 0, and as V_d increases, the difference between V_d and V_{di} gradually increases, which means that the effect on D_n is more than obvious. Therefore, considering the defect volume, the damage factor D_n is as follows:

$$D_n = 1 - \frac{V_e}{V_e + V_d - V_{di}} \cdot \frac{E_n}{E_0} \tag{8}$$

According to Figure 7c, the apparent volume and defect volume of the sample can be determined using the 3D visualization results. As shown in Equation (1), the porosity variation of the sample was calculated based on the defect volume and apparent volume [26]. Therefore, the damage index D_n of the sample was calculated using Equation (8), as presented in Figure 16. From this, it is evident that the improvement factor increases the D_n, which is due to the fact that the value of α gradually decreases to less than 1. Moreover, for the same F–T cycles, the strain rate is negatively correlated with the F–T damage, i.e., the larger the strain rate, the smaller the value of D_n.

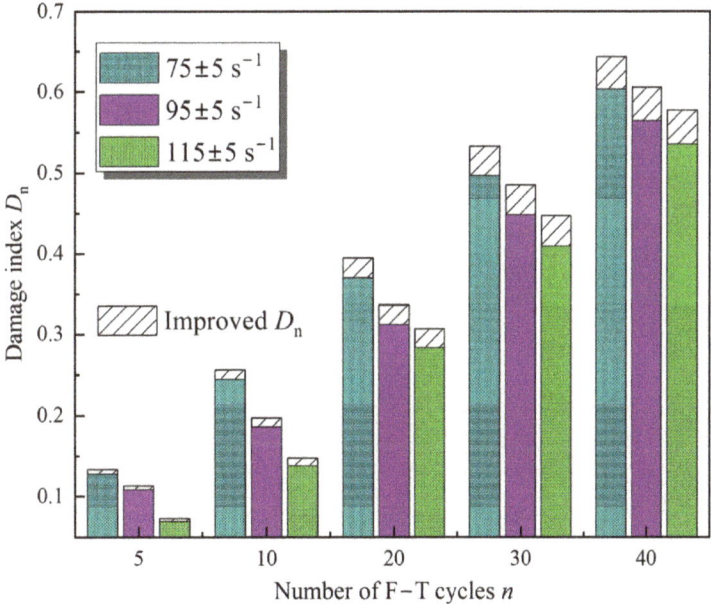

Figure 16. Variation of damage index D_n with different F–T cycles.

4.2. Damage D_m Evaluation

The following assumptions are made to characterize the evolution of damage in the rock during dynamic loading: (1) at the macroscopic level, the rock is an isotropic, homogeneous material; (2) the microunits exhibit linear elasticity before damage; and (3) the microunits strength obeys a two-parameter Weibull distribution with a probability density function of $P(F)$ [38]:

$$P(F) = \frac{m}{F_0}(\frac{F}{F_0})^{m-1} \exp[-(\frac{F}{F_0})^m] \tag{9}$$

Following the strain strength theory, the parameter F can be substituted by the strain ε [6]. Therefore, Equation (9) can be written as follows:

$$P(\varepsilon) = \frac{m}{\varepsilon_0}(\frac{\varepsilon}{\varepsilon_0})^{m-1} \exp[-(\frac{\varepsilon}{\varepsilon_0})^m] \tag{10}$$

where ε_0 is scale parameter; m refers to the shape parameter.

If the total number of microunits is N and n units fail under impact loading, the ratio n/N is defined as D_s [39]:

$$D_s = \frac{n}{N} = \frac{N\int_0^\varepsilon P(\varepsilon)d\varepsilon}{N} = 1 - \exp[-(\frac{\varepsilon}{\varepsilon_0})^m] \tag{11}$$

As a result, the stress–strain relationship under dynamic loading was determined using the strain equivalence hypothesis [6]:

$$\sigma = E_0\varepsilon(1 - D_s) \tag{12}$$

where E_0 refers to the initial elastic modulus. Considering the effect of F–T action on the elastic modulus, Equation (12) can be expressed as follows:

$$\sigma = \alpha E_n\varepsilon(1 - D_s) \tag{13}$$

Further, Equation (13) can be expressed as:

$$\sigma = E_0(1 - \frac{E_0 - \alpha E_n}{E_0})\varepsilon(1 - D_s) = E_0\varepsilon(1 - D_n)(1 - D_s) \tag{14}$$

Thus, bringing Equations (8) and (11) into Equation (14), the damage D_m evolution equation for the sample under impact loading and F–T damage is as follows:

$$D_m = D_n + D_s - D_nD_s = 1 - \frac{V_e}{V_e + V_d - V_{di}}\frac{E_n}{E_0}\exp[-(\frac{\varepsilon}{\varepsilon_0})^m] \tag{15}$$

The parameters ε_0 and m in the damage evolution model can be ascertained by fitting the stress–strain curves using Equation (14) (see Figure 17). The stress–strain curves under cyclic F–T effects were selected for the parametric study. As shown in Figure 18a, different D_n correspond to different stress–strain curves, and their peak stresses are inversely proportional to the parameter D_n. As the name implies, the higher the damage, the lower the peak stress. Figure 18b illustrates different stress–strain relationships as ε_0 is 0.0158, 0.0168, and 0.0178. The results indicate that both the peak stresses and strains increase gradually with increasing ε_0. Moreover, ε_0 clearly affects the trend of the descending branch of the stress–strain relationship. The stress–strain curves for different m are given in Figure 18c. As the m increases, the peak strength of the stress–strain curves increases, and the rate of stress decreases after the peak gradually increases.

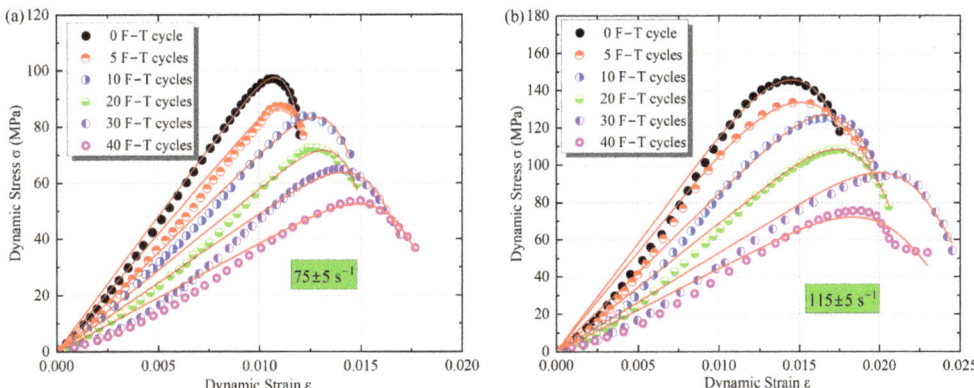

Figure 17. Measured and fitted stress–strain curves at different strain rates and F–T cycles. (**a**) $75 \pm 5\ \text{s}^{-1}$; (**b**) $115 \pm 5\ \text{s}^{-1}$.

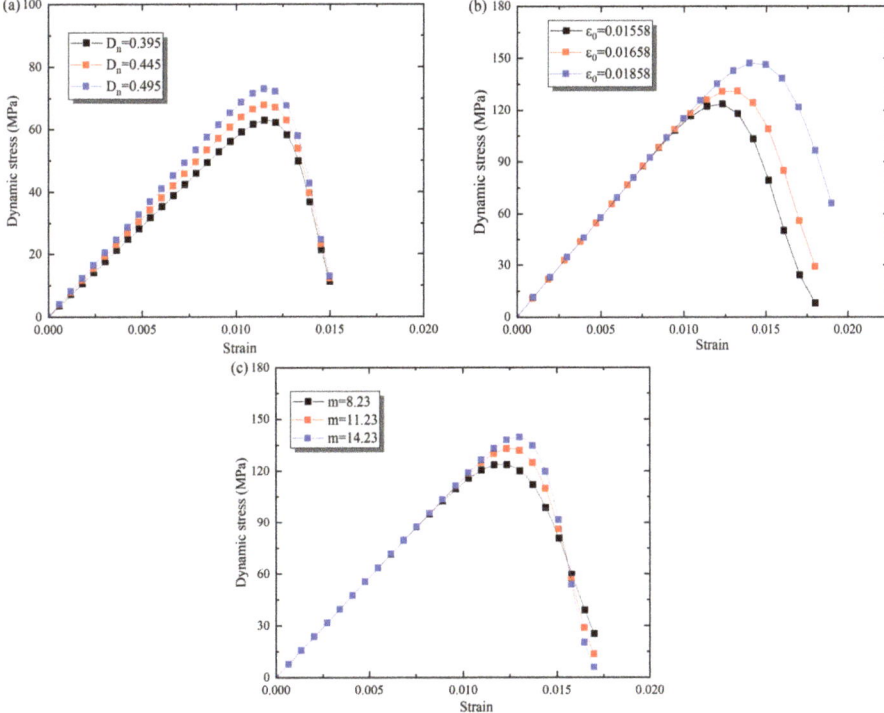

Figure 18. Analysis of parameter sensitivity in damage evolution model. (**a**) D_n; (**b**) ε_0; (**c**) m.

In addition, Figure 19 presents the variation of m and ε_0 in the damage evolution model with the F–T action and strain rates. It can be observed that the parameter ε_0 is larger under dynamic loading than under static loading, and increases with F–T damage (Figure 19a). As shown in Equation (16), the linear function can describe the change of parameter ε_0 with F–T action, and the growth rate increases with the strain rates. In contrast, the parameter m is smaller under dynamic loading than static loading and tends to decrease linearly with the F–T action (Equation (17)). Meanwhile, the decline rate of the parameter m with the F–T action is inversely correlated with the strain rates, i.e., the larger the strain rate, the slower the decline rate. After five F–T cycles, the parameter m increased by 16% from 0.001324 to 0.001559, and the slope from 0.00105 to 0.0001123 when the strain rate was increased from 75 s^{-1} to 95 s^{-1} (Figure 19b). As the F–T cycle increases, the strength of the mineral particles and cement in the sample gradually decreases, resulting in a shift in the damage of the rock from brittle to ductile [27]. The damage of the rock also shifts from brittle to ductile under impact loading [10]. The change in damage mode implies a decrease in post-peak stress drop rate; therefore, the parameter m reduces with the F–T damage and loading rate. In addition, the peak strain of the sample increases with increasing F–T damage, and there is a significant strengthening effect of strain rate, so the parameter ε_0 increases with the F–T action and strain rate.

$$\begin{cases} \varepsilon_0(10^{-5}s^{-1}, n) = 0.011 + 8.45E-5 \cdot n \ (R^2 = 0.93) \\ \varepsilon_0(75 \pm 5s^{-1}, n) = 0.013 + 1.05E-4 \cdot n \ (R^2 = 0.87) \\ \varepsilon_0(95 \pm 5s^{-1}, n) = 0.016 + 1.13E-4 \cdot n \ (R^2 = 0.95) \\ \varepsilon_0(115 \pm 5s^{-1}, n) = 0.018 + 1.46E-4 \cdot n \ (R^2 = 0.96) \end{cases} \quad (16)$$

$$\begin{cases} m(10^{-5}s^{-1}, n) = 31.61 - 0.167 \cdot n \ (R^2 = 0.91) \\ m(75 \pm 5s^{-1}, n) = 16.95 - 0.152 \cdot n \ (R^2 = 0.93) \\ m(95 \pm 5s^{-1}, n) = 11.95 - 0.131 \cdot n \ (R^2 = 0.89) \\ m(115 \pm 5s^{-1}, n) = 9.97 - 0.122 \cdot n \ (R^2 = 0.96) \end{cases} \quad (17)$$

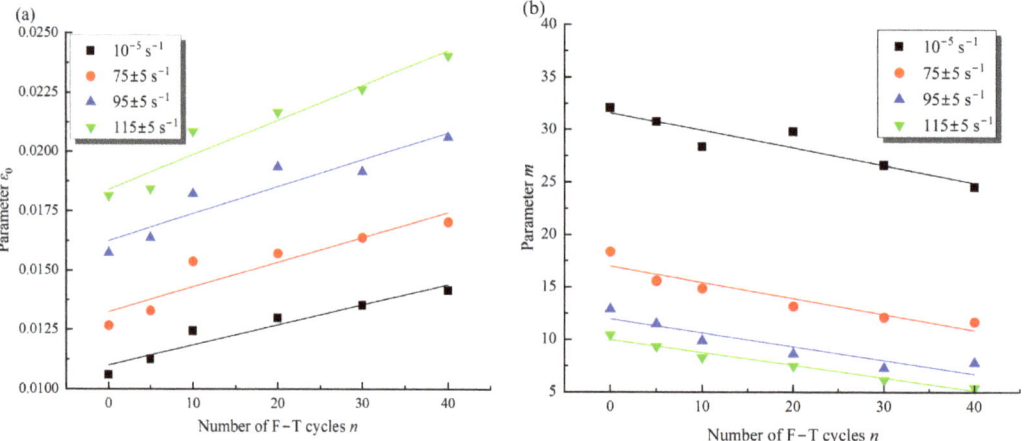

Figure 19. Variation of the parameters ε_0 (**a**) and m (**b**) with the strain rate and cyclic F–T action.

Based on the Equation (15) and the parameters ε_0, m, and α, the damage evolution of the samples under dynamic loading and F–T action was investigated. The D_m with the strain rate of the samples after 20 F–T cycles is shown in Figure 20a, and the D_m with the different F–T cycles at the same loading is displayed in Figure 20b. The evolution curve of D_m exhibits distinct phase characteristics. In the initial phase, the D_m of the samples increases slowly; as the strain increases, the D_m shows a lower convex increase; when the loading continues to increase, the damage curve gradually changes to an upper convex increase; in the final phase of the stress, the rock sample fractures, and the damage curve tends to 1.

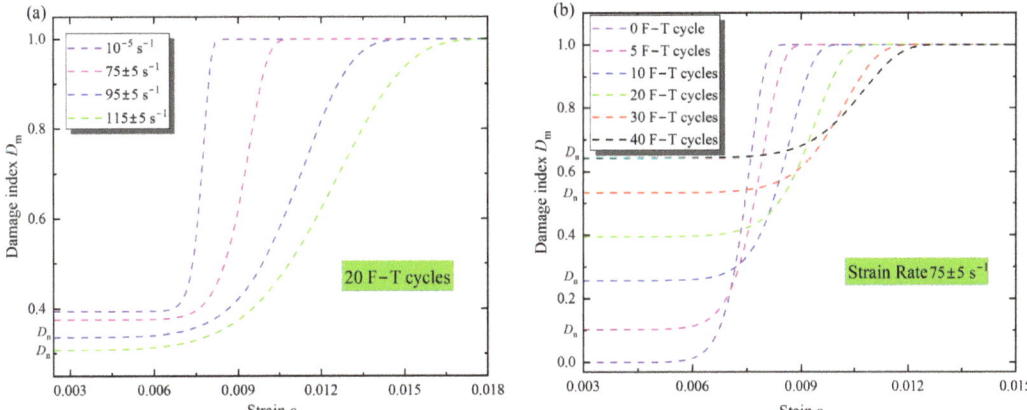

Figure 20. Damage D_m evolution with strain rate (**a**) and F–T cycles (**b**) under impact loading.

As shown in Figure 20a, the damage index D_m in the initial phase corresponds to the F–T damage D_n at different strain rates. After 20 F–T cycles, the D_n decreases with increasing strain rate, and the D_n at static loading is 0.395, which is about 1.5 times higher than that at 115 s^{-1}. As the strain increases, the D_m of the sample continues to increase, but it is always between 0 and 1. This is due to the gradual development of the pores and microcracks in the rock caused by the F–T action under the loading, which eventually aggravates the damage to the sample [20]. However, compared to static loading, there are significant differences in the D_m development curves under dynamic loading: as the strain increases, D_m develops earlier in the samples with larger strain rates, e.g., at the strain rate of 115 s^{-1}, damage begins at a strain of about 0.004 compared to about 0.007 under static loading. Also, the growth rate of D_m is greater under static loading, which means that the deformation during damage is less under static loading. As mentioned earlier, the peak strain of the sample increases with strain rate, leading to a decrease in the growth rate of D_m with strain in Figure 17a. At the same time, the damage mode of the samples changes from brittle to ductile under the impact loading, which is reflected in a decrease in the decrease rate of post-peak stress [20]. Therefore, the D_m of the sample under F–T action develops earlier under the impact loading, but the growth rate gradually decreases with the strain rate.

In Figure 20b, the initial strain corresponds to the different D_n, and it can be observed that the D_n gradually increases with the cyclic F–T action. At the same strain rate, the evolution curves of D_m show different characteristics: The larger the D_n at the initial loading, the longer the process of slow increase of D_m. This is due to the fact that the number of pores and microcracks (throats) in the sample increases continuously with the F–T action, resulting in an increase in the defect closure phase under dynamic loading [17]. With the strain increases, the pores and throats caused by the F–T action, as well as the original defects, begin to develop, so that the D_m of the sample increases in a lower convex shape [2]. As the loading continues to increase, the evolution curves of D_m gradually shift to an upper convex shape. This can be attributed to the increase in plastic strain before the peak stress with the F–T cycles, leading to a decrease in the growth rate of damage with strain. In general, the development curves of D_m changed from a sharp increase to a slow increase with strain and then to stabilization 1. The larger the strain rate, the lower the initial F–T damage, and the earlier the damage increases with strain, but the growth rate decreases with increasing strain rate; the initial damage increases gradually with the F–T action under dynamic loading, but the growth rate of damage decreases with increasing initial damage. It is worth noting that dynamic loading reduces the D_n, but does not increase the growth rate of the D_m.

5. Conclusions

The influence of microstructure on the dynamic mechanical behavior and damage evolution of frozen–thaw sandstone was investigated using the split Hopkinson pressure bar system, the MTS816 system, and the computed tomography (CT) system, revealing that microstructural change in cold regions is one of the key factors that determine the mechanical behavior of the rock under different impact loads. On this basis, a constitutive model was developed considering impact loads and F–T effects, which incorporates the influence of microscopic parameters and can well describe the stress–strain behavior under different impact loads, and provide meaningful information for understanding the dynamic hazards of rocks in cold regions.

In this study, cyclic F–T action reshapes the microstructure of the yellow sandstone, leading to an exponential increase in porosity, permeability, pore parameters, throat parameters, and fractal dimension of the samples. After 40 F–T cycles, the porosity, permeability, and fractal dimension of samples increased from 2.56, 11.3%, and 0.332 μm^2 to 2.67, 16.7% and 0.741 μm^2, respectively. Moreover, the peak strength and elastic modulus under impact loading were linearly and negatively correlated with microstructure, and the decrease rate was positively correlated with strain rate. Besides, with increasing impact loading, the D_m developed earlier in the samples with strain, while the growth rate of the D_m with strain was lower; with increasing F–T damage D_n, the slow increase of damage was longer, and the growth rate of the D_m with strain was lower.

Author Contributions: Methodology, J.X. and H.P.; investigation, Z.S.; resources, J.X.; data curation, Z.S. and J.X.; writing—original draft preparation, J.X.; writing—review and editing, J.X. and H.P.; visualization, Z.S. and J.X.; supervision, H.P. All authors have read and agreed to the published version of the manuscript.

Funding: This research was funded by the National Natural Science Foundation of China (No. 51974296, No. 52061135111).

Institutional Review Board Statement: Not applicable.

Informed Consent Statement: Not applicable.

Data Availability Statement: Data available on request due to privacy restrictions.

Conflicts of Interest: The authors declare no conflict of interest.

References

1. Li, B.; Zhang, G.; Wang, G.; Qiao, J. Damage Evolution of Frozen-Thawed Granite Based on High-Resolution Computed Tomographic Scanning. *Front. Earth Sci.* **2022**, *10*, 1–12. [CrossRef]
2. Park, J.; Hyun, C.U.; Park, H.D. Changes in microstructure and physical properties of rocks caused by artificial freeze–thaw action. *Bull. Eng. Geol. Environ.* **2015**, *74*, 555–565. [CrossRef]
3. Wang, T.; Sun, Q.; Jia, H.; Ren, J.; Luo, T. Linking the mechanical properties of frozen sandstone to phase composition of pore water measured by LF-NMR at subzero temperatures. *Bull. Eng. Geol. Environ.* **2021**, *80*, 4501–4513. [CrossRef]
4. Liu, D.; Pu, H.; Sha, Z.; Xu, J. Experimental study on dynamic tensile mechanical properties of sandstone under freeze—Thaw cycles. *Coal Sci. Technol.* **2022**, *50*, 60–67.
5. Zhou, Z.; Yude, E.; Cai, X.; Zhang, J. Coupled Effects of Water and Low Temperature on Quasistatic and Dynamic Mechanical Behavior of Sandstone. *Geofluids* **2021**, *2021*, 1–12. [CrossRef]
6. Sha, Z.; Pu, H.; Xu, J.; Ni, H.; Guo, S. Effects of Accumulated Damage on the Dynamic Properties of Coal Measures Sandstone. *Minerals* **2022**, *12*, 810. [CrossRef]
7. Xu, J.; Pu, H.; Sha, Z. Mechanical behavior and decay model of the sandstone in Urumqi under coupling of freeze–thaw and dynamic loading. *Bull. Eng. Geol. Environ.* **2021**, *80*, 2963–2978. [CrossRef]
8. Li, J.; Kaunda, R.B.; Zhou, K. Experimental investigations on the effects of ambient freeze-thaw cycling on dynamic properties and rock pore structure deterioration of sandstone. *Cold Reg. Sci. Technol.* **2018**, *154*, 133–141. [CrossRef]
9. Liu, S.; Xu, J.; Liu, S.; Wang, P. Fractal study on the dynamic fracture of red sandstone after F-T cycles. *Environ. Earth Sci.* **2022**, *81*, 1–13. [CrossRef]
10. Xu, J.; Pu, H.; Sha, Z. Dynamic Mechanical Behavior of the Frozen Red Sandstone under Coupling of Saturation and Impact Loading. *Appl. Sci.* **2022**, *12*, 7767. [CrossRef]

11. Zhang, J.; Deng, H.; Deng, J.; Ke, B. Development of energy-based brittleness index for sandstone subjected to freeze-thaw cycles and impact loads. *IEEE Access* **2018**, *6*, 48522–48530. [CrossRef]
12. Song, Y.; Yang, H.; Tan, H.; Ren, J.; Guo, X. Study on damage evolution characteristics of sandstone with different saturations in freeze-thaw environment. *Chin. J. Rock Mech. Eng.* **2021**, *40*, 1513–1524.
13. Wu, N.; Liang, Z.Z.; Li, Y.C.; Li, H.; Li, W.R.; Zhang, M. Stress-dependent anisotropy index of strength and deformability of jointed rock mass: Insights from a numerical study. *Bull. Eng. Geol. Environ.* **2019**, *78*, 5905–5917. [CrossRef]
14. Deprez, M.; De Kock, T.; De Schutter, G.; Cnudde, V. A review on freeze-thaw action and weathering of rocks. *Earth Sci. Rev.* **2020**, *203*, 103143. [CrossRef]
15. Fang, Y.; Qiao, L.; Chen, X.; Yan, S.J.; Zhai, G.L.; Liang, Y.W. Experimental study of freezing-thawing cycles on sandstone in Yungang grottos. *Yantu Lixue/Rock Soil Mech.* **2014**, *35*, 2433–2442.
16. Cheng, H.; Chen, H.; Cao, G.; Rong, C.; Yao, Z.; Cai, H. Damage mechanism of porous rock caused by moisture migration during freeze-thaw process and experimental verification. *Chin. J. Rock Mech. Eng.* **2020**, *39*, 1739–1749.
17. Xu, J.; Pu, H.; Sha, Z. Effect of Freeze-Thaw Damage on the Physical, Mechanical, and Acoustic Behavior of Sandstone in Urumqi. *Appl. Sci.* **2022**, *12*, 7870. [CrossRef]
18. De Kock, T.; Boone, M.A.; De Schryver, T.; Van Stappen, J.; Derluyn, H.; Masschaele, B.; De Schutter, G.; Cnudde, V. A pore-scale study of fracture dynamics in rock using X-ray micro-CT under ambient freeze-thaw cycling. *Environ. Sci. Technol.* **2015**, *49*, 2867–2874. [CrossRef]
19. Yang, G.; Liu, H. *Microstructure and Damage Mechanical Characteristic of Frozen Rock Based on CY Image Processing*; Science Press: Beijing, China, 2016.
20. Fan, L.F.; Fan, Y.D.; Xi, Y.; Gao, J.W. Spatial Failure Mode Analysis of Frozen Sandstone Under Uniaxial Compression Based on CT Technology. *Rock Mech. Rock Eng.* **2022**, *55*, 4123–4138. [CrossRef]
21. Maji, V. An Experimental Investigation of Micro-and Macrocracking Mechanisms in Rocks by Freeze-Thaw Cycling. Ph.D. Thesis, University of Sussex, Falmer, East Sussex, UK, August 2021.
22. Huang, S.; Cai, Y.; Liu, Y.; Liu, G. Experimental and Theoretical Study on Frost Deformation and Damage of Red Sandstones with Different Water Contents. *Rock Mech. Rock Eng.* **2021**, *54*, 4163–4181. [CrossRef]
23. De Paepe, A.E.; Sierpowska, J.; Garcia-Gorro, C.; Martinez-Horta, S.; Perez-Perez, J.; Kulisevsky, J.; Rodriguez-Dechicha, N.; Vaquer, I.; Subira, S.; Calopa, M.; et al. Using μCT to investigate water migration during freeze-thaw experiments. *J. Chem. Inf. Model.* **2019**, *53*, 1689–1699.
24. Deprez, M.; De Kock, T.; De Schutter, G.; Cnudde, V. Dynamic X-ray CT to monitor water distribution within porous building materials: A build-up towards frost-related experiments. *3rd Int. Conf. Tomogr. Mater. Struct.* **2017**, *203*, 103143.
25. Yang, H.; Liu, P.; Sun, B.; Yi, Z.; Wang, J.; Yue, Y. Study on damage mechanisms of the microstructure of sandy conglomerate at Maijishan grottoes under freeze-thaw cycles. *Chin. J. Rock Mech. Eng.* **2021**, *40*, 545–555.
26. ThermoFisher Scientific. *Thermo Scientific Avizo Software 9*, 2018; Volume 9.
27. Chen, Y.; Huidong, C.; Ming, L.; Pu, H. Study on dynamic mechanical properties and failure mechanism of saturated coal-measure sandstone in open pit mine with damage under real-time low-temperature conditions. *J. China Coal Soc.* **2022**, *47*, 1168–1179.
28. Gong, F.Q.; Si, X.F.; Li, X.B.; Wang, S.Y. Dynamic triaxial compression tests on sandstone at high strain rates and low confining pressures with split Hopkinson pressure bar. *Int. J. Rock Mech. Min. Sci.* **2019**, *113*, 211–219. [CrossRef]
29. Wan, Y.; Chen, G.Q.; Sun, X.; Zhang, G.Z. Triaxial creep characteristics and damage model for red sandstone subjected to freeze-thaw cycles under different water contents. *Chin. J. Geotech. Eng.* **2021**, *43*, 1463–1472.
30. Feng, S.; Zhou, Y.; Wang, Y.; Lei, M. Experimental research on the dynamic mechanical properties and damage characteristics of lightweight foamed concrete under impact loading. *Int. J. Impact Eng.* **2020**, *140*, 103558. [CrossRef]
31. Ding, Z.; Li, X.; Tang, Q.; Jia, J. Study on correlation between fractal characteristics of pore distribution and strength of sandstone particles. *Chin. J. Rock Mech. Eng.* **2020**, *39*, 1787–1796.
32. Liu, H.; Yang, G.; Yun, Y.; Lin, J.; Ye, W.; Zhang, H.; Zhang, Y. Investigation of Sandstone Mesostructure Damage Caused by Freeze-Thaw Cycles via CT Image Enhancement Technology. *Adv. Civ. Eng.* **2020**, *2020*, 1–13. [CrossRef]
33. Khanlari, G.; Abdilor, Y. Influence of wet–dry, freeze–thaw, and heat–cool cycles on the physical and mechanical properties of Upper Red sandstones in central Iran. *Bull. Eng. Geol. Environ.* **2015**, *74*, 1287–1300. [CrossRef]
34. Scherer, G.W. Crystallization in pores. *Cem. Concr. Res.* **1999**, *29*, 1347–1358. [CrossRef]
35. Rempel, A.W. Formation of ice lenses and frost heave. *J. Geophys. Res. Earth Surf.* **2007**, *112*, 1–17. [CrossRef]
36. Cuiying, Z.; Ning, L.; Zhen, L. Multifractal characteristics of pore structure of red beds soft rock at different saturations. *J. Eng. Geol.* **2020**, *28*, 1–9.
37. Koniorczyk, M.; Bednarska, D. Influence of the mesopore's diameter on the freezing kinetics of water. *Microporous Mesoporous Mater.* **2017**, *250*, 55–64. [CrossRef]

38. Wang, Z.L.; Shi, H.; Wang, J.G. Mechanical Behavior and Damage Constitutive Model of Granite Under Coupling of Temperature and Dynamic Loading. *Rock Mech. Rock Eng.* **2018**, *51*, 3045–3059. [CrossRef]
39. Zhou, J.; Chen, X. Stress-Strain Behavior and Statistical Continuous Damage Model of Cement Mortar under High Strain Rates. *J. Mater. Civ. Eng.* **2013**, *25*, 120–130. [CrossRef]

Disclaimer/Publisher's Note: The statements, opinions and data contained in all publications are solely those of the individual author(s) and contributor(s) and not of MDPI and/or the editor(s). MDPI and/or the editor(s) disclaim responsibility for any injury to people or property resulting from any ideas, methods, instructions or products referred to in the content.

Article

Study on Influencing Factors of Ground Pressure Behavior in Roadway-Concentrated Areas under Super-Thick Nappe

Ruojun Zhu [1,2], Xizhan Yue [2], Xuesheng Liu [3], Zhihan Shi [3,*] and Xuebin Li [3]

1. College of Mining, China University of Mining and Technology, Xuzhou 221116, China
2. China Coal Xinji Energy Co., Ltd., Huainan 232001, China
3. College of Energy and Mining Engineering, Shandong University of Science and Technology, Qingdao 266590, China
* Correspondence: shizhihan2022@163.com

Abstract: During the mining activity under the super-thick nappe formed by thrust fault, the law of mine pressure behavior is complex, and it is difficult to control the deformation and failure of surrounding rock. Combined with the actual engineering conditions, the influence of different roof lithology conditions, the thickness of nappe, the mining height, the size of the barrier coal pillar, and the creep time on mine pressure behavior was studied by UDEC numerical simulation software. The results showed that with the advancement of the coal face, due to the influence of the mining of the coal face and the slip dislocation of the super-thick nappe along the thrust faults, the roof-to-floor convergence, the two-sided convergence, and the maximum concentrated stress in the roadway-concentrated areas are significantly increased. For the above five influencing factors, the greater the thickness of the nappe and the mining height, the longer the creep time, and the stronger the ground pressure behavior. The larger the size of the barrier coal pillar, the stronger the roof lithology, and the gentler the ground pressure behavior. The research results can provide some reference for monitoring the law of ground pressure behavior in roadway-concentrated areas under super-thick nappe.

Keywords: super-thick nappe; roadway-concentrated areas; influencing factors; ground pressure behavior; numerical simulation

1. Introduction

The exploitation of coal resources under complex geological conditions has always been a matter of great concern in the coal industry [1–4]. Thrust fault is a special type of geological structure. It is a large thrust fault with a dip angle of about 30° or less [5–7]. Such faults tend to cause their hanging walls to move thousands or even tens of thousands of meters over long distances along the fault plane, creating a massive rock mass structure known as a nappe [8]. The structure of the nappe will lead to the complex law of ground pressure behavior and unclear influencing factors, which will seriously affect the safety and stability of the roadway.

In recent years, scholars at home and abroad have conducted a lot of research on the law of ground pressure behavior and its influencing factors under the influence of faults, such as the use of FLAC 3D and boundary element numerical simulation methods. Batugin et al. [9–14] obtained the best supporting parameters of surrounding rock under fault conditions and the influence of different types of faults on ground pressure behavior. Sainoki et al. [15–24] obtained the relationship between mining activities and microseismical phenomena and the influence of different mining directions on the law of ground pressure behavior in coal face. Islam et al. [25], taking Barapukuria coal mine in Bangladesh as an example, found that the deformation and stress field of the fault and its surrounding rock change significantly under mining disturbance, and stress concentration occurs at the end of the fault, using a model test or a similar material test method. Zhang et al. [26–31] studied the overburden strata movement law, mine pressure characteristics, dynamic response, slip precursor information, and instability transient process characteristics before and after the activation of a thrust fault, and the related factors affecting the law of ground pressure behavior were obtained. Combined with elastic-plastic mechanics theory, Han et al. [32–35] studied the stress distribution and failure evolution process of surrounding rock under different conditions. Using the field monitoring method, Ji et al. [36,37] studied the influence of the relative position of the coal face and the fault on the law of mine pressure behavior, and the fault activation process is inversed by the microseismical data of the mine. Using a method based on a probabilistic approach to assessing the rock strength, Begalinov et al. [38] obtained the proportion of mining active fault instability under the influence of faults by studying the physical and mechanical properties of faults, which provides a basis for determining the means of support. Apart from that, Wang et al. [39] proposed a mechanical model considering the stress redistribution caused by coal face mining, calculating the stress distribution of the fracture surface and explaining the characteristics of the fault slip. Dokht et al. [40] concluded that mining activities of the mine cause the redistribution of roadway stress, which leads to the activation of faults and eventually causes small unnatural earthquakes. The above research has played an important role in the study of the law of ground pressure behavior and its influencing factors under the influence of faults and has effectively promoted the process of quantitative research on the control of roadway-surrounding rock under the influence of complex geological conditions.

However, due to the particularity of the thrust fault, the influence of the super-thick nappe on the ground pressure behavior in the concentrated area of the lower roadway is not clear, and the influencing factors are rarely studied. Therefore, this paper takes the mining of No. 360801 coal face under the super-thick nappe structure of Xinji No. 1 Mine as the engineering background. Firstly, the main factors affecting the ground pressure behavior in the roadway-concentrated areas under the super-thick nappe are determined. Then, a numerical simulation model under different geological and mining conditions is established by using the Universal Distinct Element Code (UDEC 6.0 v6.0.336) discrete element numerical simulation software, and then the law of ground pressure behavior in roadway-concentrated areas under super-thick nappe is explored. The research results can provide some reference for monitoring the law of ground pressure behavior in roadway-concentrated areas under super-thick nappe. The remainder of this manuscript is arranged as follows. Section 2 contains the location and geological conditions of the No. 360801 coal face, and then several important factors affecting the law of mine pressure behavior are analyzed. In Section 3, the numerical simulation model is established for different influencing factors. In Section 4, the specific influence of different factors on the mine pressure behavior is analyzed according to the simulation results. In Section 5, the numerical simulation results are summarized and analyzed, and conclusions are drawn.

2. Project Profile

The No. 36 district of Xinji No. 1 Mine is located in the southwest of the Xinji minefield. There is a large range of Fufeng thrust faults above the district. The overall trend of the fault is southward, and the dip angle changes greatly along the dip direction. The strike zone is 30°–80°, and the middle dip angle gradually slows down, generally to about 5°. The local section is nearly horizontal or inclined north, and the fault drop gradually decreases from south to north. The fault zone is undulating in the strike and dips with a width of 0~27.25 m. It is composed of broken brecciated limestone, mudstone, and carbonaceous mudstone. It is easy to loosen and break, and the local cementation is good. It is a typical thrust fault. Due to the influence of faults, there is a nappe structure with a thickness of 351~562 m above the district (Figure 1). The structure of the nappe is mainly composed of gneiss. The lithology is mainly granite gneiss and hornblende gneiss, showing gray mixed with a gray-black, brown, fresh surface, a scale granular crystal structure, and a gneiss structure. The No. 360801 coal face is located in the north of the west wing of the No. 36 district, the main 8# coal-mining area. The outer section of the No. 360802 machine roadway and the outer section of the No. 360803 track crossheading are in the east of the coal face, the design cut position near the auxiliary No. 13 exploration line is in the west, the track crossheading of the No. 360803 coal face is in the south, and the F10 protective coal pillar line is in the north. The layout of the No. 360801 coal face is shown in Figure 2. The 8# coal is mainly bright coal with a thickness of 2.6~3.6 m and an average thickness of 3.1 m. The immediate roof is mudstone, and the rock stability is general. The basic roof is fine sandstone with a maximum thickness of 13.5 m, and the rock layer is stable. The floor is mudstone, in a dense block with a parallel bedding development, with a thickness of 5.54~9.15 m. The specific rock-layer histogram is shown in Figure 3.

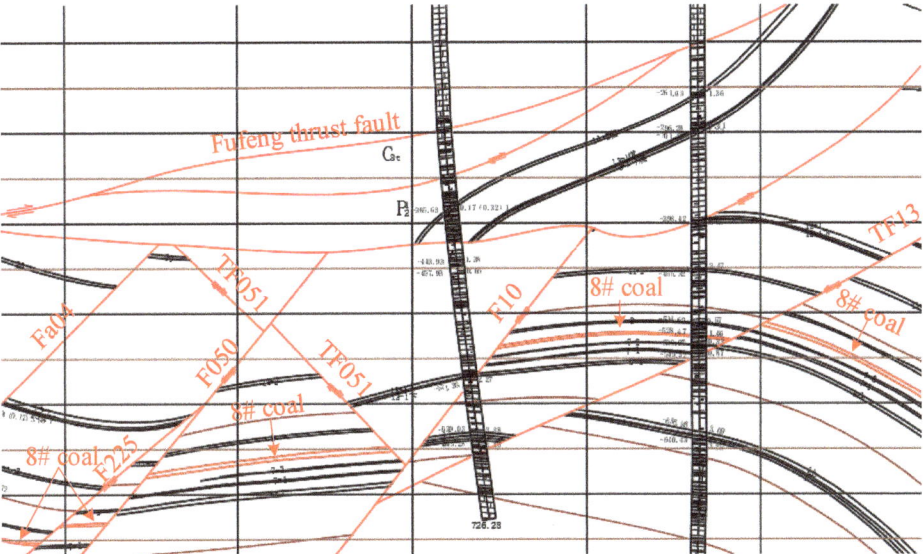

Figure 1. Nappe structure plane diagram.

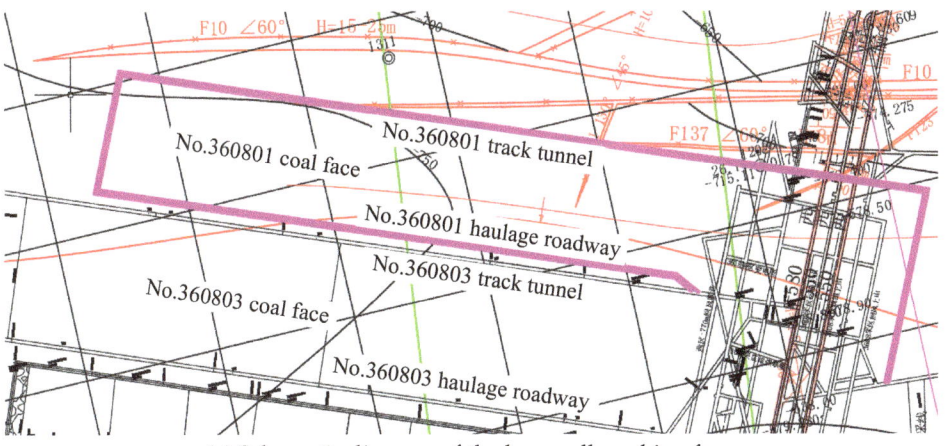

(**a**) Schematic diagram of the longwall working face.

Figure 2. *Cont.*

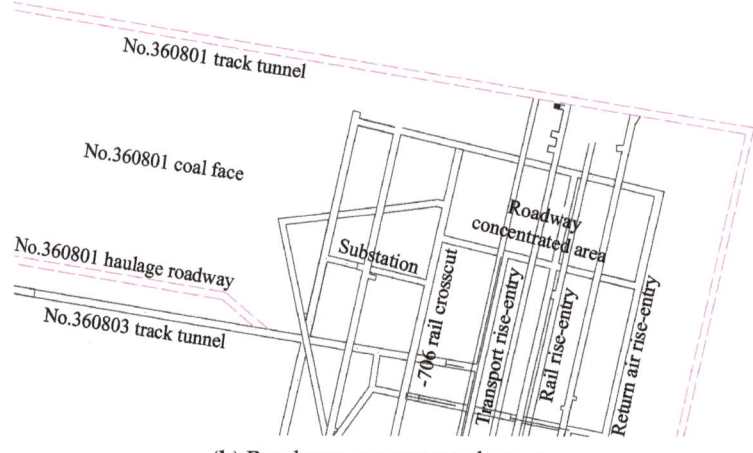

(**b**) Roadway-concentrated areas.

Figure 2. Layout plan of the No. 360801 coal face and roadway-concentrated areas.

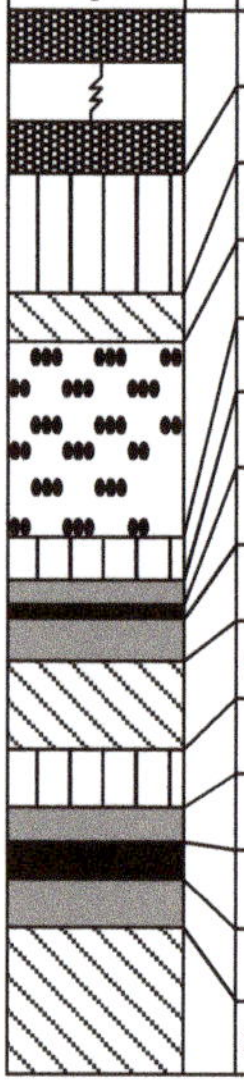

Rodman-shaped	Rock name	Average thickness/m	Rock description
	Gneiss	400	Directional distribution of dark and light minerals
	Fine sandstone	11.72	Gray, fine-grained structure, stepped fracture
	Sandy mudstone	8.16	Grey, sandy structure, massive structure
	Silicarenite	34.88	Light gray, mainly quartz, calcareous cementation
	Fine sandstone	3.86	Gray, fine-grained structure, stepped fracture
	Mudstone	2.12	Gray, blocky, broken, jagged fracture
	9# coal	1.14	Black, weak asphalt luster, blocky or scaly
	Mudstone	4.13	Gray, blocky, broken, jagged fracture
	Sandy mudstone	8.23	Grey, sandy structure, massive structure
	Fine sandstone	4.07	Light gray, medium thick layer, siliceous cementation
	Mudstone	0.89	Gray, blocky, broken, jagged fracture
	8# coal	3.1	Black, weak glass luster, scaly structure
	Mudstone	8.35	Gray, blocky, broken, jagged fracture
	Sandy mudstone	10.11	Grey, sandy structure, massive structure

Figure 3. Rock stratum histogram.

Since the thickness of the nappe is too large, the mine pressure behavior of the coal face cannot be accurately measured, and the specific influence of the change of mining conditions and geological conditions on the mine pressure cannot be confirmed. In the actual engineering survey, we found that the mine pressure behavior in many areas of the roadway-concentrated area is very strong. The −706-rail crosscuts and transport rise-entry show serious deformation and failure of the surrounding rock, poor stability, and other issues, as shown in the Figure 4a,b. Often, multiple repairs are needed to ensure normal use, seriously affecting the safety of the mine production. Therefore, it is necessary to study the influencing factors of mine pressure behavior.

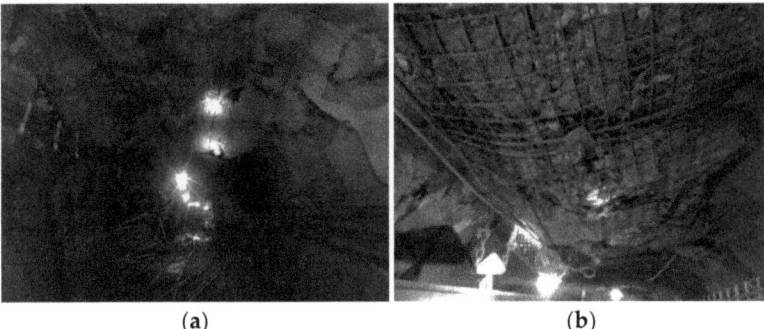

(a)   (b)

Figure 4. Deformation diagram of partial roadway-surrounding rock. (**a**) Overall damage diagram (**b**) Local damage diagram.

3. Factors and Methods

3.1. Analysis of the Main Factors

3.1.1. Geologic Factor

(1) Roof lithology

Roof lithology refers to the mechanical properties of the rock layer on the top of the roadway in a mine. When the roof lithology is different, the stability of roof strata is also different [41–43]. The deformation of the roof and the degree of rock fragmentation are quite different, which leads to the possibility of rock failure under the long-term action of a small external load or a small load. The direct roof of the 8# coal in the No. 36 district of the Xinji No. 1 Mine is mainly mudstone and sandy mudstone. The stability of rock strata is general, and the stability of rock strata is poor in the structural development area. The direct roof is scoured by the main roof of sandstone, which leads to direct contact between the main roof sandstone and the coal seam. At the same time, it is affected by the structure of the super-thick nappe, which makes the ground pressure behavior law more complicated, and then affects the safety and stability of the roadway-concentrated areas under the super-thick nappe.

(2) Thickness of nappe

Nappe is a huge rock mass structure formed by thrust faults that cause the hanging wall to move thousands or even tens of thousands of meters along the fault plane. The nappe structure will lead to inaccurate identification of the coal seam buried depth and distribution, and the inability to determine the structure development and distribution for the calculation and accurate measurement of the mine pressure caused great difficulties. When the thickness of the nappe greatly changes, the ground pressure behavior also dramatically changes. Therefore, it is particularly important to determine the relationship between the law of ground pressure behavior and the thickness of the nappe. Combined with the actual engineering conditions, the thickness of the overlying nappe in the No. 36 district of the Xinji No. 1 Mine is 351~562 m, and the thickness greatly varies. Therefore, the ground pressure behavior in different positions of the roadway-concentrated areas under the super-thick nappe is different.

3.1.2. Mining Technological Factors

(1) Mining height

Mining height refers to the distance difference between upper repeated mining and lower normal mining for inclined coal seam mining. The mining height of the coal face is directly related to the law of mine pressure behavior, and it is also the fundamental factor of the deformation and failure of the overlying strata. In combination with the engineering practice, in the No. 36 district of the Xinji No. 1 Mine, with the progress of the 8# coal-mining work, the stress of the rock mass around the roadway is redistributed, resulting in a significant change in the law of ground pressure behavior in the roadway-concentrated areas under the super-thick nappe. If the mining height of the coal face is adjusted, the roadway stress will be redistributed, and the actual ground pressure behavior law in the roadway-concentrated areas is extremely complex.

(2) Rise-entry barrier coal pillar size

The barrier coal pillar refers to the part of the coal body that is specially left underground and is not mined, and the purpose is to protect the above-mentioned protected objects inside the rock stratum and on the surface from mining. Due to the excavation of roadways and the mining of coal seams, the stability of roadways in roadway-concentrated areas under super-thick nappe is reduced, making the district within their influence range vulnerable to damage. To protect the rise-entry of the district and maintain the stability of roadways, it is necessary to setup a barrier coal pillar. However, with the change of the size of the rise-entry barrier coal pillar, the ground pressure behavior of roadway-concentrated areas also changes, and under the combined action of it and super-thick nappe, the influence law of roadway-concentrated areas under the nappe is more complicated.

3.1.3. Time Factor

The phenomenon that the deformation of rock increases with time under the continuous action of an external force whose size and direction do not change is called creep. The roof rack of the roadway is subjected to external force for a long time, and its deformation is increasing, and the stability is reduced. If its creep cannot be stabilized at a certain limit value, the breaking characteristics will be different from the creep failure of instantaneous failure, thus changing the ground pressure behavior in the roadway-concentrated areas. Therefore, the creep time also has an important influence on the ground pressure behavior in the roadway-concentrated areas under the super-thick nappe.

3.2. Construction of the Numerical Model

To study the influence of different factors on the ground pressure behavior in the roadway-concentrated areas under the super-thick nappe, based on the actual geological conditions of the No. 360801 coal face in the Xinji No. 1 Mine, the −706-rail crosscut, rail rise-entry, transport rise-entry, and return air rise-entry in this district are the most frequently used roadways and serve as the main transportation and return air tasks in the district. At the same time, the above four roadways are located in or through the tendency center of the No. 360801 coal face, and the research is representative. Therefore, the above four roadways were selected as the main research objects. Establishment of a numerical simulation model was performed using UDEC discrete element software [44]. Under the influence of different roof lithology conditions, the thickness of nappe, the mining height, the size of the barrier coal pillar, and the creep time, the law of ground pressure behavior in roadway-concentrated areas caused by mining was analyzed. The numerical simulation foundation model was established according to actual engineering geological conditions. The thickness of the nappe was 400 m, the mining height was 3 m, and the size of the barrier coal pillar was 190 m. The model size was length × height = 1000 m × 450 m. The boundary condition of the model was to apply the X-axis horizontal constraint on the right boundary of the model and the Y-axis vertical constraint on the bottom of the model. The vertical load was applied on the top of the model to simulate the weight of the overlying strata. The self-weight stress generated by the actual buried depth was used as the load on the top-end face. The upper self-weight stress was applied in the horizontal direction multiplied by the lateral pressure coefficient, λ, to simulate the initial ground stress. The specific numerical model is shown in Figure 5. The mechanical parameters of the whole rock stratum of the model are shown in Table 1.

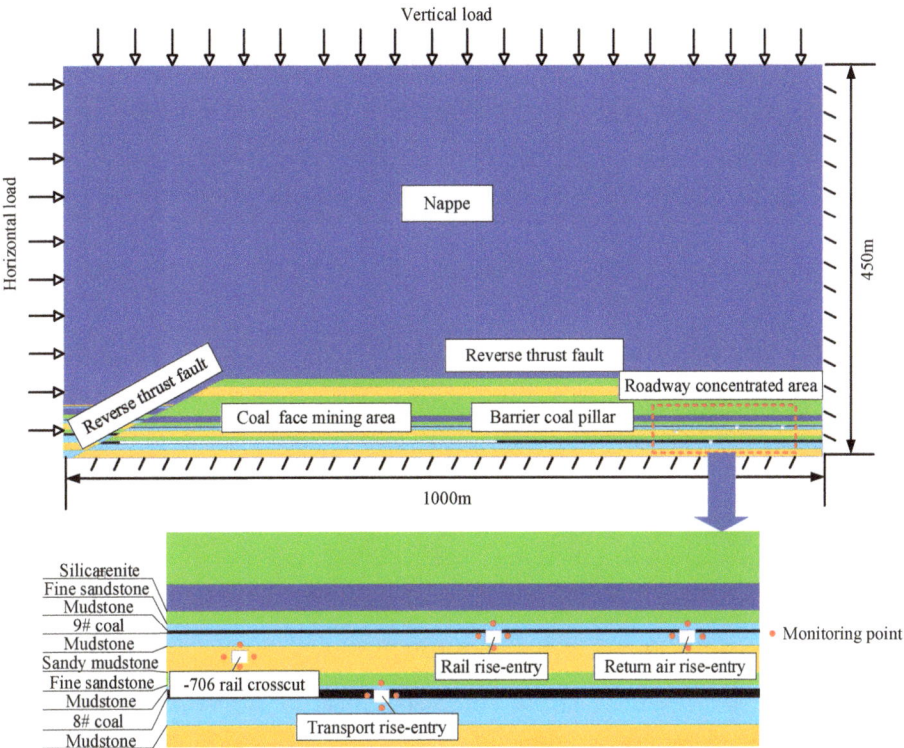

Figure 5. Overall numerical model and roadway-concentrated areas diagram.

Table 1. Numerical model of rock physical and mechanical parameters table.

	Rock Name	Thickness	Density/kg $\times$ m^{-3}	Tensile Strength/MPa	Elastic Modulus/GPa	Cohesion/MPa	The Angle of Internal Friction/°	Poisson'd Ratio
1	Gneiss	400	2763	11.2	44.8	15	40	0.25
2	Fine sandstone	12	2532	5.38	37.15	3.2	42	0.27
3	Sandy mudstone	8	2562	2.6	12.08	2.45	40	0.25
4	Silicarenite	35	2423	4.9	21.75	21	75	0.29
5	Fine sandstone	4	2532	5.38	37.15	3.2	42	0.27
6	Mudstone	2	2582	2.0	10.37	1.2	32	0.28
7	No. 9 coal seam	1	1401	0.3	2.79	0.8	29	0.32
8	Mudstone	4	2582	2.0	10.37	1.2	32	0.28
9	Sandy mudstone	8	2562	2.6	12.08	2.45	40	0.25
10	Fine sandstone	4	2532	5.38	37.15	3.2	42	0.27
11	Mudstone	1	2582	2.0	10.37	1.2	32	0.28
12	No. 8 coal seam	3	1378	0.4	1.32	0.8	29	0.31
13	Mudstone	8	2582	2.0	10.37	1.2	32	0.28
14	Sandy mudstone	10	2562	2.6	12.08	2.45	40	0.25

To study the influence of different factors on the ground pressure behavior of roadway-surrounding rock, the numerical model under different conditions was established by adjusting the above basic model. At the same time, in the simulation process, the corresponding stress and displacement monitoring points were set at the top and bottom plates of the above four roadways and on the left side (near the side of the No. 360801 coal face) to monitor the roof-to-floor convergence, two-sided convergence, the maximum concentrated stress, and its position changes, to clarify the law of ground pressure behavior in the roadway-concentrated areas under the influence of different factors.

3.3. Simulation Scheme Design under Different Influence Factors

3.3.1. Geologic Factor

(1) Roof lithology

To deeply analyze the influence of roof lithology on ground pressure behavior in roadway-concentrated areas under super-thick nappe, under the condition of ensuring other conditions are unchanged, the roof lithology of the No. 360801 coal face was set as very soft, soft, softer, and stiff. Different lithology is to be achieved by the assignment command in numerical simulation. The stress and displacement changes of roadway-surrounding rock after coal face mining and the influence law of roof lithology on ground pressure behavior were obtained. The specific scheme is shown in Table 2.

Table 2. Simulation scheme under different roof lithology conditions.

Name	Simulation Scheme	Specific Parameters					
		Density/kg·m^{-3}	Bulk Modulus/GPa	Shear Modulus/GPa	Internal Friction Angle/°	Cohesion/MPa	Tensile Strength/MPa
Roof lithology	Very soft	1378	2.8	1.51	32	0.3	0.945
	Soft	2582	4.23	2.3	40	0.3	2.4
	Softer	2532	8.64	5.69	38	2.1	4.5
	Stiff	2532	12.64	8.69	36	5.2	11.5

(2) Thickness of nappe

To ensure other conditions are unchanged, the specific influence of the thickness variation of the nappe on the ground pressure behavior of the roadway was determined. The thickness of the nappe was set to 100, 250, 400, and 550 m, respectively. By observing the changes of stress and displacement of the surrounding rock of the roadway after mining in the coal face, the influence of the different thicknesses of nappe on the ground pressure behavior was explored.

3.3.2. Mining Technologic Factors

(1) Mining height

To explore the influence of mining height on the ground pressure behavior in the roadway-concentrated areas under the super-thick nappe, the other conditions were also guaranteed. The mining height was set to 1, 3, 5, and 7 m, respectively. We analyzed and compared the changes of stress and displacement of roadway-surrounding rock after coal face mining to explore the law of ground pressure behavior in roadway-concentrated areas.

(2) Rise-entry barrier coal pillar size

Under the premise of ensuring other conditions remained unchanged, the size of the rise-entry barrier coal pillar was changed. It was set to 150, 190, 230, and 270 m, respectively. The stress and displacement changes of the surrounding rock of the roadway after coal face mining and the influence law of the size of the rise-entry barrier coal pillar on ground pressure behavior were obtained.

3.3.3. Time Factor

To determine the influence of the creep time on the mine pressure behavior, the principle of the only variable was followed. The creep time was set to 30, 60, 90, and 120 days. The stress and displacement changes of roadway-surrounding rock after coal face mining and the influence law of the creep time on the ground pressure behavior were obtained.

4. Results

4.1. Appearance Law of Ground Pressure Behavior in Roadway-Concentrated Areas under Actual Engineering Conditions

To determine the change of ground pressure behavior in roadway-concentrated areas during the ground pressure behavior of the No. 360801 coal face, the advancing distance of the coal face was set as 100, 200, 300, 400, and 512 m, respectively (the size of the rise-entry barrier coal pillar was 190 m). The roof-to-floor convergence, two-side convergence, maximum concentrated stress, and their positions of the four roadways of the −706-rail crosscut, rail rise-entry, transport rise-entry, and return air rise-entry were monitored and compared. The specific results are shown in Figures 6 and 7.

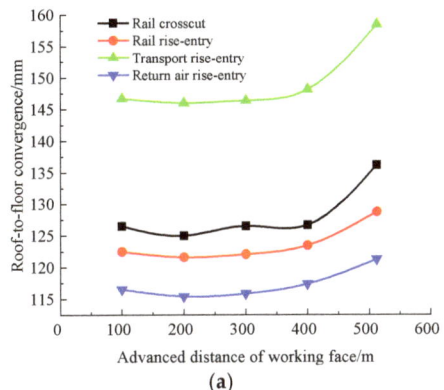
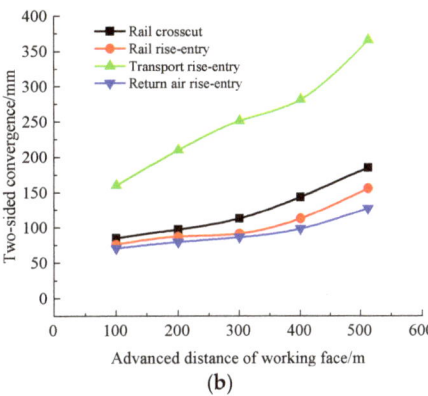

Figure 6. Monitoring results of convergence in roadway-concentrated areas during the advancing process of the No. 360801 coal face. (**a**) Roof-to-floor convergence, (**b**) Two-sided convergence.

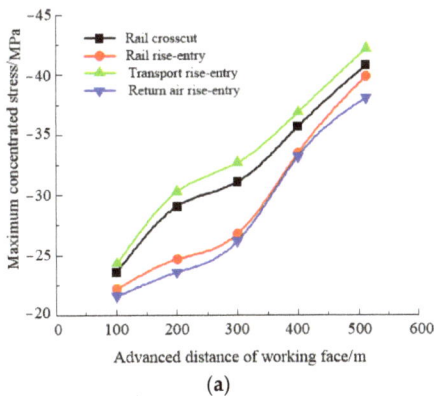
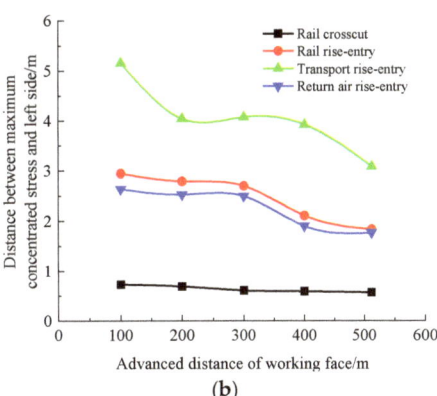

Figure 7. Maximum concentrated stress and its position monitoring results in the advancing process of the No. 360801 coal face. (**a**) Maximum concentrated stress, (**b**) Position of maximum concentrated stress.

It can be seen from Figure 6 that with the increase of the advancing distance of the coal face, for the roof-to-floor convergence, each roadway did not change before the advancing distance was less than 300 m. When the advancing distance was greater than 300 m, there was a significant increasing trend, and the two-sided convergence showed a gradually increasing trend from the beginning. In general, when the coal faces advances to 512 m, the −706-rail crosscut, rail rise-entry, transport rise-entry, and return air rise-entry the roof-to-floor convergence increased by 7.66%, 5.11%, 8%, and 4.06%, respectively, compared with 100 m. The two-sided convergence increased by 117.32%, 103.8%, 128.6%, and 59.65%, respectively. Combined with the two above sets of data, it can be seen that with the increase of the advancing distance of the coal face, the influence on the two-sided convergence was greater than that of the roof-to-floor convergence, but the basic trend remained unchanged.

It can be seen from Figure 7 that with the continuous advancement of the coal face, the maximum concentrated stress in the roadway-concentrated areas gradually increased, and the position of the stress concentration was constantly approaching the left side of the roadway. The maximum concentrated stress of the transport rise-entry roadway and its distance from the left side showed the maximum values of the four roadways, and the maximum concentrated stress of the return air rise-entry was the smallest. The closest distance between the maximum concentrated stress position and the left side of the roadway was the −706-rail crosscut, and the position of the maximum concentrated stress did not change significantly with the continuous advancement of the coal face.

In summary, with the advance of the coal face, the roof-to-floor convergence, the two-sided convergence, the maximum concentrated stress, and its location in the roadway-concentrated areas significantly changed. The reason is mainly due to the mining of the coal face and the slip dislocation of the super-thick nappe along the thrust fault caused by mining.

4.2. Influence Law of Roof Lithology

According to the simulation scheme in Table 2, the roof-to-floor convergence, two-sided convergence, maximum concentrated stress, and its position of the four roadways of the −706-rail crosscut, rail rise-entry, transport rise-entry, and return air rise-entry in the roadway-concentrated areas after the No. 360801 coal face mining were monitored and compared. The specific results are shown in Figures 8 and 9.

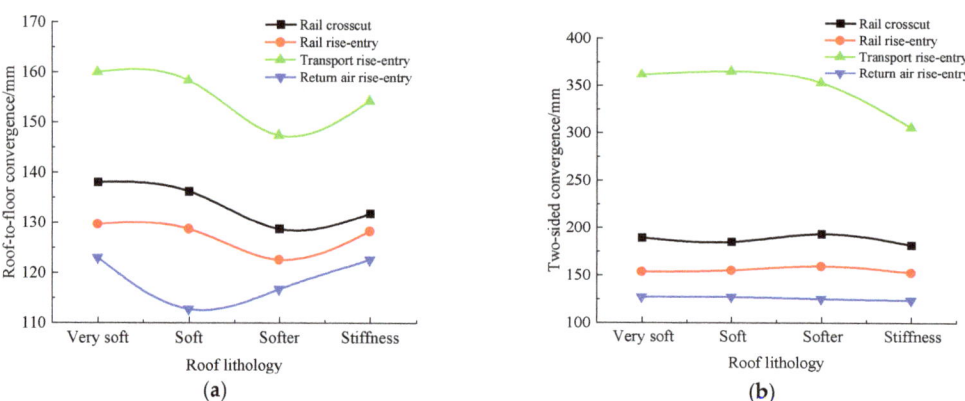

Figure 8. Monitoring results of convergence in roadway-concentrated areas under different roof lithology conditions. (**a**) Roof-to-floor convergence, (**b**) Two-sided convergence.

It can be seen from Figure 8 that in the roof-to-floor convergence, affected by the roof lithology, each roadway showed a trend of decreasing first and then increasing. As for the two-sided convergence, the −706-rail crosscut, rail rise-entry, and return air rise-entry were less affected by the lithology of the roof, and only the transport rise-entry showed a decreasing trend. Overall, when the roof lithology was stiff, the −706-rail crosscut, rail rise-entry, transport rise-entry, and return air rise-entry roof-to-floor convergence compared to the lithology was very soft, and decreased by 4.49%, 1.1%, 3.63%, and 0.3%, while the two-sided convergence was reduced by 4.01%, 0.7%, 15.37%, and 2.61%, respectively.

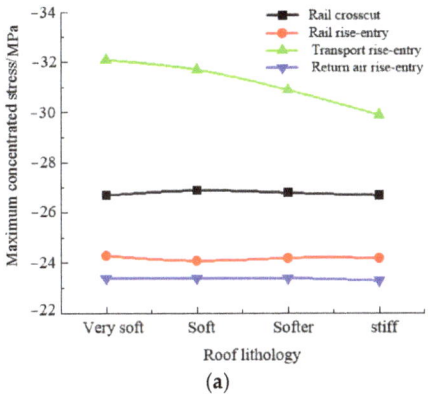
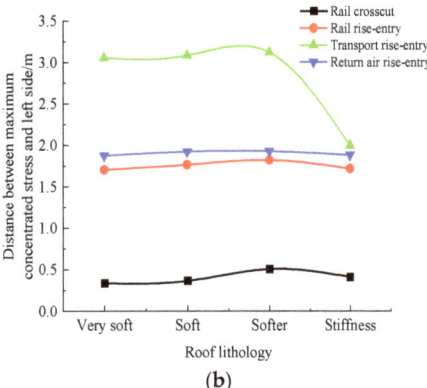

Figure 9. Maximum concentrated stress and its location monitoring results under different roof lithology conditions. (**a**) Maximum concentrated stress, (**b**) Position of maximum concentrated stress.

It can be seen from Figure 9 that with the continuous enhancement of roof lithology, for the maximum concentrated stress of the three roadways of the −706-rail crosscut, rail rise-entry, and return air rise-entry, the distance between their position and the left side of the roadway did not change much. The maximum concentrated stress of transport rise-entry showed a decreasing trend, but the distance between the maximum concentrated stress and the left side of the roadway was unchanged when the roof lithology was very soft, soft, and softer. Only when the roof lithology changes from softer to stiff will the distance between the maximum concentrated stress and the left side of the roadway be greatly reduced.

4.3. Influence Law of Nappe Thickness

According to the simulation scheme, the roof-to-floor convergence, two-sided convergence, maximum concentrated stress, and its position of the four roadways of the −706-rail crosscut, rail rise-entry, transport rise-entry, and return air rise-entry in the roadway-concentrated areas after mining of the No. 360801 coal face with different thicknesses of nappe were monitored and compared. The specific results are shown in Figures 10 and 11.

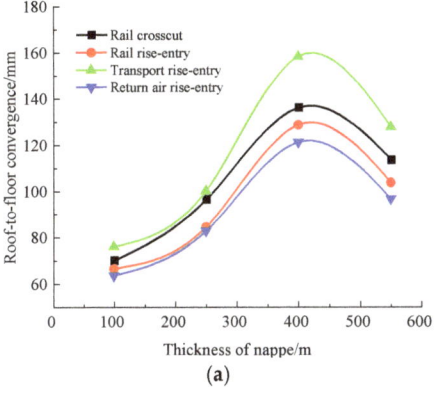
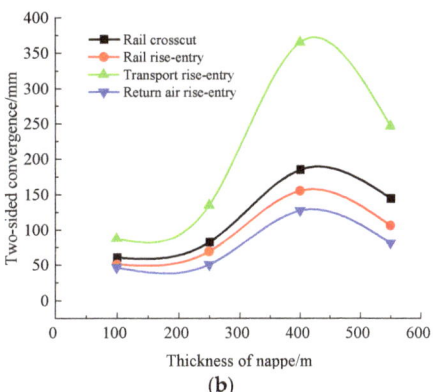

Figure 10. Monitoring results of convergence in roadway-concentrated areas under the different thicknesses of nappe conditions. (**a**) Roof-to-floor convergence, (**b**) Two-sided convergence.

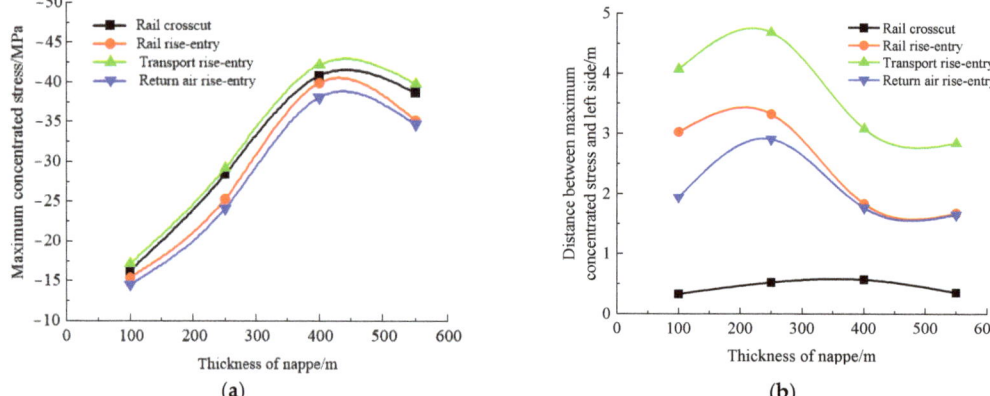

Figure 11. Maximum concentrated stress and its location monitoring results under the different thicknesses of nappe conditions. (**a**) Maximum concentrated stress, (**b**) Position of maximum concentrated stress.

It can be seen from Figure 10 that the thickness of the nappe had the same influence on the changing trend of the roof-to-floor convergence and the two-sided convergence in the roadway-concentrated areas under the super-thick nappe, both showing a trend of increasing first and then decreasing, and both changes were significant when the thickness of the nappe was 400 m. In general, when the thickness of the nappe was 550 m, the roof-to-floor convergence of the −706-rail crosscut, rail rise-entry, transport rise-entry, and return air rise-entry was increased by 61.89%, 55.7%, 67.88%, and 52.1%, respectively, compared with the thickness of 100 m, while the two-sided convergence was increased by 138.45%, 107.29%, 180.87%, and 75.05%, respectively.

It can be seen from Figure 11 that the thickness of the nappe had the same influence on the variation trend of the maximum concentrated stress and the distance between the maximum concentrated stress and the left side of the roadway-concentrated areas under the super-thick nappe, which increased first and then decreased. The maximum concentrated stress of the four roadways of the −706-rail crosscut, rail rise-entry, transport rise-entry, and return air rise-entry obviously changed when the thickness of the nappe was 400 m. The distance between the maximum concentrated stress of the three roadways of the rail rise-entry, transport rise-entry, and return air rise-entry with the left side of the roadway obviously changed under the condition where the thickness of the nappe was 250 m, and only the −706-rail crosscut changed when the thickness of the nappe was 400 m.

4.4. Influence Law of Mining Height

According to the simulation scheme, the roof-to-floor convergence, two-sided convergence, maximum concentrated stress, and its position of the four roadways of the −706-rail crosscut, rail rise-entry, transport rise-entry, and return air rise-entry in the roadway-concentrated areas after mining of the No. 360801 coal face with different mining heights were monitored and compared. The specific results are shown in Figures 12 and 13.

It can be seen from Figure 12 that the influence of the mining height on the roof-to-floor convergence in the roadway-concentrated areas under the super-thick nappe and the two-sided convergence, except for the −706-rail crosscut roadway, was the same, showing a trend of increasing and then decreasing, and both obviously changed when the mining height was 3 m. The two-sided convergence of the −706-rail crosscut was gradually increasing. In general, when the mining height was 7 m, the roof-to-floor convergence of the −706-rail crosscut, rail rise-entry, transport rise-entry, and return air rise-entry was reduced by 10.15%, 10.09%, 12.97%, and 6.35%, respectively, compared to that when the mining height was 1 m. The two-sided convergence increased by 120.8%, 73.29%, 130.08%, and 46.68%, respectively.

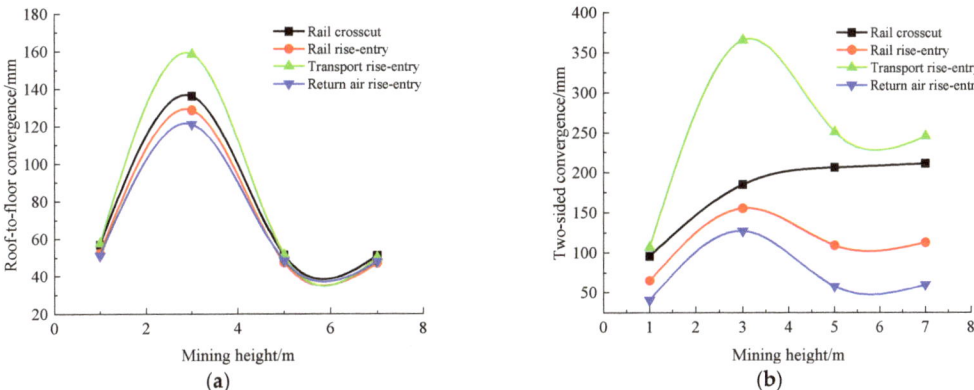

Figure 12. Monitoring results of convergence in roadway-concentrated areas under different mining height conditions. (**a**) Roof-to-floor convergence, (**b**) Two-sided convergence.

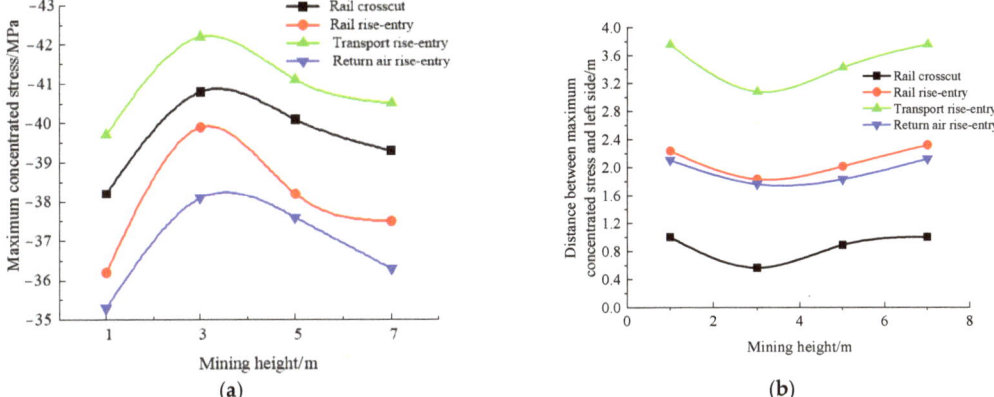

Figure 13. Maximum concentrated stress and its location monitoring results under different mining height conditions. (**a**) Maximum concentrated stress, (**b**) Position of maximum concentrated stress.

It can be seen from Figure 13 that the continuous increase of mining height, −706-rail crosscut, rail rise-entry, transport rise-entry, and return air rise-entry maximum concentrated stress showed a trend of increasing first and then decreasing, and when the mining height was 3 m, the changes were significant. The distance between the maximum concentrated stress and the left side of the roadway decreased first and then increased, and obviously changed when the mining height was 3 m.

4.5. Influence Law of Rise-Entry Barrier Coal Pillar Size

According to the simulation scheme, the roof-to-floor convergence, two-sided convergence, maximum concentrated stress, and its position of the four roadways of the −706-rail crosscut, rail rise-entry, transport rise-entry, and return air rise-entry in the roadway-concentrated areas after mining of the No. 360801 coal face with different sizes of the rise-entry barrier coal pillar were monitored and compared. The specific results are shown in Figures 14 and 15.

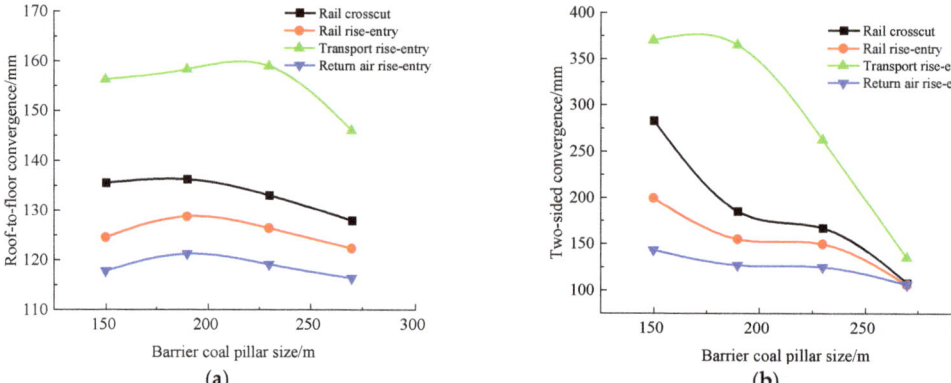

Figure 14. Monitoring results of convergence in roadway-concentrated areas under different sizes of rise-entry barrier coal pillar conditions. (**a**) Roof-to-floor convergence, (**b**) Two-sided convergence.

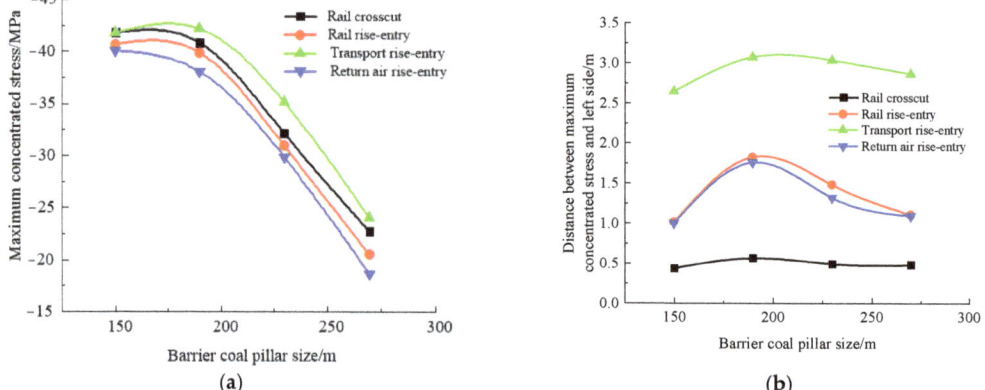

Figure 15. Maximum concentrated stress and its location monitoring results under different sizes of rise-entry barrier coal pillar conditions. (**a**) Maximum concentrated stress, (**b**) Position of maximum concentrated stress.

It can be seen from Figure 14 that the influence of the size of the rise-entry barrier coal pillar on the roof-to-floor convergence in the roadway-concentrated areas under the super-thick nappe showed a trend of increasing first and then decreasing, while the two-sided convergence showed a trend of gradually decreasing. In general, when the size of the rise-entry barrier coal pillar increased to 270 m, the roof-to-floor convergence of the −706-rail crosscut, rail rise-entry, transport rise-entry, and return air rise-entry decreased by 5.61%, 1.76%, 6.55%, and 1.25%, respectively, compared to that of 150 m, while the two-sided convergence decreased by 61.93%, 46.74%, 63.5%, and 25.9%, respectively.

It can be seen from Figure 15 that with the increase of the size of the rise-entry barrier coal pillar, the maximum concentrated stress of the four roadways showed a decreasing trend, and the decreasing speed suddenly increased when the size of the rise-entry barrier coal pillar was greater than 190 m. The distance between the maximum concentrated stress and the left side of the roadway increased first and then decreased, and obviously changed when the size of the rise-entry barrier coal pillar was 190 m.

4.6. Influence Law of Creep Time

According to the simulation scheme, the roof-to-floor convergence, two-sided convergence, maximum concentrated stress, and its position of the four roadways of the −706-rail crosscut, rail rise-entry, transport rise-entry, and return air rise-entry in the roadway-concentrated areas after mining of the No. 360801 coal face with different creep times were monitored and compared. The specific results are shown in Figures 16 and 17.

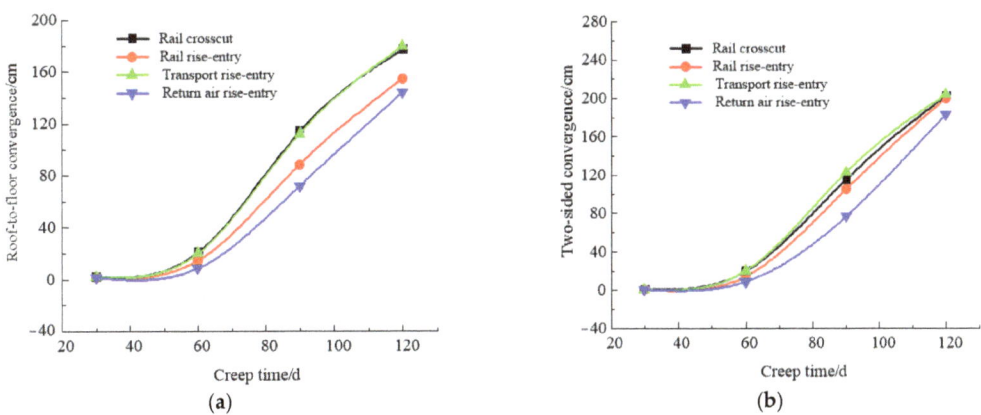

Figure 16. Monitoring results of convergence in roadway-concentrated areas under different creep time conditions. (**a**) Roof-to-floor convergence, (**b**) Two-sided convergence.

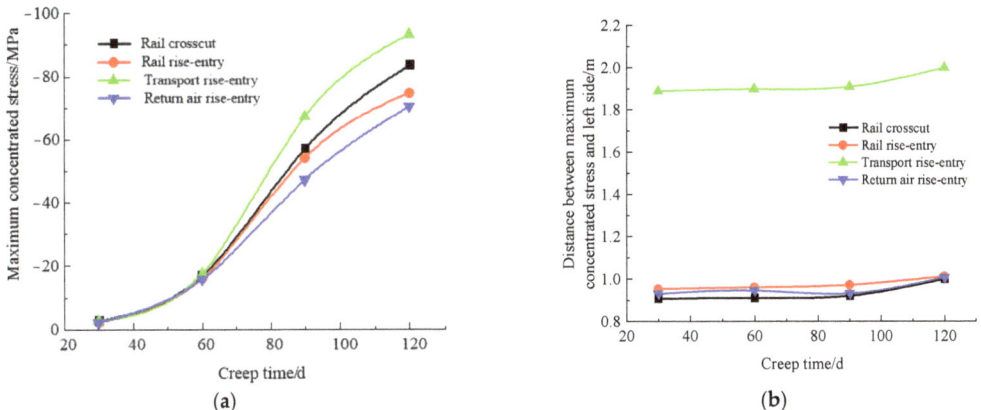

Figure 17. Maximum concentrated stress and its location monitoring results under different creep time conditions. (**a**) Maximum concentrated stress, (**b**) Position of maximum concentrated stress.

It can be seen from Figure 16 that the creep time showed a gradual increase in the trend of roof-to-floor convergence in the roadway-concentrated areas under the super-thick nappe. In general, when the creep time was 120 days, the −706-rail crosscut, rail rise-entry, transport rise-entry, and return air rise-entry roof-to-floor convergence compared to the creep time of 30 days increased by 78.21, 74.4, 80.56, and 73.95 times, respectively. The two-sided convergence increased by 164.1, 160.9, 167.1, and 151.1 times, respectively. The reason is that due to the influence of the rock layer relationship, the transport rise-entry was the closest to the coal seam roof, followed by the −706-rail crosscut, so the impact on these two roadways was greater than the rail rise-entry and return air rise-entry.

It can be seen from Figure 17 that the creep time had a serious influence on the maximum concentrated stress in the roadway-concentrated areas under the super-thick nappe, showing a gradual increasing trend, while the change range of the transport rise-entry was larger than that of the other three roadways, and the change range of the return air rise-entry was the smallest. The distance between the maximum concentrated stress and the left side of the roadway was almost unchanged before the creep time of 90 days, and when the creep time exceeded 90 days, there was a slightly obvious increasing trend.

4.7. Comparative Analysis of Factors

Based on the above numerical simulation results, it can be seen that for the roof-to-floor convergence in roadway-concentrated areas, with the increase of the thickness of the nappe, the mining height, and the size of the rise-entry barrier coal pillar, they all showed a trend of increasing first and then decreasing. With the increase of roof lithology conditions, the opposite trend was shown, and the creep time increased. For the two-sided convergence and the maximum concentrated stress, it increased first and then decreased with the increase of the thickness of the nappe and the mining height, decreased with the increase of the roof lithology and the size of the rise-entry barrier coal pillar, and showed the opposite trend with the increase of the creep time. For the maximum concentrated stress and the distance from the side, it increased first and then decreased with the increase of the thickness of the nappe and the size of the rise-entry barrier coal pillar. It showed an opposite trend with the increase of the mining height, the increase of the roof lithology, and the increase of the creep time.

By comparing the influence of roof lithology, the thickness of nappe, the mining height, the size of the rise-entry barrier coal pillar, and the creep time on the overall ground pressure behavior in the roadway-concentrated areas, it can be seen that the creep time had the greatest influence on the ground pressure behavior in the roadway-concentrated areas, followed by the thickness of the nappe, the mining height, and the size of the rise-entry barrier coal pillar, and the roof lithology had the least influence on the ground pressure behavior of each roadway. At the same time, by comparing the influence degree of the ground pressure behavior of each roadway in the roadway-concentrated areas, it was found that the above influencing factors had the most significant influence on the transport rise-entry, followed by the −706-rail crosscut, the rail rise-entry, and the return air rise-entry, which was the least affected by the various factors.

5. Conclusions

(1) According to the actual geological mining conditions of the No. 36 district in the Xinji No. 1 Mine, the influencing factors of ground pressure behavior in roadway-concentrated areas under super-thick nappe were analyzed from three aspects: geological factors, mining technology factors, and time factors. The main influencing factors were determined as roof lithology, the thickness of the nappe, the mining height, the size of the rise-entry barrier coal pillar, and the creep time.

(2) With the advancement of the No. 360801 coal face, the roof-to-floor convergence, the two-sided convergence, and the maximum concentrated stress of each roadway in the roadway-concentrated areas gradually increased, the distance between the position of the maximum concentrated stress and the side gradually decreased, and the deformation of the two-sided convergence was significantly greater than that of the roof-to-floor convergence. Among them, the two-sided convergence and the maximum concentrated stress can reach 365.4 mm and 42.2 MPa.

(3) By comparing the influence of various factors on the ground pressure behavior in roadway-concentrated areas, it can be concluded that: the greater the thickness of nappe and the greater the mining height, the longer the creep time, the stronger the mine pressure behavior, the greater the size of the rise-entry barrier coal pillar, and the stronger the roof lithology, and the ground pressure behavior tends to be gentle.

(4) By comparing the degree of mine pressure behavior in each roadway under the influence of different factors, it can be found that: The intensity of mine pressure behavior in each roadway was mainly affected by the strata. The closer the roadway

was to the coal seam roof, the stronger the mine pressure behavior, while the farther away from the coal seam roof, the gentler the mine pressure behavior. Transport rise-entry was the closest to the coal seam roof, followed by the rail crosscut, so the impact on these two roadways was greater than on the rail rise-entry and the return air rise-entry.

Author Contributions: Conceptualization, R.Z. and X.L. (Xuesheng Liu); Methodology, R.Z. and X.Y.; Software, R.Z., Z.S. and X.L. (Xuebin Li); Validation, X.Y. and Z.S.; Formal analysis, X.Y.; Investigation, X.L. (Xuebin Li); Data curation, X.L. (Xuesheng Liu); Writing – original draft, Z.S.; Writing – review & editing, X.L. (Xuesheng Liu) and X.L. (Xuebin Li). All authors have read and agreed to the published version of the manuscript.

Funding: This research was funded by the National Natural Science Foundation of China (No. 52174122), Outstanding Youth Fund of Shandong Natural Science Foundation (No. ZQ2022YQ49) and Taishan Scholar Young Expert in Shandong Province.

Institutional Review Board Statement: Not applicable.

Informed Consent Statement: Not applicable.

Data Availability Statement: Not applicable.

Acknowledgments: We thank Cao Anye of China University of Mining and Technology for his support and help with this paper.

Conflicts of Interest: The authors declare no conflict of interest.

Nomenclature

Terminology	Representative meaning
Thrust fault	A large reverse fault with a dip angle of about 30° or less.
Nappe	A huge rock mass structure formed by thrust faults that cause the hanging wall to move thousands or even tens of thousands of meters along the fault plane.
Ground pressure behavior	Mine pressure phenomenon shown by surrounding rock movement and support force under the mine pressure.
Barrier coal pillar	It refers to the part of the coal body that is specially left underground and is not mined, and the purpose is to protect the above-mentioned protected objects inside the rock stratum and on the surface from mining.
Creep	The phenomenon that the deformation of rock increases with time under the continuous action of an external force whose size and direction do not change.
Numerical simulation	By means of electronic computers, combined with the concept of finite element or finite volume, through numerical calculation and image display, with the purpose of studying engineering and physical problems and even various problems in nature.
Gneiss	Rocks formed by deep metamorphism of magmatic or sedimentary rocks.
Immediate roof	Refers to the rock strata directly above the ore layer and can fall down in time during caving.
Main roof	The roof located on the direct roof or coal seam, also known as the old roof. Usually large thickness and rock strength, difficult to collapse rock.
Rail crosscut	Transport is the main task in the inner track of the connecting roadway through the coal seam.
Rail rise-entry	Inclined roadway upward from the transport layer.
Transport rise-entry	Roadways for transporting ores when mining deposits above the mining level.
Return air rise-entry	A roadway extending upward from a mining level for ventilation.
Roof-to-floor convergence	Under the support of hydraulic support, the mining space is affected by mining, the reflection of rock deformation, and movement.

References

1. Kayabasi, A.; Gokceoglu, C. Coal mining under difficult geological conditions: The Can lignite open pit (Canakkale, Turkey). *Eng. Geol.* 2012, *135*, 66–82. [CrossRef]
2. Zhao, T.; Zhang, Z.; Tan, Y.; Shi, C.; Wei, P.; Li, Q. An innovative approach to thin coal seam mining of complex geological conditions by pressure regulation. *Int. J. Rock Mech. Min. Sci.* 2014, *71*, 249–257. [CrossRef]
3. Kang, H. Support technologies for deep and complex roadways in underground coal mines: A review. *Int. J. Coal Sci. Technol.* 2014, *1*, 261–277. [CrossRef]
4. Yao, N.; Yi, W.; Yao, Y.; Song, H.; Li, W.; Peng, T.; Sun, X. Progress of drilling technologies and equipments for complicated geological conditions in underground coal mines in China. *Coal Field Geol. Explor.* 2020, *48*, 1–7.
5. Elizalde, C.; Griffith, W.A.; Miller, T. Thrust fault nucleation due to heterogeneous bedding plane slip: Evidence from an Ohio coal mine. *Eng. Geol.* 2016, *206*, 1–17. [CrossRef]
6. Liu, X.; Fan, D.; Tan, Y.; Ning, J.; Song, S.; Wang, H.; Li, X. New detecting method on the connecting fractured zone above the coal face and a case study. *Rock Mech. Rock Eng.* 2021, *54*, 4379–4391. [CrossRef]
7. Jing, F.; Wei, Q.; Wang, C.; Yao, S.-L.; Zhang, Y.; Han, R.-J.; Wei, X.-Z.; Li, Z.-C. Analysis of rock burst mechanism in extra-thick coal seam controlled by huge thick conglomerate and thrust fault. *J. China Coal Soc.* 2014, *39*, 1191–1196.
8. Zhao, Y.; Wang, H.; Jiao, Z.; Zhang, X. Experimental study of the activities of reverse fault induced by footwall coal mining. *J. China Coal Soc.* 2018, *43*, 914–922.
9. Batugin, A.; Wang, Z.; Su, Z.; Sidikovna, S.S. Combined support mechanism of rock bolts and anchor cables for adjacent roadways in the external staggered split-level panel layout. *Int. J. Coal Sci. Technol.* 2021, *8*, 659–673. [CrossRef]
10. Tan, Y.; Fan, D.; Liu, X.; Song, S.; Li, X.; Wang, H. Numerical investigation on failure evolution of surrounding rock for super-large section chamber group in deep coal mine. *Energy Sci. Eng.* 2019, *7*, 3124–3146. [CrossRef]
11. Liu, X.; Fan, D.; Tan, Y.; Song, S. Failure evolution and instability mechanism of surrounding rock for close-distance parallel chambers with super-large section in deep coal mines. *Int. J. Geomech.* 2021, *21*, 04021049. [CrossRef]
12. Wu, Q.; Jiang, L.; Wu, Q. Study on the law of mining stress evolution and fault activation under the influence of normal fault. *Acta Geodyn. Geomater.* 2017, *14*, 357–369. [CrossRef]
13. Jiao, Z.; Jiang, Y.; Zhao, Y.; Hu, H. Study of dynamic mechanical response characteristics of working face passing through reverse fault. *J. China Univ. Min. Technol.* 2019, *48*, 54–63.
14. Jiao, Z.; Zhao, Y.; Jiang, Y.; Wang, H.; Lu, Z.G.; Wang, X.Z. Fault damage induced by mining and its sensitivity analysis of influencing factors. *J. China Coal Soc.* 2017, *42* (Suppl. S1), 36–42.
15. Sainoki, A.; Schwartzkopff, A.K.; Jiang, L.; Mitri, H.S. Numerical Modeling of Complex Stress State in a Fault Damage Zone and Its Implication on Near-Fault Seismic Activity. *J. Geophys. Res. Solid Earth* 2021, *126*, e2021JB021784. [CrossRef]
16. Li, Z.; Wang, C.; Shan, R.; Yuan, H.; Zhao, Y.; Wei, Y. Study on the influence of the fault dip angle on the stress evolution and slip risk of normal faults in mining. *Bull. Eng. Geol. Environ.* 2021, *80*, 3537–3551. [CrossRef]
17. Chen, X.; Li, W.; Yan, X. Analysis on rock burst danger when fully-mechanized caving coal face passed fault with deep mining. *Saf. Sci.* 2012, *50*, 645–648. [CrossRef]
18. Zhang, X.; Zhou, F.; Zou, J. Numerical Simulation of Gas Extraction in Coal Seam Strengthened by Static Blasting. *Sustainability* 2022, *14*, 12484. [CrossRef]
19. Cao, A.; Jing, G.; Ding, Y.; Liu, S. Mining-induced static and dynamic loading rate effect on rock damage and acoustic emission characteristic under uniaxial compression. *Saf. Sci.* 2019, *116*, 86–96. [CrossRef]
20. Cao, A.; Hu, Y.; Li, B. Research and application on anchor cable reinforcement of distressed zone in sidewall along roadway under dynamic disturbance. *Coal Sci. Technol. Mag.* 2021, *42*, 39–46.
21. Zhang, F.; Cui, L.; An, M.; Elsworth, D.; He, C. Frictional stability of Longmaxi shale gouges and its implication for deep seismic potential in the southeastern Sichuan Basin. *Deep. Undergr. Sci Eng.* 2022, *1*, 3–14. [CrossRef]
22. Kong, P.; Jiang, L.; Shu, J.; Wang, L. Mining stress distribution and fault-slip behavior: A case study of fault-influenced longwall coal mining. *Energies* 2019, *12*, 2494. [CrossRef]
23. Wu, Q.; Wu, Q.; Yuan, A.; Wu, Y. Analysis of mining effect and fault stability under the influence of normal faults. *Geotech. Geol. Eng.* 2021, *39*, 49–63. [CrossRef]
24. Jiao, Z.; Wang, L.; Zhang, M.; Wang, J. Numerical Simulation of Mining-Induced Stress Evolution and Fault Slip Behavior in Deep Mining. *Adv. Mater. Sci. Eng.* 2021. [CrossRef]
25. Islan, M.R.; Shinjo, R. Mining-induced fault reactivation associated with the main conveyor belt roadway and safety of the Barapukuria coal mine in Bangladesh: Constraints from BEM simulations. *Int. J. Coal Geol.* 2009, *79*, 115–130. [CrossRef]
26. Zhang, D.; Duan, Y.; Du, W.; Chai, J. Experimental Study on Physical Similar Model of Fault Activation Law Based on Distributed Optical Fiber Monitoring. *Shock. Vib.* 2021. [CrossRef]
27. Wang, A.; Pan, Y.; Li, Z.; Liu, C.S.; Han, R.J.; Lv, X.F.; Lu, H.Q. Similar experimental study of rock burst induced by mining deep coal seam under fault action. *Rock Soil Mech.* 2014, *35*, 2486–2492.
28. Luo, H.; Li, Z.; Wang, A.; Xiao, Y.-H. Study on the evolution law of stress field when approaching fault in deep mining. *J. China Coal Soc.* 2014, *39*, 322–327.
29. Wang, H.; Jiang, Y.; Xue, S.; Mao, L.; Lin, Z.; Deng, D.; Zhang, D. Influence of fault slip on mining-induced pressure and optimization of roadway support design in fault-influenced zone. *J. Rock Mech. Geotech. Eng.* 2016, *8*, 660–671. [CrossRef]

30. Cai, W.; Dou, L.; Wang, G.; Hu, Y. Mechanism of fault reactivation and its induced coal burst caused by coal mining activities. *J. Min. Saf. Eng.* **2019**, *36*, 1193–1202.
31. Babets, D.; Sdvyzhkova, O.; Shashenko, O.; Kravchenko, K.; Cabana, E.C. Implementation of probabilistic approach to rock mass strength estimation while excavating through fault zones. *Min. Miner. Depos.* **2019**, *13*, 72–83. [CrossRef]
32. Han, Z.; Li, D.; Li, X. Dynamic mechanical properties and wave propagation of composite rock-mortar specimens based on SHPB tests. *Int. J. Min. Sci. Technol.* **2022**, *32*, 793–806. [CrossRef]
33. Cao, A.; Chen, F.; Liu, Y.; Dou, L.M.; Wang, C.B.; Yang, X.; Bai, X.Q.; Song, S.K. Response characteristics of rupture mechanism and source parameters of mining tremors in frequent coal burst area. *J. China Coal Soc.* **2022**, *47*, 722–733.
34. Fan, D.; Liu, X.; Tan, Y.; Li, X.; Lkhamsuren, P. Instability energy mechanism of super-large section crossing chambers in deep coal mines. *Int. J. Min. Sci. Technol.* **2022**, *32*, 1075–1086. [CrossRef]
35. Luo, S.; Gong, F. Evaluation of rock burst proneness considering specimen shape by storable elastic strain energy. *Deep Underground Sci. Eng.* **2022**, 1–15.
36. Ji, H.; Ma, H.; Wang, J.; Zhang, Y.H.; Cao, H. Mining disturbance effect and mining arrangements analysis of near-fault mining in high tectonic stress region. *Saf. Sci.* **2012**, *50*, 649–654. [CrossRef]
37. Potvin, Y.; Jarufe, J.; Wesseloo, J. Interpretation of seismic data and numerical modeling of fault reactivation at El Teniente, Reservas Norte sector. *Trans. Inst. Min. Metall.* **2015**, *119*, 175–181.
38. Begalinov, A.; Almenov, T.; Zhanakova, R.; Bektur, B. Analysis of the stress deformed state of rocks around the haulage roadway of the Beskempir field (Kazakhstan). *Min. Miner. Depos.* **2020**, *14*, 28–36. [CrossRef]
39. Wang, H.; Shi, R.; Song, J.; Tian, Z.; Deng, D.; Jiang, Y. Mechanical model for the calculation of stress distribution on fault surface during the underground coal seam mining. *Int. J. Rock Mech. Min. Sci.* **2021**, *144*, 104765. [CrossRef]
40. Dokht, R.M.H.; Smith, B.; Kao, H.; Visser, R.; Hutchinson, J. Reactivation of an intraplate fault by mine-blasting events: Implications to regional seismic hazard in Western Canada. *J. Geophys. Res. Solid Earth* **2020**, *125*. [CrossRef]
41. Pan, C.; Xia, B.; Zuo, Y.; Yu, B.; Ou, C. Mechanism and control technology of strong ground pressure behavior induced by high-position hard roofs in extra-thick coal seam mining. *Int. J. Min. Sci. Technol.* **2022**, *32*, 499–511. [CrossRef]
42. Tan, Y.L.; Liu, X.S.; Shen, B.; Ning, J.G.; Gu, Q.H. New approaches to testing and evaluating the impact capability of coal seam with hard roof and/or floor in coal mines. *Geomech. Eng.* **2018**, *14*, 367–376.
43. Liu X, S.; Tan Y, L.; Ning J, G.; Lu, Y.W. Gu, Q.H. Mechanical properties and damage constitutive model of coal in coal-rock combined body. *Int. J. Rock Mech. Min. Sci.* **2018**, *110*, 140–150. [CrossRef]
44. Małkowski, P.; Niedbalski, Z.; Balarabe, T. A statistical analysis of geomechanical data and its effect on rock mass numerical modeling: A case study. *Int. J. Coal Sci. Technol.* **2021**, *8*, 312–323. [CrossRef]

Disclaimer/Publisher's Note: The statements, opinions and data contained in all publications are solely those of the individual author(s) and contributor(s) and not of MDPI and/or the editor(s). MDPI and/or the editor(s) disclaim responsibility for any injury to people or property resulting from any ideas, methods, instructions or products referred to in the content.

Article

Mechanical and Microcrack Evolution Characteristics of Roof Rock of Coal Seam with Different Angle of Defects Based on Particle Flow Code

Qinghai Deng [1], Jiaqi Liu [1], Junchao Wang [2] and Xianzhou Lyu [1,*]

1 College of Earth Science and Engineering, Shandong University of Science and Technology, Qingdao 266590, China
2 Longshou Mine of Jinchuan Group Co., Ltd., Jinchuan 737100, China
* Correspondence: lyuxianzhou0608@sdust.edu.cn

Abstract: The creation of the natural ceiling rock of the coal seam is rife with fractures, holes, and other flaws. The angle of the defects has a significant influence on the mechanical characteristics and crack evolution of coal seam roof rock. Multi-scale numerical simulation software PFC2D gets adapted to realize the crack propagation and coalescence process in the roof rock of a coal seam with different angles of defects under uniaxial compression. The effect of flaw angles on the micro and macro mechanical characteristics of rock is also discovered. The results show that: (1) the defect angle has influence on the stress-strain, elastic modulus, peak strength, peak strain, acoustic emission (AE) and strain energy of roof rock of coal seam. When the defect angles are different, the starting position of the roof rock in a coal seam fracture is different. The quantity of microcracks firstly reduces with an increase in defect angles before gradually increasing. At the same fault angle, the cracks are mostly tensile ones and only a few shear ones. (2) When the defect angle is less than 90°, tensile and shear fractures are mostly localized at the defect's two tips and propagate along the loading direction. When the defect angle is 90°, the tensile and shear cracks are not concentrated at the tip of the defect. (3) As the defect angles increase, the elastic strain energy rises initially and then falls, and the dissipated energy and total input energy both increase continuously. The elastic strain energy is greatest at the highest strength. The study provides a certain reference for the use of various analysis methods in practical engineering to evaluate the safety and stability of rock samples with pre-existing defects.

Keywords: rock mechanics; microcrack evolution; defects angles; failure modes; multi-scale characterization; particle flow code

Citation: Deng, Q.; Liu, J.; Wang, J.; Lyu, X. Mechanical and Microcrack Evolution Characteristics of Roof Rock of Coal Seam with Different Angle of Defects Based on Particle Flow Code. *Materials* 2023, 16, 1401. https://doi.org/10.3390/ma16041401

Academic Editor: Alessandro Pirondi

Received: 4 January 2023
Revised: 31 January 2023
Accepted: 4 February 2023
Published: 7 February 2023

Copyright: © 2023 by the authors. Licensee MDPI, Basel, Switzerland. This article is an open access article distributed under the terms and conditions of the Creative Commons Attribution (CC BY) license (https://creativecommons.org/licenses/by/4.0/).

1. Introduction

In mining engineering, defective rock mass is a type of complicated engineering medium, as is water conservancy engineering, transportation engineering, and underground tunnel engineering [1–3]. The complex structure of rock makes its physical properties inhomogeneous and anisotropic, which makes the research of rock masses with defects more difficult [4–7]. In previous studies on fractured rocks, cracks are mostly placed in the middle of rock specimens [8–12]. When the rock specimen's top or lower half is fractured, what effects will the angle change have on its mechanics and crack evolution characteristics.

At present, the research work on rocks with defects is mainly based on indoor rocks and includes the material model test or numerical simulation test. Through the laboratory test of rock with defects, we can find some influence rules such as inclination angle and number of prefabricated cracks, which have effect on the axial compressive strength, crack initiation, propagation and failure mode. And the numerical simulation software can only change one factor of the defect when other conditions remain unchanged, so that the research on the mechanical characteristics of defective rocks can be validated [13–17]. By controlling the numerical software program, the stress condition and changes of cracks near the defect can be obtained, which cannot be obtained experimentally. Hence, further analysis of the crack model's stress variation properties during uniaxial compression can be done. For example, Yang et al. used PFC2D to study the tensile strength and fracture propagation of intermittently double-fractured rocks and discovered the effect of cracks on these properties. [18]. Su et al. conducted laboratory experiments to investigate how longitudinal fractures affect the mechanical characteristics of sandstone and obtained some useful research results [2]. Through indoor experiments, Yang et al. studied the law and mechanism of fracture propagation for various inclination defects. The test results have important guidance and reference significance for underground engineering construction design. [19]. Cao et al. have done a lot of research on fractured rock with pores, including laboratory tests, numerical simulation, and different influencing factors [20,21]. Li et al. investigated marble with holes' dynamic failure properties under impact stress. The complete process of crack germination, propagation, penetration, and destruction was captured using a high-speed camera. [22]. The prefabricated hole specimen's dynamic compressive strength, mechanism of failure, and crack propagation characteristics under impact stress were also explored.

It can be found that there are many kinds of research methods for rocks with cracks, cavities, and other defects. There are various research methods, including the indoor rock mechanics test, the numerical simulation test, the uniaxial compression test, triaxial compression test, the impact dynamic load test, and also the cyclic loading and unloading test, and so on. Hence, it is possible to obtain the rock's mechanical, deformational, AE, failure, and energy properties, and so on. In addition, the effects of different crack combinations, water, heat, chemistry, and other environments on it are studied, and many useful results are obtained [23–29]. However, most of the defects in these studies are concentrated in the middle of the rock specimen or symmetrically distributed in the middle of the rock, and there are also a few studies on the distribution of shear cracks and tensile cracks in rock failure. However, in practical rock engineering, the distribution of cracks is not uniform. Hence, according to engineering practice, it is required to investigate the mechanical features of defects in a specific section. As fracture initiation and propagation are energy-driven processes, various analysis methods must be applied to assess the characteristics of rock samples with different defects.

In view of this, the numerical model of samples with various angle defects on one side of the roof of the coal seam is established using the PFC2D software, and its uniaxial compression is explored. The process of crack propagation and the number of microcracks are traced by using fish language [30–32]. Therefore, further analysis on the strain energy, strength, deformation, and acoustic emission with various crack inclinations is carried out, and the distribution characteristics of shear and tension cracks are also obtained.

2. Methodology

2.1. Particle Flow Code

How particles interact with one another can be described by the particle contact constitutive model. Among them, the most commonly used are the contact bond model and the parallel bond model [30]. Infinitesimal, linear elastic, and interfacial bearing capacity characteristics can be provided by the contact bond model, whether the surface is bonded or frictional, as shown in Figure 1a. It can transmit force and moment at the same time. While the contact area of the contact bond model is point that can not transmit moment. The bond loses its function, and the spring will break, if the tensile or shear stress between the particles is greater than the normal or tangential bond strength. This means that the contact bond model is usually suitable for soil. While the parallel bond model can efficiently simulate the bonding between rock particles because it can regard the bonding between particles as a group of parallel springs with the functions of tension, shear, and torque, as shown in Figure 1b [31]. Hence, the parallel bond model is widely used in building PFC2D rock models and simulations [18,21,30].

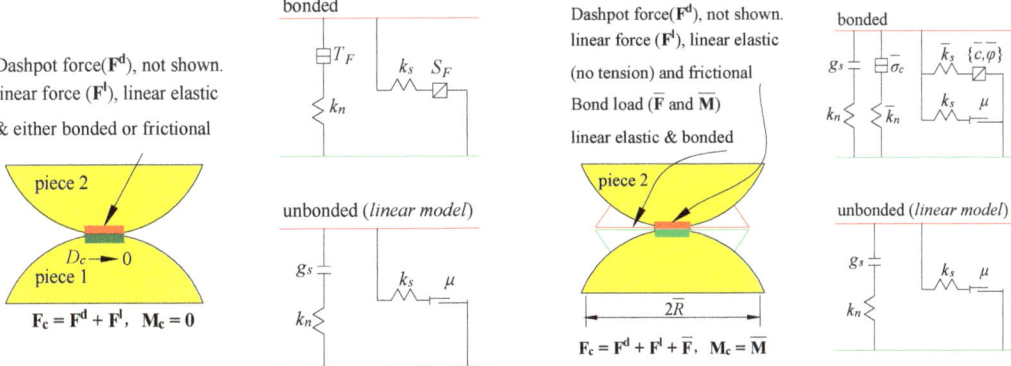

(a) Contact bond model   (b) Parallel bond model

Figure 1. Bonded-particle model and its micro-mechanical behavior.

2.2. Parameter Calibration of Roof Rock of Coal Seam

The purpose of meso-parameter calibration is to make the macro-mechanical properties of rock match the established PFC2D model. The meso-parameters include two parts: particle meso-parameters and meso-parameters of the contact constitutive model. The meso-parameters are usually calibrated by the "trial and error" method [18,21,30–32]. The most popular technique for calibrating the meso-parameters of the rock PFC2D model is "trial and error," which involves gradually modifying the meso-parameters until the numerical test results are consistent with the rock's large-scale mechanical properties. Figure 2 describes the parameter checking process of the "trial and error" method for the PFC model (version 5.0) [30–32].

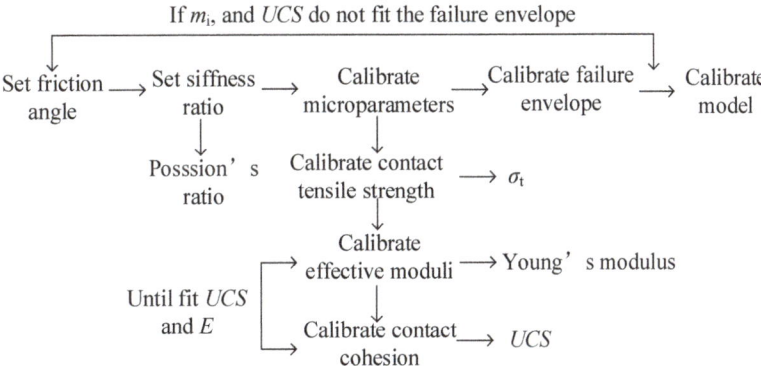

Figure 2. Parameter checking process of "trial and error" method for PFC model.

The parameters provided by previous studies were used to conduct numerical tests because of the limitations of laboratory testing. Through the method of "trial and error" with repeated check comparisons, we obtained the physical mechanical characteristics for the PFC model, as shown in Table 1. Certainly, they were close to the macroscopic mechanical parameters of the real rock.

Table 1. Micro mechanical parameters of coal and rock for numerical simulation.

Contact Parameter	Rock	Meaning
R_{min}	0.2	Minimum particle size
R_{max}/R_{min}	1.5	Ratio of maximum particle size to minimum particle size
E_c (GPa)	1.8	Effective modulus of particles
K_n/K_s	1.5	Ratio of the contact stiffness between the normal direction and the tangential bond of particles
$\bar{E}$ (GPa)	2.4	Bond effective modulus
$\bar{K}_n/\bar{K}_s$	1.5	Ratio of normal to tangential bonding contact stiffness
σ_b (MPa)	16	Average and standard deviation of normal bond strength
c_b (MPa)	20	Mean and standard deviation of cohesive force
ϕ (°)	42	Bond internal friction angle
$\bar{\mu}$	0.5	Linear friction coefficient of particles

2.3. Numerical Models Roof Rock of Coal Seam with Different Defects Angle

Five rock models with varying angles of defects and one model without defects were built to explore the mechanical properties and the crack evolution law (Figure 3). The values in Table 1 were used to create these models as intact rocks first, and then defects were deleted before the simulation run. A certain thickness of grain element is deleted to simulate fracture. This is the most commonly used method at present and is mainly used to simulate discontinuous and unclosed fractures [18,21,30]. This is because after deleting the particles, the particles on both sides have been separated without any interaction. The loading rate needs to be low enough to guarantee the quasistatic loading condition. Hence, the loading rate in this study is set at 0.05 mm/s.

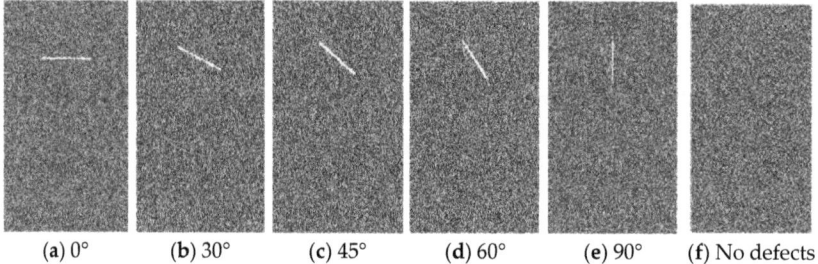

Figure 3. Numerical roof rock of coal seam specimens with different angle of defect and no defects.

3. Numerical Simulation and Results

3.1. Strength and Deformation Properties

Uniaxial compressive strength (UCS), elastic modulus (E), and stress-strain curves of rock samples with various defect angles are shown in Figure 4a,b. The defect angle has influence on the stress-strain curve, UCS and E of rock. Three stages, referred to as the elastic stage, plastic stage, and failure stage, may be distinguished between the stress-strain of rocks with various defect angles. Although the PFC rock model's particles are rigid and have no initial damage, the numerical rock model's stress-strain curve does not have a closed stage at the initial elastic stage similar to that of real rock. That is due to the fact that the PFC model is relatively homogeneous, which makes it beneficial to study the influence of defects from different angles on rock mechanical properties.

UCS increases with the increase of angle, but the increase is different with different angles. When the crack angle is below 45°, the UCS increases from 37.30 MPa at defect angle 0° to 37.91 MPa at defect angle 45° by 1.6%, then increases from 37.91 MPa at defect angle 30° to 78.1 MPa at defect angle 90° by 106%. Similarly, E has a similar trend, while its strain at maximal strength reduces initially, then increases as the angle increases.

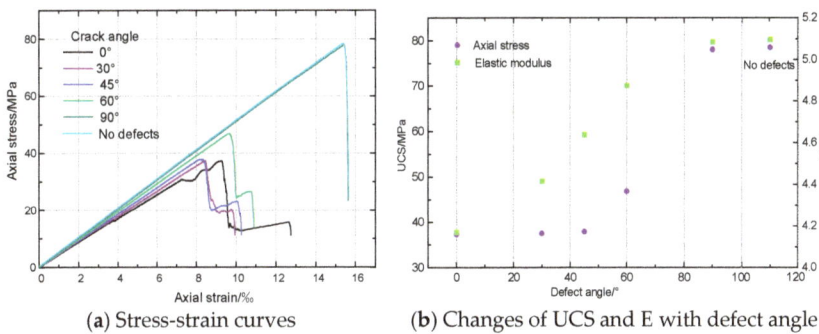

(a) Stress-strain curves   (b) Changes of UCS and E with defect angle

Figure 4. Stress-strain curves, UCS and E of rock with different defect angle.

Figure 5 shows the peak strain and failure forms of rock specimens with different defect angles. It shows that when the angle rises, the peak strain initially reduces and then increases. The maximum is reached when the angle is 90°, and the value is 15.43‰. This is 85.68% higher than the minimum value (8.31‰) when the angle is 45°, and the peak strain value of rock without defect is 13.45‰.

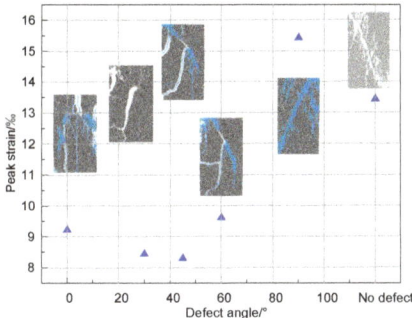

Figure 5. Peak strain of rock with different defect angle.

3.2. Laws of Crack Evolution

Figure 6 shows the relationship between the axial strain and the number of cracks in rock samples with various defect angles. Three stages can be distinguished in the fracture development under uniaxial compression. Taking the relationship between the rock crack number and strain at an angle of 90° as an example, with continuous loading, the number of cracks can be divided into three stages: I, II, and III; the zero crack stage, the crack slowly propagating stage, and the crack rapidly growing stage, respectively. In addition, the number of cracks in rock without defects can also be divided into three stages with loading, but it is completely different from the rock with cracks. However, the evolution of microcracks in rock with defects has a stage of basically unchanged microcracks after the rapid increase of microcracks in the third stage, and then continues to appear the stage of rapid increase of microcracks.

To better understand the impact of defect angles and the development of the micro crack of rock samples, we list the number of microcracks when rock failure occurs with varied defect angles in Figure 7. It shows that the number of microcracks during rock failure initially decreases and subsequently increases as the defect angle rises. The relationship between the crack numbers and defect angles can be described by a nonlinear relationship, where the crack number during rock failure decreases from 5571 in defect angle 0° to 3594 in defect angle 30° by 35.5%, then increases from 3594 in defect angle 30° to 9040 in defect angle 90° by 151.5%.

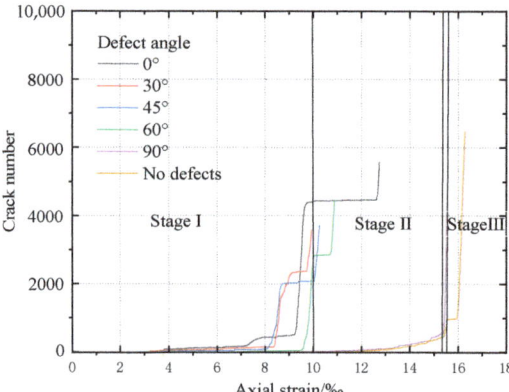

Figure 6. Number of cracks with axial strain of rock with different angle of defects.

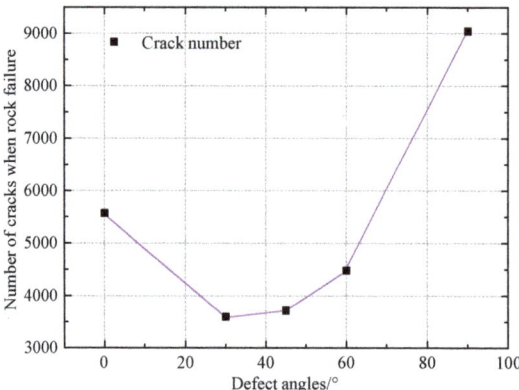

Figure 7. Number of cracks with different defect angles when rock failure.

3.3. Acoustic Emission Characteristics

As we all know, each crack in the PFC2D BPM creates an AE pulse. When the defect angle in rock samples is different, the AE characteristics under load are also different. And the AE event calculation of rock sample failure can be simulated by counting the number of cracks during uniaxial compression.

The stress-strain-AE event curves for various defect angles in rocks are shown in Figure 8. It can be seen that the AE event characteristics are strongly connected to the stress-strain relationship. The variation of AE events with strain can also be divided into three stages. In elastic stage, AE events rarely appear. The frequency of AE events rises when the stress-strain developed into plastic stage. When there is a failure, the AE events reach a peak and then rapidly decrease. It shows that the rock with different angles of cracking has been greatly damaged at this stage.

To sum up, the number of rock cracks significantly affects the number of AE events. The maximum AE events. The maximum AE events first decrease and then increase with the increase of the angle. The minimum value is 458 when the angle is 45°, and it is 2753 when the angle is 90°. In addition, when the angle is less than 90°, the AE events in the third stage are significantly different. When the angle is less than 90°, the third stage is relatively long, and there is a second peak of AE events. The AE events of rocks without defects are roughly similar to those of rocks with 90° defects, and there are basically no large AE events before the peak strength.

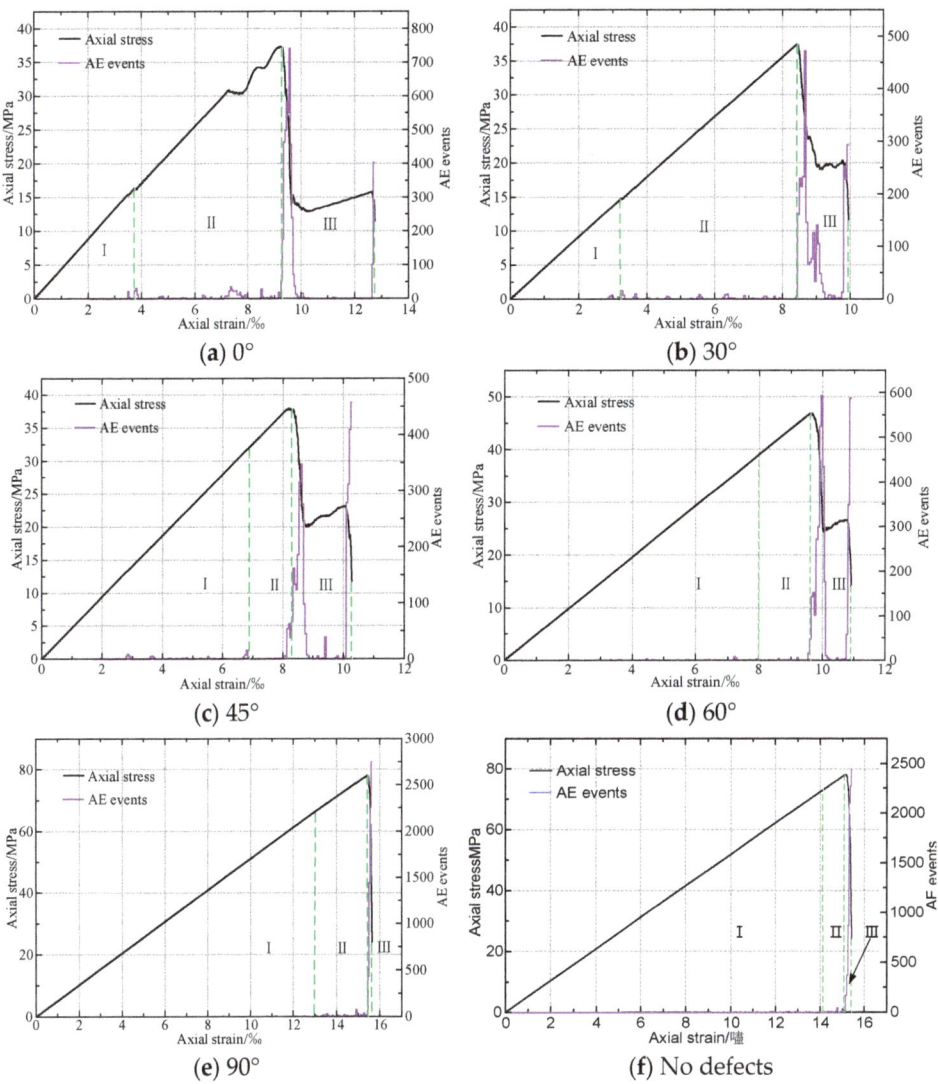

Figure 8. Stress-strain-AE events characteristics of rock with different angle of crack.

3.4. Crack Initiation and Distribution

In the process of simulation, the microcracks caused by intergranular bonding failure can be recognized as tensile microcracks and shear microcracks, which are represented by blue lines and green lines, respectively. The crack initiation of rocks with defects at different angles is shown in Figure 9. We can see that as the defect angles varied, the initial fracture position also varied. When the defect angle is 0°, the failure starts at the center, which is close to the lowest half of the defect. When the angle is 30, 45, and 60°, the fracture initiation occurs at the defect tip. The difference is that at 30° and 45°, the fracture starts at the lower part of the defect tip. When the angle is 60°, the fracture starts at the upper part of the defect tip. When the angle is 90°, the rock failure appears at the lower end of the whole rock, far away from the defect instead of appearing near the defect.

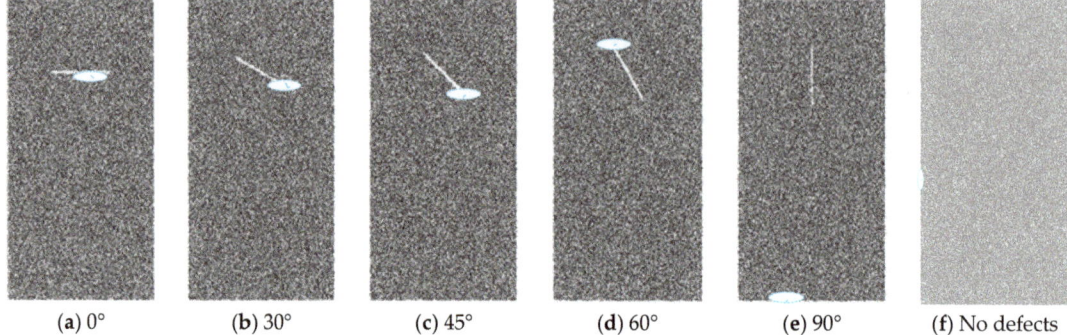

Figure 9. Crack initiation of rocks with defects at different angles.

Figure 10 shows the distribution of tension and shear cracks in samples with defects at various angles. It shows that at the same defect angle, the cracks of rock specimens are mainly tensile cracks, while shear cracks are relatively few. At different defect angles, tensile and shear crack counts initially drop and then subsequently rise. When the defect angle is 30°, the number of tensile cracks and reducing cracks is the least. We can also see that when the angle of the defect is less than 90°, the tensile crack and shear crack are mainly concentrated at the two tips of the defect and propagate along the loading direction. While the defect angle is 90°, the tensile and shear cracks are not concentrated at the defect tip. However, similar to the rock without defects, the microcracks are mainly concentrated on a fracture surface at about 45° in the horizontal direction. In Figure 4, the UCS and E of the 90° defect are much larger, which shows that they have little effect on the mechanical characteristics of rock. Accordingly, there are noticeably more tensile cracks than shear cracks when the defect angle is 90°.

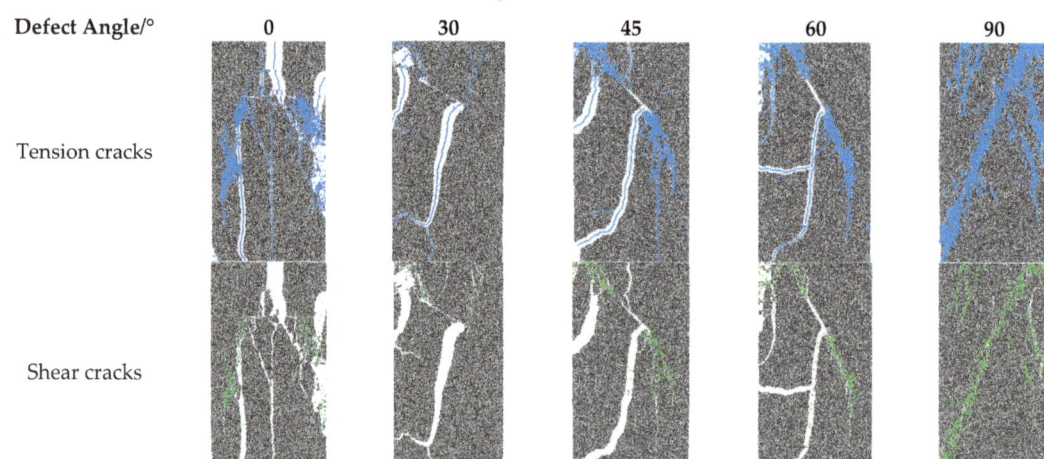

Figure 10. Distribution of tensile and shear cracks when rock failure.

3.5. Strain Energy Evolution Characteristics

According to previous studies, rock failure is driven by energy release. Hence, a more accurate reflection of the rock failure law can be achieved if we can thoroughly understand the energy transfer and transformation during the process of rock loading until failure [33,34]. The stress-strain curve of a rock mass element is given in Figure 11. The area U^d represents the energy consumed by the element when damage and plastic deformation occur. The releasable strain energy kept in the cell is represented by the darkened region U^e. $\bar{E}$ is the unloading elastic modulus [34].

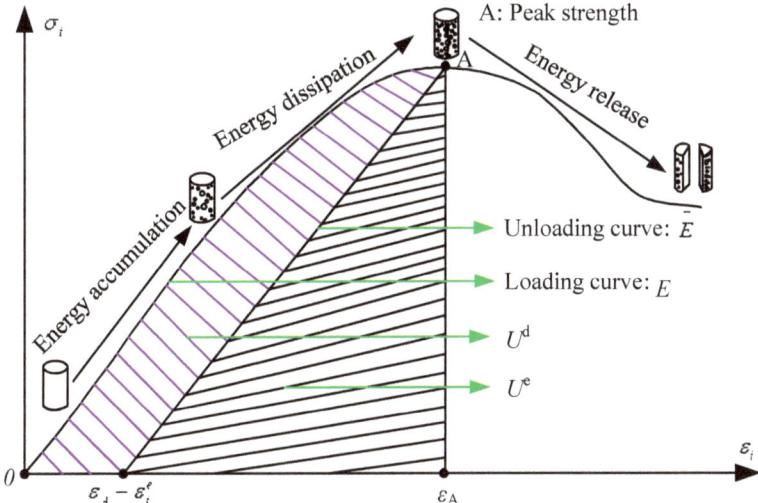

Figure 11. Relationship between dissipated energy and releasable strain energy in rock.

We assume that a unit volume rock mass element generates deformation when external force loading without experiencing any heat exchange with the environment. U is the total input energy that is produced by the external force work. According to the first law of thermodynamics [33,34]:

$$U = U^d + U^e \tag{1}$$

where, U^d is the dissipative energy, while U^e is the elastic strain energy.

$$U = \int \sigma_1 d\varepsilon_1 = \sum_{i=1}^{n} \frac{1}{2}(\sigma_{1i} + \sigma_{1i-1})(\varepsilon_{1i} + \varepsilon_{1i-1}) \tag{2}$$

$$U^e = \frac{1}{2\bar{E}}\left[\sigma_1^2 + \sigma_2^2 + \sigma_3^2 - 2\bar{\nu}(\sigma_1\sigma_2 + \sigma_2\sigma_3 + \sigma_1\sigma_3)\right] \tag{3}$$

where, $\bar{\nu}$ and $\bar{E}$ represent the mean values of the Poisson's ratio and the unloading elastic modulus respectively.

The mathematical method used to determine the elastic strain energy that generate form triaxial compression is Equation (3). When the rock is under uniaxial compression ($\sigma_2 = \sigma_3 = 0$), Equation (3) becomes:

$$U^e = \frac{\sigma_1^2}{2\bar{E}} \tag{4}$$

The change trend of the total input energy U, elastic strain energy U^e, and dissipative energy U^d of rocks with different defect angles is shown in Figure 12. Figure 12a shows that the elastic strain energy increases first and subsequently decreases with loading, reaching its maximum value when the rock is destroyed. With increasing loading, the total input energy tends to rise. At the end of the test, the total input energy can reach the maximum value. However, the dissipated energy does not change at first and then increases rapidly with the loading.

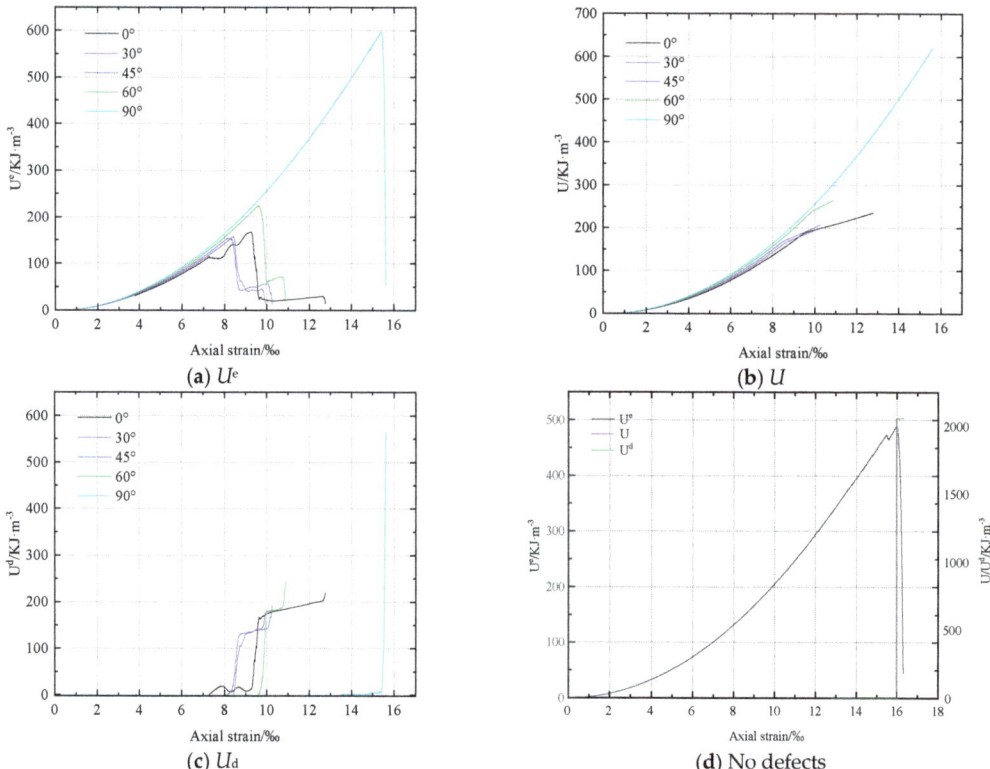

Figure 12. Variation diagram of total input energy U, elastic strain energy U^e and dissipative energy U^d of rock with strain at different defect angles and no defect.

At the same time, we also calculated the total input energy U_A, elastic strain energy U_A^e, dissipative energy U_A^d and the proportion of U_A^e and U_A^d to U_A at the rock's maximal strength with various defect angles is shown in Table 2 and Figure 13.

Table 2. U_A, U_A^e, U_A^d and the proportion at the peak strength of rock with different defect angles and no defects.

Defect Angles/°	U_A/KJ·m^{-3}	U_A^e/KJ·m^{-3}	U_A^d/KJ·m^{-3}	U_A^e/U_A/%	U_A^d/U_A/%
0	180.47	166.60	13.87	92.3	7.7
30	159.70	159.06	0.64	99.6	0.4
45	156.23	154.95	1.28	99.2	0.8
60	226.44	225.53	0.91	99.6	0.4
90	607.04	599.50	7.54	98.8	1.2
No defects	583.50	489.68	93.82	83.9	16.1

Table 2 and Figure 13 both show that when defect angles rise, U_A at the peak strength initially reduces and then subsequently increases. The minimum value is 156.23 KJ·m^{-3} at 45° and the maximum value is 607.04 KJ·m^{-3} at 90°. The difference between the minimum value and the maximum value is 74.3%. The value of U_A at peak strength also drops and subsequently increases with the increase in defect angles. The minimum value is 154.95 KJ·m^{-3} at 45°, and the maximum value is 599.50 KJ·m^{-3} at 90°. The difference between the minimum value and the maximum value is 74.2%. However, the dissipation energy U_A^d at the peak strength is unstable with the increase of the defect angles. The minimum value is 0.64 KJ·m^{-3} at 30° and the maximum value is 13.87 KJ·m^{-3} at 0°, which have a 99.5% difference. In addition, it can be seen from Table 2 that the elastic strain energy at the peak strength of rock without defects is relatively small, while the dissipated energy is relatively large.

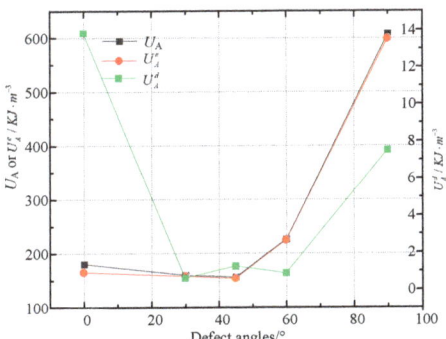

Figure 13. U_A, U_A^e, and U_A^d at the peak strength with different defect angles.

4. Conclusions

(1) The stress-strain curve, UCS, and E of rock are all significantly affected by the defect angle. The values of UCS become larger as the angle increases; however, the increase varies depending on the angle. E also presents such a trend. When the angle values increase, the strain values at peak strength drop first and then increase.

(2) The number of microcracks decreases first and then increases as the defect angle increases. The crack number when the roof rock of a coal seam failure decreases from 5571 at defect angle 0° to 3594 at defect angle 30° by 35.5%, then increases from 3594 at defect angle 30° to 9040 at defect angle 90° by 151.5%. With the increase of the defect angle, the maximum AE events first decrease and then increase. When the angle is less than 90°, the AE events in the third stage are significantly different. And when the angle is less than 90°, the third stage is relatively long, and there is a second peak of AE events.

(3) The starting position of rock failure is different at different angles of defects. When the defect angle is 0°, the failure starts in the middle of the defect, close to its bottom section. When the angle is 30, 45, or 60°, the fracture initiation occurs at the defect tip. When the defect is 90°, the rock failure occurs at the lower end of the whole rock far away from the defect, while not near the defect. Tensile fractures predominate at the same defect angle, but shear cracks are few.

(4) During loading, the elastic strain energy increases first and then decreases, reaching its peak value when the rock is broken. U_A^d does not change at first and then increases rapidly with the loading. At the peak strength stage, U_A initially decreases and then increases. U_A^e at the peak strength stage also decreases first and then increases. However, the dissipation energy at the peak strength is unstable with the increase of defect angles.

Author Contributions: Q.D.: methodology, formal analysis, writing—reviewing and editing, funding acquisition. J.L.: visualization, validation, data curation, validation. J.W.: funding acquisition, project administration, investigation, visualization. X.L.: writing—original draft preparation, conceptualization, software, resources. All authors have read and agreed to the published version of the manuscript.

Funding: This work is supported by the Natural Science Foundation of Shandong Province ZR2021QE187.

Institutional Review Board Statement: Not applicable.

Informed Consent Statement: Not applicable.

Data Availability Statement: The data used to support of this study are included within the article.

Conflicts of Interest: The authors declare no conflict of interest.

References

1. Yang, S.Q. Crack coalescence behavior of brittle sandstone samples containing two coplanar fissures in the process of deformation failure. *Eng. Fract. Mech.* **2011**, *78*, 3059–3081. [CrossRef]
2. Su, H.J.; Jing, H.W.; Zhao, H.H.; Wang, Y.C. Experimental study on the influence of longitudinal fissure on mechanics characteristic of sandstone. *J. Min. Saf. Eng.* **2014**, *31*, 644–649.
3. Cao, P.; Wang, H.; Jin, J.; Cao, R.; Fan, W. Experimental study of the cracks failure of sandstone containing hole and fissure under seepage water pressure. *J. China Univ. Min. Technol.* **2018**, *47*, 240–246.
4. Huang, Y.H.; Yang, S.Q. Particle flow simulation of macro- and meso-mechanical behavior of red sandstone containing two pre-existing non-coplanar fissures. *Chin. J. Rock Mech. Eng.* **2014**, *33*, 1644–1653.
5. Jiang, M.J.; Chen, H.; Zhang, N. Distinct element numerical analysis of crack evolution in rocks containing pre-existing double flaw. *Rock Soil Mech.* **2014**, *35*, 3259–3268.
6. Liu, H.W.; Yang, C. Micro-analysis of uniaxial compression of cracked rock containing open or closing fissure based on PFC. *Water Resour. Power* **2016**, *34*, 131–135.
7. Xu, Y.; Ren, F.; Ahmed, Z.; Wang, K.Y.; Wang, Z.H. Mechanical characteristics and damage evolution law of sandstone with prefabricated cracks under cyclic loading. *Arab. J. Sci. Eng.* **2021**, *46*, 10641–10653. [CrossRef]
8. Wan, W.; Liu, J.; Zhao, Y.; Fan, X. The effects of the rock bridge ligament angle and the confinement on crack coalescence in rock bridges: An experimental study and discrete element method. *Comptes Rendus Mec.* **2019**, *347*, 490–503. [CrossRef]
9. Xu, Y.; Ren, F.; Ahmed, Z.; Wang, K.Y. Mechanical and fatigue damage evolution properties of cracked sandstone under cyclic loading. *Proc. Pak. Acad. Sci.* **2020**, *57*, 59–72.
10. Zhou, X.P.; Bi, J.; Qian, Q.H. Numerical simulation of crack growth and coalescence in rock-like materials containing multiple pre-existing flaws. *Rock Mech. Rock Eng.* **2015**, *48*, 1097–1114. [CrossRef]
11. Aliabadian, Z.; Sharafisafa, M.; Tahmasebinia, F.; Shen, L.M. Experimental and numerical investigations on crack development in 3D printed rock-like specimens with pre-existing flaws. *Eng. Fract. Mech.* **2020**, *241*, 107396. [CrossRef]
12. Zhao, Z.H.; Liu, H.; Gao, X.J.; Feng, Y.H. Meso-macro damage deterioration of weakly cemented red sandstone under the coupling effect of high-humidity and uniaxial loading. *Eng. Fail. Anal.* **2023**, *143*, 106911. [CrossRef]
13. Jiang, M.J.; Zhang, N.; Shen, Z.F.; Chen, H. DEM analyses of crack propagation in flawed rock mass under uniaxial compression. *Rock Soil Mech.* **2015**, *36*, 3293–3300.
14. Luo, K.; Zhao, G.D.; Zeng, J.Z.; Zhang, X.X.; Pu, C.Z. Fracture experiments and numerical simulation of cracked body in rock-like materials affected by loading rate. *Chin. J. Rock Mech. Eng.* **2018**, *37*, 1833–1842.
15. Zhang, B.; Guo, S.; Yang, X.Y.; Li, Y.; Xu, X.J.; Yang, L. Hydraulic fracture propagation of rock-like material with X-type flaws. *J. China Coal Soc.* **2019**, *44*, 2066–2073.
16. Zhang, K.; Liu, X.H.; Li, K.; Wu, W.Y. Investigation on the correlation between mechanical characteristics and fracturing fractal dimension of rocks containing a hole and multi-flaws. *Chin. J. Rock Mech. Eng.* **2018**, *37*, 149–158. [CrossRef]
17. Li, A.; Ma, Q.; Lian, Y.; Ma, L.; Mu, Q.; Chen, J.B. Numerical simulation and experimental study on floor failure mechanism of typical working face in thick coal seam in Chenghe mining area of Weibei, China. *Environ. Earth Sci.* **2020**, *79*, 118. [CrossRef]
18. Yang, S.Q.; Huang, Y.H.; Liu, X.R. Analysis of tensile strength and crack propagation particle flow of discontinuous double fracture rock. *J. China Univ. Min. Technol.* **2014**, *43*, 220–226.
19. Yang, W.; Wei, B.; Xu, Y. Study on the crack growth law and mechanism of the rock mass with defect combination. *Geotech. Geol. Eng.* **2021**, *39*, 1319–1327. [CrossRef]
20. Cao, P.; Liu, T.; Pu, C.; Lin, H. Crack propagation and coalescence of brittle rock-like specimens with pre-existing cracks in compression. *Eng. Geol.* **2015**, *187*, 113–121. [CrossRef]
21. Cao, P.; Zhong, Y.F.; Li, Y.J.; Liu, J. Investigations on direct shear tests and pfc2D numerical simulations of rock-like materials with a hole and prefabricated cracks. *Key Eng. Mater.* **2014**, *627*, 477–480. [CrossRef]
22. Li, D.Y.; Cheng, T.J.; Zhou, T.; Li, X.B. Experimental study on dynamic mechanical failure characteristics of marble with holes under impact loading. *Chin. J. Rock Mech. Eng.* **2015**, *34*, 249–260.
23. Zhao, X.D.; Li, Y.H.; Yuan, R.P.; Yang, T.H.; Zhang, J.Y.; Liu, J.P. Study on crack dynamic propagation process of rock samples based on acoustic emission location. *Chin. J. Rock Mech. Eng.* **2007**, *26*, 944–950.
24. Luo, Y. Influence of water on mechanical behavior of surrounding rock in hard-rock tunnels: An experimental simulation. *Eng. Geol.* **2020**, *277*, 105816. [CrossRef]
25. Li, P.; Rao, Q.H.; Li, Z.; Jing, J. Thermal-hydro-mechanical coupling stress intensity factor of brittle rock. *Chin. J. Nonferrous Met.* **2014**, *24*, 499–508. [CrossRef]
26. Huang, X.; Shi, C.; Ruan, H.N.; Zhang, Y.P.; Zhao, W. Stable crack propagation model of rock based on crack strain. *Energies* **2022**, *15*, 1885. [CrossRef]
27. Sharafisafa, M.; Aliabadian, Z.; Tahmasebinia, F.; Shen, L.M. A comparative study on the crack development in rock-like specimens containing unfilled and filled flaws. *Eng. Fract. Mech.* **2021**, *241*, 107405. [CrossRef]

28. Ma, Z.; Cheng, S.; Gong, P.; Hu, J.; Chen, Y. Particle flow code simulation of the characteristics of crack evolution in rock-like materials with bent cracks. *Geofluids* **2021**, *2021*, 8889025. [CrossRef]
29. Ge, J.J.; Xu, Y.; Huang, W.; Wang, H.B.; Yang, R.Z.; Zhang, Z.Y. Experimental study on crack propagation of rock by blasting under bidirectional equal confining pressure load. *Sustainability* **2021**, *13*, 12093. [CrossRef]
30. Potyondy, D.O.; Cundall, P.A. A bonded-particle model for rock. *Int. J. Rock Mech. Min. Sci.* **2004**, *41*, 1329–1364. [CrossRef]
31. Itasca. *PFC (Particle Flow Code) Version 5.0*; Itasca Consulting Group Inc.: Minneapolis, MN, USA, 2014.
32. Castro-Filgueira, U.; Alejano, L.R.; Arzúa, J.; Ivars, D.M. Sensitivity analysis of the micro-parameters used in a PFC analysis towards the mechanical properties of rocks. *Procedia Eng.* **2017**, *191*, 488–495. [CrossRef]
33. Hou, Z.K.; Gutierrez, M.; Ma, S.Q.; Almrabat, A.; Yang, C.H. Mechanical behavior of shale at different strain rates. *Rock Mech. Rock Eng.* **2019**, *52*, 3531–3544. [CrossRef]
34. Xie, H.P.; Peng, R.D.; Ju, Y.; Zhou, H.W. On energy analysis of rock failure. *Chin. J. Rock Mech. Eng.* **2005**, *24*, 2603–2608.

Disclaimer/Publisher's Note: The statements, opinions and data contained in all publications are solely those of the individual author(s) and contributor(s) and not of MDPI and/or the editor(s). MDPI and/or the editor(s) disclaim responsibility for any injury to people or property resulting from any ideas, methods, instructions or products referred to in the content.

Article

A New Digital Analysis Technique for the Mechanical Aperture and Contact Area of Rock Fractures

Yong-Ki Lee [1,*,†], Chae-Soon Choi [1,†], Seungbeom Choi [2] and Kyung-Woo Park [1]

1. Disposal Performance Demonstration Research Division, Korea Atomic Energy Research Institute (KAERI), 111 Daedeok-daero, 989 Beon-gil, Yuseong-gu, Daejeon 34057, Republic of Korea
2. Disposal Safety Evaluation Research Division, Korea Atomic Energy Research Institute (KAERI), 111 Daedeok-daero, 989 Beon-gil, Yuseong-gu, Daejeon 34057, Republic of Korea
* Correspondence: yklee12@kaeri.re.kr
† These authors contributed equally to this work.

Abstract: In this study, a new digital technique for the analysis of the mechanical aperture and contact area of rock fractures under various normal stresses is proposed. The technique requires point cloud data of the upper and lower fracture surfaces, pressure film image data of the fracture, and normal deformation data of the fracture as input data. Three steps of algorithms were constructed using these input data: (1) a primary matching algorithm that considers the shape of the fracture surfaces; (2) a secondary matching algorithm that uses pressure film images; and (3) a translation algorithm that considers the normal deformation of a fracture. The applicability of the proposed technique was investigated using natural fracture specimens sampled at an underground research facility in Korea. In this process, the technique was validated through a comparison with the empirical equation suggested in a previous study. The proposed technique has the advantage of being able to analyze changes in the mechanical aperture and contact area under various normal stresses without multiple experiments. In addition, the change in the contact area on the fracture surface according to the normal stress can be analyzed in detail.

Keywords: rock fractures; mechanical aperture; contact area; digital analysis; pressure film

1. Introduction

A growing interest in engineering projects that utilize underground spaces, such as deep mining, enhanced geothermal energy systems, geological disposal of CO_2, and high-level radioactive waste disposal, has highlighted the importance of understanding the hydro-mechanical behavior of rock masses. This is because the groundwater and rock masses inevitably interact in underground spaces. As a result, the importance of rock fractures, which are mechanically weak planes and primary flow paths of fluids, has also emerged. Rock fractures comprise various parameters, including roughness, aperture, and contact area, which constitute the void geometry governing the hydro-mechanical behavior of the rock mass [1]. The mechanical aperture and contact area are the main parameters of the void geometry. The mechanical aperture is a parameter defined as the average height difference between the upper and lower fracture surfaces, and the contact area is a parameter that refers to the area where the upper and lower fracture surfaces come into contact with each other and transfer stress. The size of the fracture surfaces vary, and the contact area is thus generally defined as the ratio of the total surface area. Due to the importance of these two parameters for the hydro-mechanical behavior of rock masses, various studies have been conducted to measure them.

First, studies related to the measurement of the mechanical aperture are divided into surface topography, injection, and casting approaches. The surface topography approach was first applied by Gentier [2] to measure the mechanical aperture by digitizing the shape of the fracture with a profiler moving across the surface. Since then, precise surface measurements have been made possible thanks to the development of laser scanning devices, which remain the most commonly used method. However, a matching algorithm for the upper and lower fracture surfaces is required to derive the mechanical aperture, and the validity of this algorithm must be supported. The injection approach involves the application of epoxy resin (or metal) via injection onto the fracture surface, which is then hardened under normal stress. Next, the mechanical aperture is measured using this hardened resin. Since Gale [3] first applied this approach, it has been utilized in subsequent studies [4,5]. After cutting, the specimen containing the hardened resin as well as the thickness and cross section of the resin are analyzed using optical microscopy. As a result, this approach has the disadvantage of damaging the specimens [6]. In addition, the precision of the measurements is affected by the physical properties of the resin. Alternatively, the casting approach measures the mechanical aperture by creating transparent replicas of the fracture aperture space and then analyzing the images taken. This approach has been utilized in several studies [7–9]; however, it has disadvantages, such as damaging the specimen and a relatively low precision [6]. Approaches using X-ray CT have also been studied [10], which have the advantage of directly measuring the mechanical aperture of fractures. However, despite their considerable cost, these approaches have a lack of precision compared with the surface topography approach, which uses recently developed laser scanning devices.

Next, studies related to the contact area have mainly introduced methods of inserting a specific material into the fracture. Early studies measured the contact area by inserting a deformable plastic material between the fracture surfaces and applying a normal load [11]. Similarly, approaches have been introduced that insert a layer of low-viscosity casting epoxy [12] or a layer of white tempera with a black ink film [13]. However, these approaches have the disadvantage of overestimating the contact area because of the thickness of the insertion material. Recently, several studies have focused on measuring the contact area by performing compression tests, in which a pressure film is inserted instead of these materials. A pressure film has the advantage of enabling a detailed analysis as it can consider even the stress level acting on the contact area [14]. An approach that uses electrical current resistance has also been proposed [15], but this has a fatal disadvantage in that the measurement precision cannot be guaranteed, owing to the high current resistance of the rocks. In addition, techniques for analyzing the contact area of fractures using point cloud data have been proposed with the recent development of laser scanning devices [16,17]. These techniques are characterized by the fact that the matching algorithm of the upper and lower fracture surfaces affects the measurement result of the contact area, as in the case of the mechanical aperture.

In summary, using point cloud data obtained with a laser scanning device is the preferred approach for measuring the mechanical aperture and contact area. The use of a pressure film is preferred to measure the contact area. However, as the approaches in previous studies have been limited to measuring the mechanical aperture and contact area under a specific normal stress, only results under the normal stress applied in the measurement process can be analyzed. In other words, multiple measurements are required to analyze the changes in the mechanical aperture and contact area under various normal stresses.

In this context, this study proposes a new digital technique that can derive the mechanical aperture and contact area of fractures under various normal stresses by using point cloud data with both the pressure film data and normal deformation data of the fracture. The proposed technique consists of three steps: (i) a primary matching algorithm based on the shape of the fracture surfaces; (ii) a secondary matching algorithm using pressure film images; and (iii) a translation algorithm that considers the normal deformation of fractures. The applicability of the proposed technique was investigated using natural fracture specimens sampled at an underground research facility in Korea. In this process, the technique was validated through a comparison with the empirical equation suggested in a previous study.

2. Methodology of the Digital Analysis Technique

To construct a digital technique with which to analyze the mechanical aperture and the contact area of the rock fractures under various normal stresses, we attempted to utilize the advantages of the surface topography approach using a laser scanning device and the method using a pressure film. In addition, an algorithm using the normal deformation of rock fractures was applied to analyze the mechanical aperture and contact area according to the change in normal stress without additional experiments. The proposed technique derives the results in three steps. This chapter presents the method of preparing the input data for applying the technique and the algorithms used for each step.

2.1. Preparation of Input Data

The digital analysis technique requires three types of input data: (i) point cloud data for fracture surfaces; (ii) image data from a pressure film compression test; and (iii) normal deformation data from compression tests on intact and fractured rock specimens. The input data can be prepared using laser scanning and compression tests.

First, point cloud data for fracture surfaces can be obtained using a commonly used laser scanning device. The higher the precision of the laser scanning device, the more advantageous it is. In particular, higher precision is required for the z-axis, which is the elevation direction. For the x- and y-axes, it is recommended to ensure measurement intervals of less than 0.5 mm. Three-dimensional point cloud data can be obtained by measuring the upper and lower fracture surfaces using a laser scanning device. Figure 1 shows an example of the point cloud data measured on the upper and lower surfaces. Outliers were observed at the fracture boundaries owing to the characteristics of laser scanning; therefore, a removal process was required during analysis.

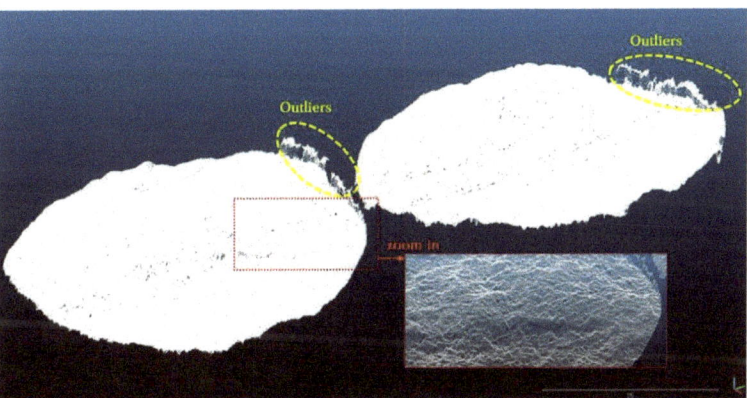

Figure 1. Example of point cloud data of fracture surfaces acquired using a laser scanning device.

Second, the pressure film image data can be obtained by performing a compression test on a rock fracture in which a pressure film is inserted. The pressure film loaded to a specific normal stress is photographed using a digital camera to obtain the image data. After the pressure film is inserted between the fractures, care must be taken to ensure that the upper and lower surfaces are joined in conformance. In general, there is no problem unless the surface is significantly distorted because the fracture is well joined as the normal load begins to be applied. The applied normal stress should not be too high so that sufficient normal deformation of the fracture occurs. Considering the properties of general rock fractures, normal stress in the range of 3–5 MPa is appropriate. In addition, the pressure film should be capable of covering the low normal stress range. Figure 2 shows an example of a compression test using a pressure film and the image data obtained from it.

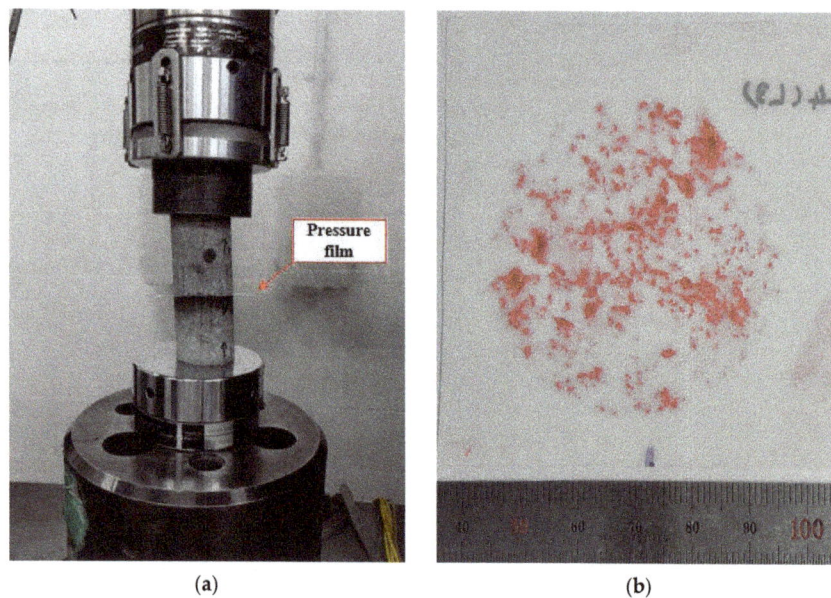

Figure 2. Example of pressure film image data acquisition: (**a**) pressure film compression test; (**b**) pressure film image data.

Third, normal deformation data for rock fractures can be obtained from compression tests on intact and fractured rock specimens [18]. After measuring the normal displacement of each specimen under various normal stress conditions, the normal deformation data of the fracture is derived by calculating the difference between the displacements of the fractured rock and intact rock under the same normal stress. Figure 3 shows a schematic representation of the data analysis to derive the normal deformation data for the rock fracture.

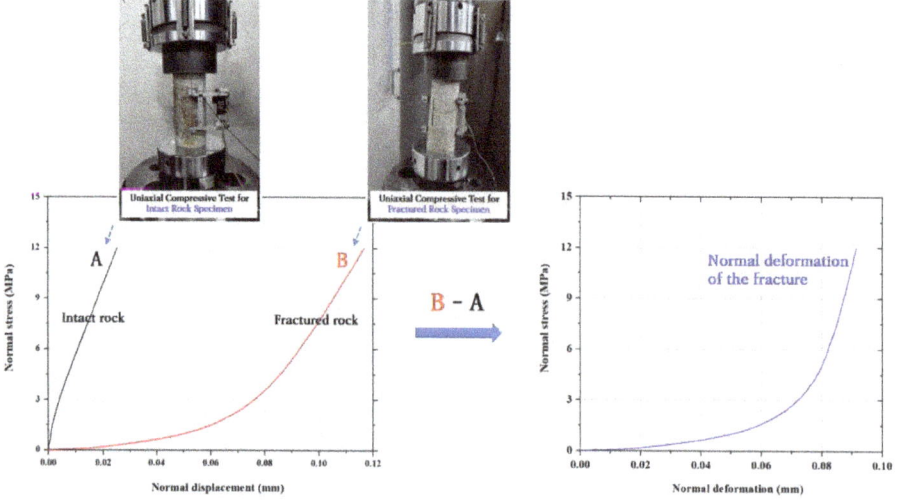

Figure 3. Schematic of data analysis to derive the normal deformation data of the rock fracture.

2.2. First Step: Primary Matching According to the Shape of Fracture Surfaces

Because the point cloud data is obtained by measuring the separated upper and lower surfaces of the rock fracture, it is necessary to match them to one fracture. An algorithm that matches the upper and lower surfaces considering the surface shape was applied as the first step. In this step, the iterative closest point (ICP) algorithm [19], widely used for matching different point cloud data, was mainly used. The ICP algorithm estimates the correspondences between two point cloud data and then aligns them until the distance error is below the threshold. In estimating correspondences, pairs of points with the closest distance from each point cloud data are used. Assuming that there are different point cloud data P_1 and P_2, the flow of the ICP algorithm is briefly summarized as follows:

- The correspondences of P_1 to P_2 are constructed by searching for the points of P_2 closest to each point of P_1.
- A rotation matrix (R) and translation vector (t) are derived based on the correspondences.
- P_1 is updated to $\overline{P_1}(= R \times P_1 + t)$, then aligned with P_2.
- The above process is repeated until the point-to-point distance error between the $\overline{P_1}$ and P_2 is smaller than the threshold.

The first step starts with clustering the measured point cloud data based on distance and dividing it into two sets of point cloud data composed of points corresponding to the upper and lower fracture surfaces. Then, the points corresponding to the upper surface are rotated to match the approximate position, and the ICP algorithm is applied. Next, the outliers are removed by setting a region of interest so that outliers are not included. The ICP algorithm is reapplied to the two point cloud data, from which outliers have been removed. This is because, in the first matching using the ICP algorithm, the matching accuracy is lowered owing to the influence of outliers. A schematic of the primary matching algorithm is shown in Figure 4.

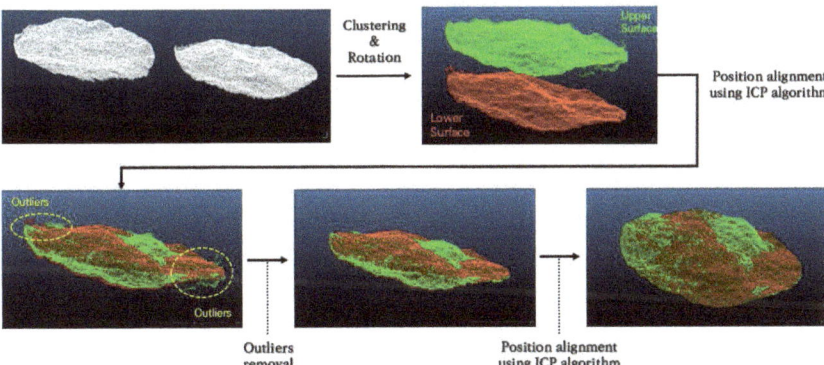

Figure 4. Schematic of an algorithm for primary matching according to the shape of the fracture surfaces.

2.3. Second Step: Secondary Matching Using Pressure Film Image

The matched point cloud data in the first step is not under normal stress conditions because only the shape of the fracture surfaces is considered. This study used pressure film images to match the upper and lower fracture surfaces under specific normal stress conditions. As described in Section 2.1, because the pressure film image data is obtained by applying a load up to a specific normal stress, the contact area at that normal stress appears in the image. Therefore, if the point cloud data are adjusted such that the contact area coincides with the pressure film image data, the point cloud data can represent the state of fracture under the normal stress condition applied to the pressure film. In this study, a secondary matching algorithm was constructed to adjust the point cloud data to the pressure film image. Figure 5 shows a schematic of the secondary matching algorithm, which is summarized as follows:

- The pressure film image obtained by applying specific normal stress is converted into numerical data via image processing. Because only the identification of the contact area is of interest, the presence or absence of contact is determined without considering the intensity of the image.
- The contact area of the point cloud data is converted into numerical data. If the coordinates of the upper surface are located lower than those of the lower surface, it is determined as a contact state, and the coordinates of the upper surface are changed to be identical to those of the lower surface.
- Interpolation is performed such that the numerical data of the pressure film and point cloud data are located on the exact grid coordinates.
- The matching ratio is calculated by comparing the two numerical datasets. The matching ratio refers to the percentage of grids that match contact/non-contact to the total number of grids.
- The above process is repeated while the point cloud data of the upper surface is rotated and translated about the x-, y-, and z-axes, and the matching ratio is calculated under all conditions.
- Subsequently, the condition where the matching ratio is minimized is assigned as the state under the normal stress applied in the pressure film compression test.

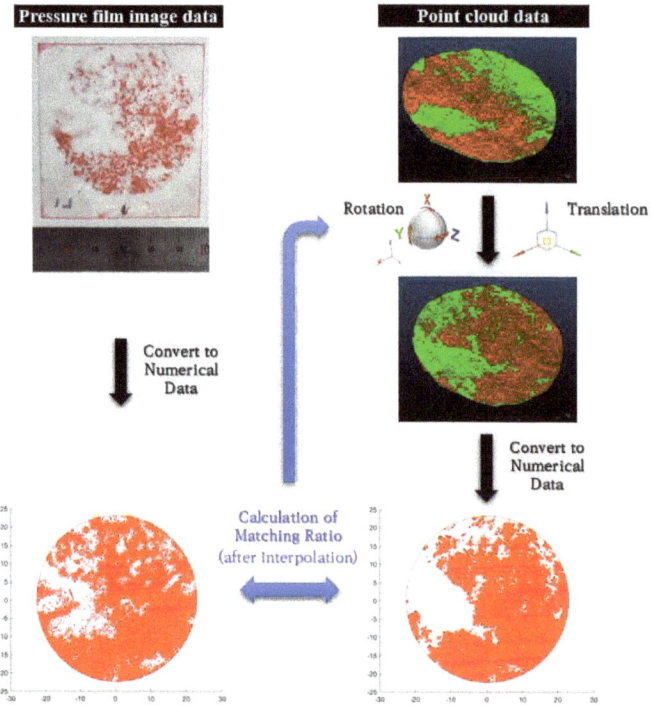

Figure 5. Schematic of an algorithm for secondary matching using the pressure film image.

Because the point cloud data is already primarily matched, it is not necessary to include a wide range of rotations and translations in the secondary matching algorithm. For each axis, it is sufficient to apply ±3° rotation and ±3 mm translation. As the z-axis, which is the elevation direction, is affected by the normal deformation of the fracture, it is recommended to apply a range of about ±5 mm along the z-axis. The smaller the rotation and translation interval within the range, the higher the reliability of the result. Therefore, considering the computational efficiency, it is better to set the interval as small as possible. In general, reasonable results were obtained at a rotation interval of about 0.1° and a translation interval of about 0.01 mm.

2.4. Third Step: Translation Considering the Normal Deformation of Fracture

The point cloud data matched to the pressure film image shows the results under the normal stress condition applied in the pressure film compression test. The normal deformation of the fracture caused by normal stress should be considered to analyze the change in the mechanical aperture and contact area under various normal stress conditions. In this study, a translation algorithm using the normal deformation data of rock fractures was applied to calculate the mechanical aperture and contact area under various normal stress conditions. The algorithm is as follows:

- The normal stress applied in the pressure film compression test is assigned as the reference normal stress. Then, the difference in the normal displacement of the fracture caused by the change from the reference normal stress to the target normal stress is calculated using the normal deformation data.
- The point cloud data on the upper surface of the fracture are translated in the z-axis direction using the difference in the normal displacement. If the coordinates of the upper surface are located below the coordinates of the lower surface during translation (the elevation difference is negative), it is determined to be a contact, and the coordinates of the upper surface are set to the same value as the coordinates of the lower surface. When the elevation difference changes from negative to positive due to translation, the upper surface coordinates are restored by considering the original shape.
- After the translation, the mechanical aperture and contact area of the fracture are calculated. The average value of the elevation differences at each grid point is derived as the mechanical aperture, and the ratio of the number of contacted grids to the total number of grids is derived as the contact area ratio.

Figure 6 shows an example of the application of the translation algorithm. The left side of the figure shows a schematic of the algorithm, and the right side shows the results derived from this. The example results in the figure correspond to a normal stress of 0 MPa as the initial condition, 4 MPa as the reference condition, and 12 MPa as the condition of approximately 500 m depth. The blue and green points show the changes in the contact area according to each condition.

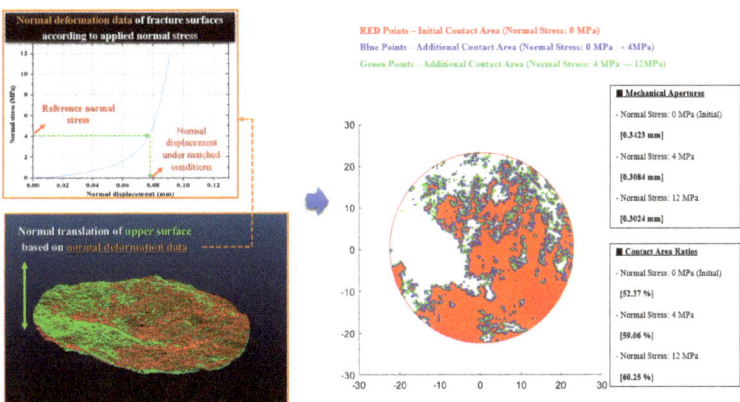

Figure 6. Example of the application result of the translation algorithm considering a normal deformation of fracture.

3. Applicability of the Proposed Technique

The applicability of the proposed technique was examined using natural fracture specimens sampled from the KAERI underground research tunnel (KURT), an underground research facility in Korea. KURT was built to develop disposal technologies for high-level radioactive waste and to verify the performance of the disposal system. Natural fracture specimens for applicability examination were sampled from a deep borehole with various rock types. In this section, the results of the applicability examination with the experimental procedure are presented, including information on the sampled specimens. In addition, the analysis of the change in the mechanical aperture and contact area according to the normal stress of the sampled specimens using the proposed technique is also presented.

3.1. Experimental Procedure

A total of 12 natural fracture specimens were sampled from deep boreholes (DB-2) in KURT to examine the applicability of the proposed technique. The DB-2 borehole is located in Daejeon, Korea, as shown in Figure 7.

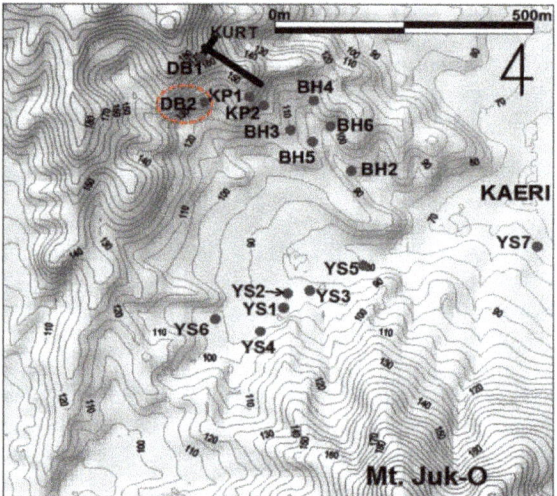

Figure 7. Location map for the DB-2 borehole.

To analyze the applicability at different rock types and depths, granite, altered rock, and fine-grained andesite dyke specimens were sampled at low and high depths. Because KURT is related to the disposal of high-level radioactive waste, it was divided into low and high depths based on a disposal depth of 500 m. The types of sampled specimens and their notations used in this study were as follows:

- Granite at a depth of 250 m: GL.
- Granite at a depth of 840 m: GH.
- Altered rock at a depth of 320 m: AR.
- Fine-grained andesite dyke at a depth of 750 m: FAD.

As shown in Figure 8, three specimens were sampled for each type, and intact rock specimens were prepared to obtain normal deformation data of the rock fracture. In addition, the physical and mechanical properties were analyzed using additional intact rock specimens, and the results are listed in Table 1.

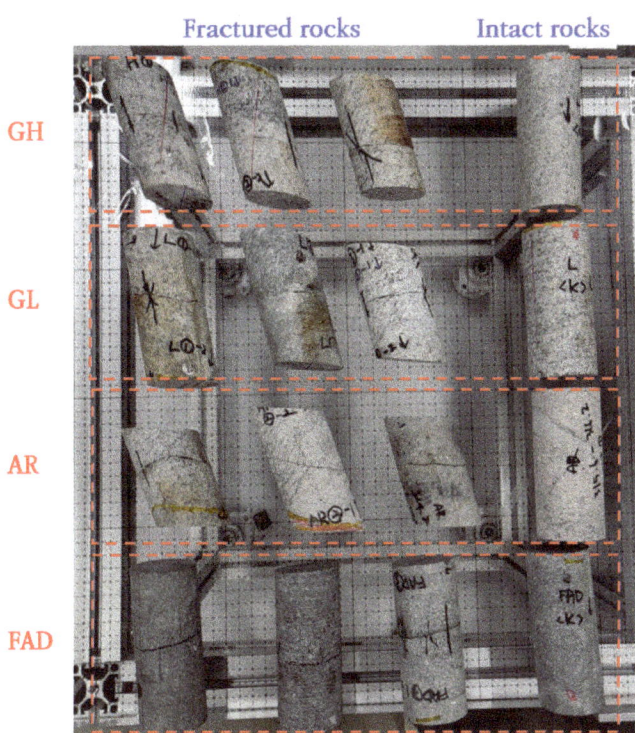

Figure 8. Intact and fractured rock specimens sampled to investigate the applicability of the proposed technique.

Table 1. Physical and mechanical properties of the specimens used in this study.

	GL	GH	AR	FAD
Dry unit weight (kg/m^3)	2637.68	2604.12	2602.46	2743.17
Porosity (%)	0.91	0.51	0.70	0.51
P-wave velocity (m/s)	3276.19	3783.51	4029.59	3593.14
S-wave velocity (m/s)	2484.88	2566.43	2723.80	2783.54
Uniaxial compressive strength (MPa)	184.73	161.46	129.42	135.61
Young's modulus (GPa)	57.30	71.40	47.00	50.40
Poisson's ratio	0.21	0.31	0.05	0.36

The laser scanning device for acquiring point cloud data of the fracture surfaces used the equipment shown in Figure 9a, manufactured by SPOS Corporation. This equipment includes the LMI technologies' Gocator 2530 sensor, enabling elevation measurement with a resolution of 0.001 mm at intervals of 0.03–0.05 mm. The MTS 816 system (as shown in Figure 9b), which is widely used for rock testing, was used as the compression test equipment to acquire the normal deformation data of the fracture as well as the physical and mechanical properties of the specimens.

The pressure film used for acquiring the pressure film image data was Fujifilm Corporation's Prescale super low. This film is a two-sheet type, and a detailed classification is possible up to normal stress of 2.5 MPa. This type of film showed normal stresses exceeding 2.5 MPa with the same intensity. However, because the proposed technique does not consider intensity, this film that covered low normal stresses in detail was suitable. The pressure film compression test applied a load up to a normal stress of 4 MPa using this film.

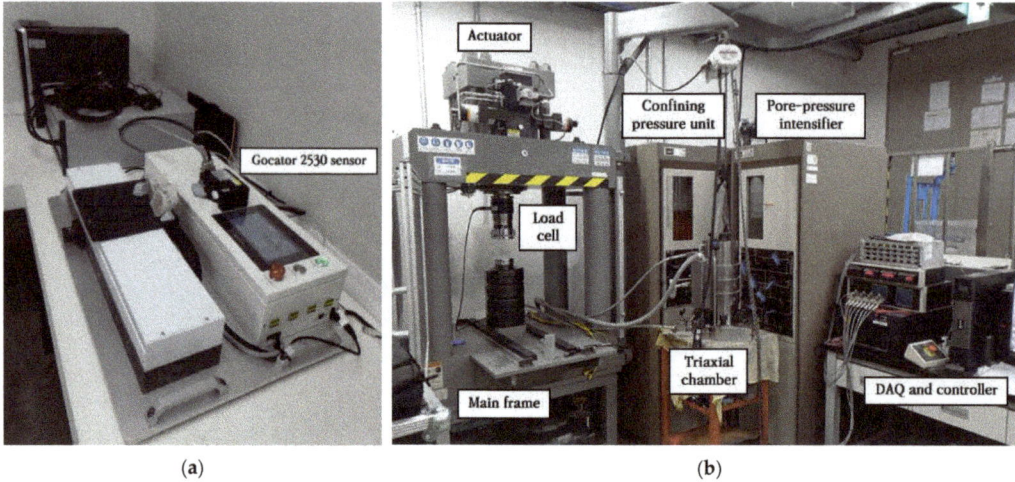

Figure 9. Test equipment used to investigate the applicability of the proposed technique: (**a**) 3-D laser scanning device; (**b**) MTS 816 system.

3.2. Validation of the Proposed Technique

The applicability of the proposed technique was validated using the prepared specimens. The matching algorithm (first and second steps) was examined using the matching ratio with the pressure film image data, which were related to the validation of the contact area analysis. The result of the mechanical aperture derived by applying the entire algorithm was compared with the empirical equation of a previous study for validation.

Figure 10 shows the results obtained by applying the matching algorithm to the 12 natural fracture specimens. The numerical data of the matching result (point cloud matching data) and numerical data of the pressure film image (pressure film data) are shown together in the figure.

The analysis results for the matching ratio and the contact area are listed in Table 2. The contact area was analyzed using the contact area ratio, which is the ratio of the contact area to the area of the entire fracture surface. This result corresponded to the reference normal stress condition of 4 MPa applied in the pressure film compression test.

Table 2. Analysis result of matching ratio and contact area for natural fracture specimens.

Sample Name	Contact Area Ratio in Point Cloud Matching Data (%)	Contact Area Ratio in Pressure Film Data (%)	Difference in Contact Area Ratio (%)	Matching Ratio (%)
GL-1	59.06	48.08	10.98	66.16
GL-2	25.19	29.58	4.39	64.66
GL-3	24.63	22.75	1.88	66.44
GH-1	45.55	38.83	6.72	62.23
GH-2	28.51	32.25	3.74	61.74
GH-3	22.13	26.54	4.41	63.38
AR-1	17.03	16.33	0.70	77.29
AR-2	14.10	14.84	0.74	81.45
AR-3	12.51	13.30	0.79	78.14
FAD-1	26.23	26.35	0.12	63.62
FAD-2	35.74	24.06	11.68	70.75
FAD-3	20.07	19.50	0.57	75.94
Average			3.89%	69.32%

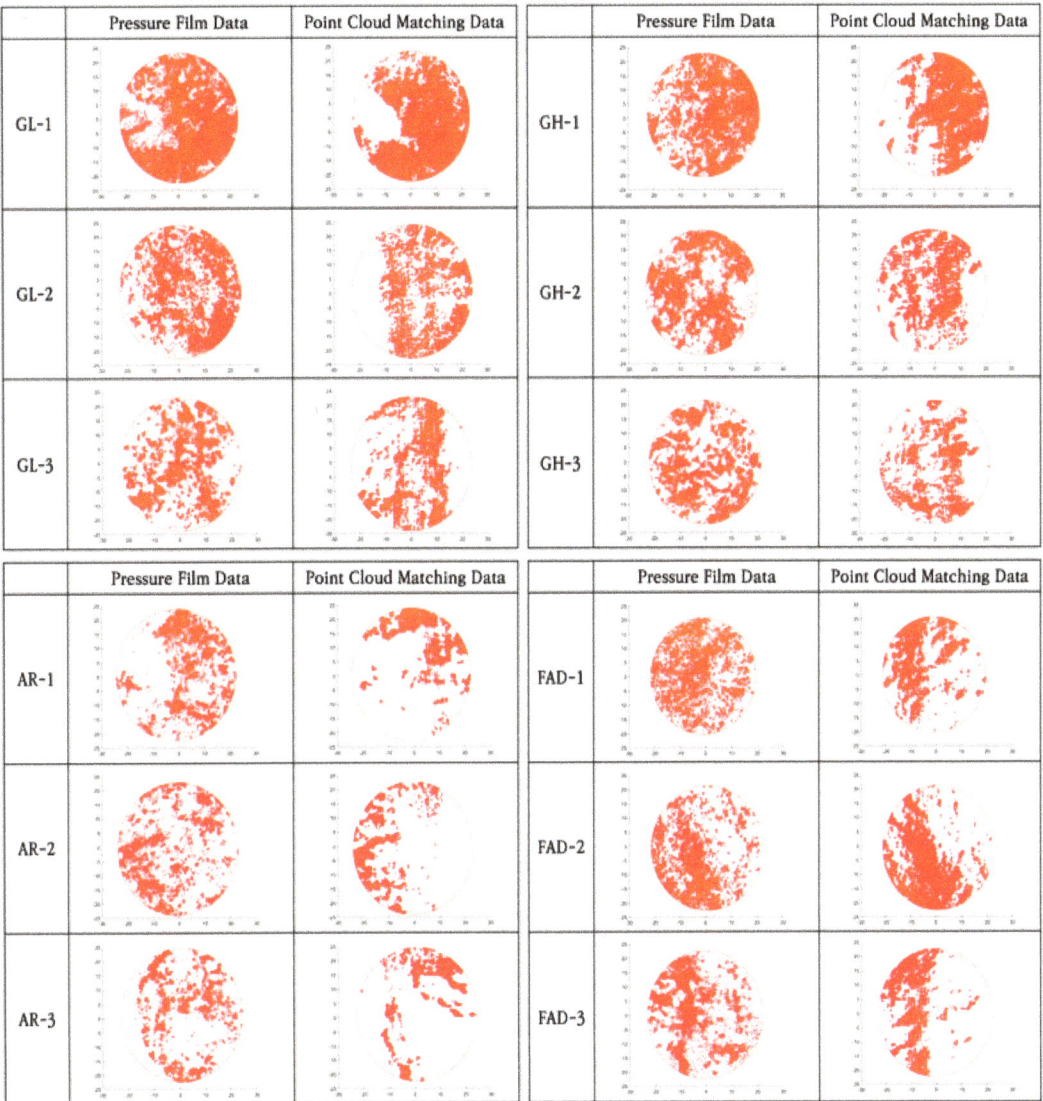

Figure 10. Comparison result of numerical data through the application of the matching algorithm.

The average matching ratio was approximately 70%, and the lowest value was approximately 62%. However, the distribution of the contact area was similar, as shown in Figure 10, and the contact area ratio exhibited an average difference of 3.89%, as shown in Table 2. These results indicate that there were no significant problems with the application of the matching algorithm.

The mechanical aperture derived using the proposed technique was compared with an empirical equation from a previous study. Most studies have aimed to measure the mechanical aperture under a specific normal stress; therefore, they have mainly dealt with the initial mechanical aperture (e_0) under zero normal stress. Therefore, the empirical formula for the initial mechanical aperture proposed by Bandis et al. [18] was used for comparative analysis. This empirical equation is defined in Equation (1), and has been widely used because it is composed of general fracture parameters:

$$e_0 = \frac{JRC}{5}\left(0.2\frac{\sigma_c}{JCS} - 0.1\right) \quad (1)$$

where JRC, σ_c, and JCS are the joint roughness coefficient, uniaxial compressive strength, and joint wall compressive strength, respectively.

JRC was calculated through its relationship with Z_2 (the root mean square of the first derivative of the profile), as in Equation (2) suggested by Tse and Cruden [20]. Profiles were extracted at intervals of 0.1 mm, and the average Z_2 value was used. For the calculation interval of Z_2, 0.5 mm belonging to the appropriate interval range suggested in the previous study [21] was applied. Table 3 lists the calculation results of the JRC of the upper and lower surfaces of the specimens, and the average value was used as the JRC of the specimen.

$$JRC = 32.2 + 32.47 \log(Z_2) \quad (2)$$

where

$$Z_2 = \sqrt{\frac{1}{L}\int_0^L \left(\frac{dy}{dx}\right)^2 dx} = \sqrt{\frac{1}{L}\sum_{i=1}^{N-1}\frac{(y_{i+1}-y_i)^2}{x_{i+1}-x_i}} \quad (3)$$

Table 3. Joint roughness coefficient values of natural fracture specimens.

Sample Name	Upper Surface	Lower Surface	Average
GL-1	16.29	17.57	16.93
GL-2	13.47	14.16	13.82
GL-3	11.48	11.88	11.68
GH-1	10.30	10.28	10.29
GH-2	15.21	15.73	15.47
GH-3	11.30	9.33	10.32
AR-1	8.82	9.04	8.93
AR-2	10.13	9.54	9.84
AR-3	10.28	10.40	10.34
FAD-1	15.19	13.59	14.39
FAD-2	15.21	16.70	15.96
FAD-3	11.73	10.94	11.34

The JCS of unweathered rocks is known to be equal to the uniaxial compressive strength [22], and so σ_c/JCS was substituted with one for the GL, GH, and FAD specimens of fresh surfaces. The AR specimens are altered rocks with surface weathering. Therefore, JCS was measured using a Schmidt hammer, and, as a result, a value of 1.54 was substituted into σ_c/JCS.

The initial mechanical apertures derived by applying the empirical equation and digital analysis technique proposed in this study are presented in Table 4. In the digital analysis technique, the results under a normal stress condition of 0 MPa were derived by considering the normal deformation from the reference normal stress condition of 4 MPa. The results show an average difference of 0.0319 mm and an error rate of 10.68%.

Figure 11 compares the two results and shows that the digital analysis technique can reasonably derive the initial mechanical aperture. However, the digital analysis method showed a slight underestimation of the initial mechanical aperture compared with the empirical equation. This is because the normal deformation of the fracture was slightly affected by the insertion of the pressure film. Better results could be obtained if the effect of the pressure film on the normal deformation of the fracture was considered.

Table 4. Initial mechanical apertures derived using the empirical equation and digital analysis technique.

Sample Name	Derived Initial Mechanical Aperture (mm)		Difference in Initial Mechanical Aperture (mm)	Error Rate Compared to Empirical Equation (%)
	Empirical Equation	Digital Analysis Technique		
GL-1	0.3386	0.3423	0.0037	1.09
GL-2	0.2763	0.2481	0.0282	10.21
GL-3	0.2336	0.2331	0.0005	0.21
GH-1	0.2058	0.1607	0.0451	21.91
GH-2	0.3094	0.3024	0.0070	2.26
GH-3	0.2063	0.1579	0.0484	23.46
AR-1	0.3709	0.4089	0.0380	10.24
AR-2	0.4085	0.3492	0.0593	14.52
AR-3	0.4295	0.3682	0.0613	14.27
FAD-1	0.2878	0.3019	0.0141	4.90
FAD-2	0.3191	0.2501	0.0690	21.62
FAD-3	0.2267	0.2189	0.0078	3.44
Average			0.0319 mm	10.68%

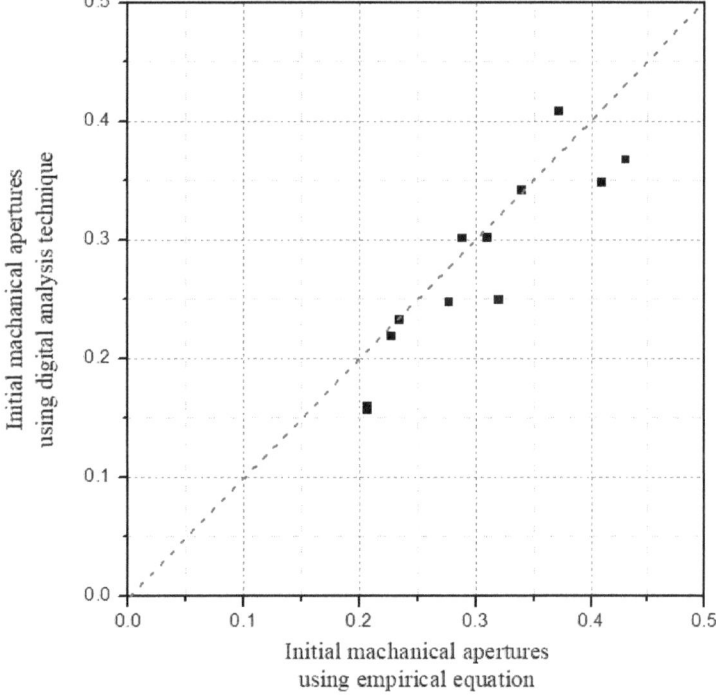

Figure 11. Result of the comparison between initial mechanical apertures derived using the empirical equation and digital analysis technique.

3.3. Application to Natural Fractures at the KURT Site

The proposed technique was applied to the prepared specimens to examine the applicability in the analysis of the mechanical aperture and contact area according to the normal stress variation. In general rock fractures, the change in the mechanical aperture and contact area becomes small because the normal deformation rapidly decreases above specific normal stress. This tendency has also been observed in previous studies [23,24]. The natural fracture specimens at the KURT site sampled in this study showed a rapid decrease in normal deformation above 4 MPa.

Figure 12 shows the results of analyzing the change in the mechanical aperture and contact area under various normal stresses for the GL1 specimen, which showed similar results to other specimens. Table 5 summarizes the mechanical aperture and contact area ratio when the normal stresses of 0 MPa, 4 Mpa, and 12 Mpa were applied. The mechanical aperture and contact area ratio of all the specimens did not show significant differences when 4 Mpa and 12 Mpa were applied. This result indicates that the normal deformation change was not substantial when normal stress conditions of over 4 Mpa were applied.

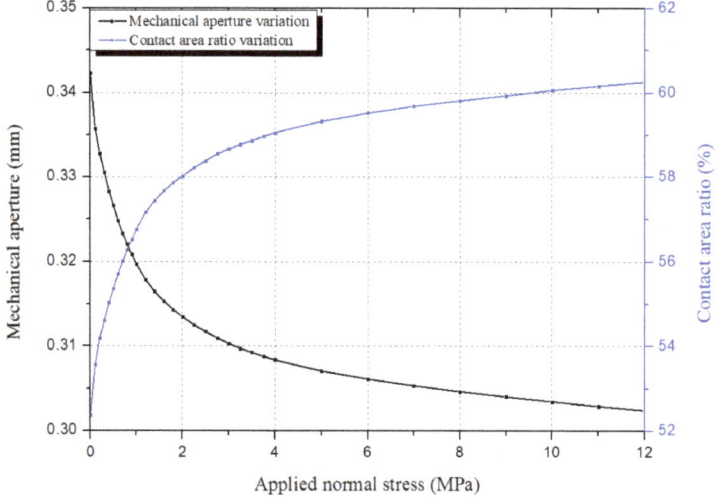

Figure 12. Variation of the mechanical aperture and contact area ratio under various normal stress conditions for the GL1 specimen.

Table 5. Mechanical apertures and contact area ratios under various normal stresses for natural fracture specimens.

Sample Name	Mechanical Aperture (mm)			Contact Area Ratio (%)		
	$\sigma_n =$ 0 MPa	$\sigma_n =$ 4 MPa	$\sigma_n =$ 12 MPa	$\sigma_n =$ 0 MPa	$\sigma_n =$ 4 MPa	$\sigma_n =$ 12 MPa
GL1	0.3423	0.3084	0.3024	52.37	59.06	60.25
GL2	0.2481	0.228	0.2209	21.29	25.19	26.69
GL3	0.2331	0.1835	0.1655	17.15	24.63	27.98
GH1	0.1607	0.145	0.1392	41.8	45.55	46.97
GH2	0.3024	0.2291	0.2251	17.39	28.51	29.28
GH3	0.1579	0.1491	0.1425	20.02	22.13	23.85
AR1	0.4089	0.3858	0.3815	15.04	17.03	17.43
AR2	0.3492	0.32	0.315	11.18	14.1	14.68
AR3	0.3682	0.3357	0.334	9.95	12.51	12.68
FAD1	0.3019	0.2511	0.2392	20.23	26.23	27.88
FAD2	0.2501	0.1984	0.1894	28.67	35.74	37.11
FAD3	0.2189	0.1992	0.1929	17.59	20.07	20.89

Figure 13 shows the change in the contact area under various normal stresses in more detail. This figure shows the changes in the contact area as the normal stress condition increases to 0 MPa, 4 MPa, and 12 MPa (red, blue, and green points, respectively). As the normal stress increases, the area near the initial contact area gradually changed to the contact state.

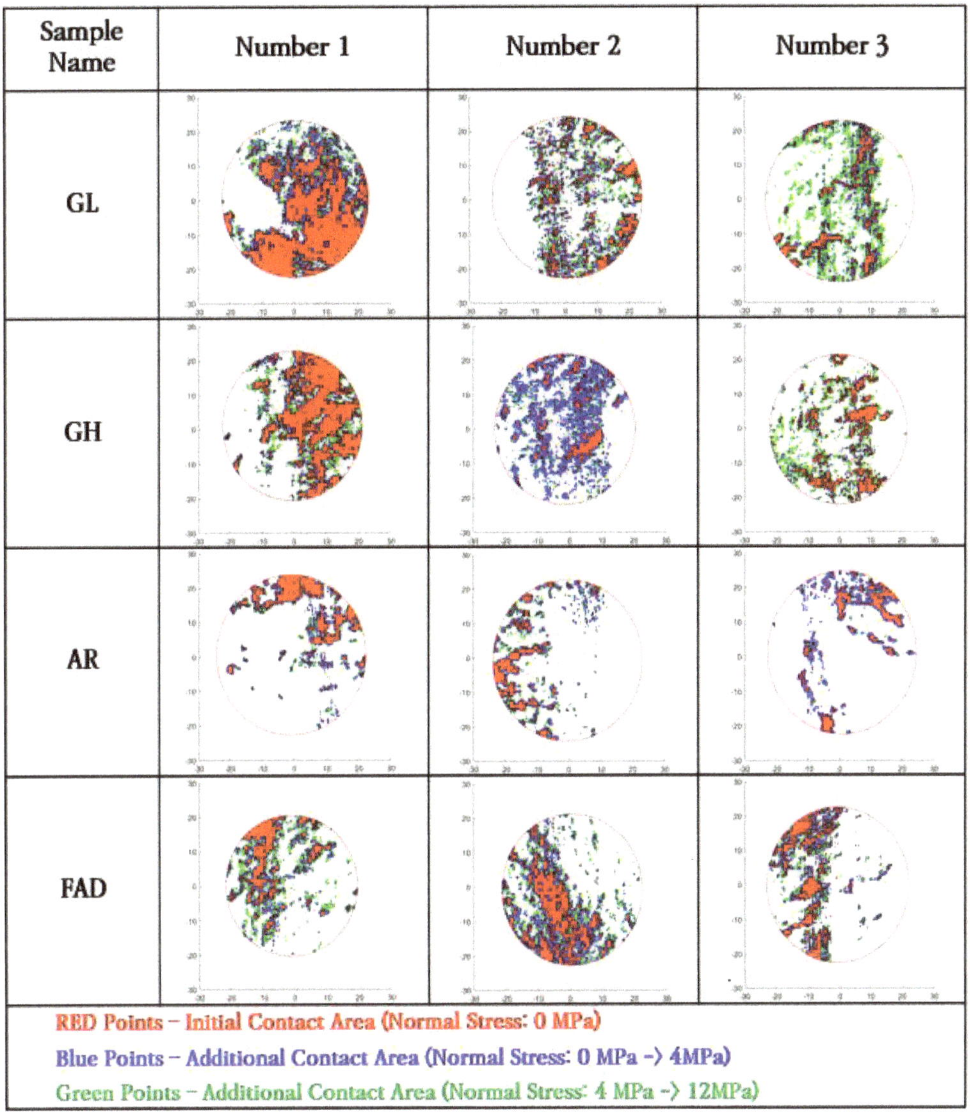

Sample Name	Number 1	Number 2	Number 3
GL			
GH			
AR			
FAD			
RED Points – Initial Contact Area (Normal Stress: 0 MPa)			
Blue Points – Additional Contact Area (Normal Stress: 0 MPa -> 4MPa)			
Green Points – Additional Contact Area (Normal Stress: 4 MPa -> 12MPa)			

Figure 13. Changes in the contact area with increasing normal stress for natural fracture specimens.

These results indicate that the mechanical aperture and contact area, under various normal stress conditions in rock fractures, can be derived without additional experiments using the proposed digital analysis technique. In addition, the proposed technique has the advantage of being able to analyze the change in the contact area on the fracture surface in detail, as shown in Figure 13.

4. Conclusions

This study proposes a new digital analysis technique for the analysis of the mechanical aperture and contact area of rock fractures under various normal stress conditions. The proposed technique includes matching algorithms for point cloud data using pressure film image data and a translation algorithm using normal deformation data of the fracture. The applicability of the proposed technique was examined using natural fracture specimens from the KURT site, an underground research facility in Korea.

The main findings of this study are summarized as follows:

(1) A new algorithm was proposed to match the point cloud data of the upper and lower surfaces using the pressure film image data of rock fractures. This consisted of a primary matching algorithm using the ICP algorithm and a secondary matching algorithm using pressure film image data. The proposed matching algorithm was validated using the results of the contact area and the matching ratio of natural fracture specimens.

(2) A new algorithm was proposed to analyze the mechanical aperture and contact area according to the change in normal stress using the normal deformation data of rock fractures. The proposed algorithm was validated by deriving the initial mechanical aperture using matched point cloud data under reference normal stress and then comparing the results with the empirical equation of a previous study.

(3) A digital analysis technique was proposed by synthesizing the above algorithms. The types and acquisition methods of the input data required to apply the proposed technique were presented. In addition, the entire algorithm was organized sequentially.

(4) To examine its applicability, the mechanical aperture and contact area under various normal stresses were analyzed by applying digital analysis techniques to natural fracture specimens at the KURT site, an underground research facility in Korea. As the normal stress increased, the change in the mechanical aperture and contact area was found to decrease rapidly, and the area adjacent to the initial contact area gradually switched to the contact state.

This study is significant because it proposed a new technique for analyzing the mechanical aperture and contact area, which are considered important parameters in the hydro-mechanical behavior of rock mass. As in the cubic law [25], the aperture is a parameter that is generally required in the hydro-mechanical coupled analysis. The proposed technique has the main advantage of being able to analyze changes in the mechanical aperture and contact area under various normal stress conditions without additional experiments. In addition, the change in the contact area on the fracture surface according to the normal stress can be analyzed in detail.

However, when applying the proposed technique, noise due to the sensitivity of the pressure film affects the contact area and matching ratio results. In addition, the slight deformation of the pressure film affected the derivation of the mechanical aperture based on normal deformation. Therefore, the proposed technique should be supplemented in the future by selecting an appropriate pressure film or applying a correction according to the type of pressure film.

The technique and application results of this study are expected to be utilized in various underground space-related projects that require analyzing rock mass properties from borehole cores.

Author Contributions: Conceptualization, Y.-K.L. and C.-S.C.; methodology, Y.-K.L.; validation, Y.-K.L. and S.C.; formal analysis, Y.-K.L. and C.-S.C.; writing—original draft preparation, Y.-K.L.; writing—review and editing, C.-S.C., S.C. and K.-W.P.; visualization, Y.-K.L. and C.-S.C.; supervision, K.-W.P. All authors have read and agreed to the published version of the manuscript.

Funding: This work was supported by the Institute for Korea Spent Nuclear Fuel (iKSNF) and the National Research Foundation of Korea (NRF) grant funded by the Korea government (Ministry of Science and ICT, MSIT) (No.2021M2E1A1085200).

Institutional Review Board Statement: Not applicable.

Informed Consent Statement: Not applicable.

Data Availability Statement: Not applicable.

Conflicts of Interest: The authors declare no conflict of interest.

References

1. Olsson, R.; Barton, N. An improved model for hydromechanical coupling during shearing of rock joints. *Int. J. Rock Mech. Min. Sci.* **2001**, *38*, 317–329. [CrossRef]
2. Gentier, S. Morphologie et Comportement Hydromécanique d'une Fracture Naturelle dans un Granite Sous Contrainte Normale. Ph.D. Thesis, Univ. d'Orléans, Orléans, France, 1986.
3. Gale, J. Comparison of coupled fracture deformation and fluid flow models with direct measurements of fracture pore structure and stress-flow properties. In Proceedings of the 28th US Symposium on Rock Mechanics, Tucson, AZ, USA, 29 June–1 July 1987.
4. Hakami, E.; Larsson, E. Aperture measurements and flow experiments on a single natural fracture. *Int. J. Rock Mech. Min. Sci. Geomech. Abstr.* **1996**, *33*, 395–404. [CrossRef]
5. Archambault, G.; Gentier, S.; Riss, J.; Flamand, R. The evolution of void spaces (permeability) in relation with rock joint shear behavior. *Int. J. Rock Mech. Min. Sci.* **1997**, *34*, 14-e1. [CrossRef]
6. Chen, Y.; Liang, W.; Lian, H.; Wang, J.; Nguyen, V.P. Experimental study on the effect of fracture geometric characteristics on the permeability in deformable rough-walled fractures. *Int. J. Rock Mech. Min. Sci.* **2017**, *98*, 121–140. [CrossRef]
7. Gentier, S.; Billaux, D.; van Vliet, L. Laboratory testing of the voids of a fracture. *Rock Mech. Rock Eng.* **1989**, *22*, 149–157. [CrossRef]
8. Pyrak-Nolte, L.; Myer, L.; Cook, N.; Witherspoon, P. Hydraulic and mechanical properties of natural fractures in low-permeability rock. In Proceedings of the 6th ISRM Congress, Montreal, QC, Canada, 30 August–3 September 1987.
9. Yeo, I.; Freitas, M.; Zimmerman, R. Effect of shear displacement on the aperture and permeability of a rock fracture. *Int. J. Rock Mech. Min. Sci.* **1998**, *35*, 1051–1070. [CrossRef]
10. Keller, A. High Resolution, Non-Destructive Measurement and Characterization of Fracture Apertures. *Int. J. Rock Mech. Min. Sci.* **1998**, *35*, 1037–1050. [CrossRef]
11. Iwai, K. Fundamental Studies of Fluid Flow through a Single Fracture. Ph.D. Thesis, University of California, Berkeley, CA, USA, 1976.
12. Stesky, R.M.; Hannan, S.S. Growth of contact area between rough surfaces under normal stress. *Geophys. Res. Lett.* **1987**, *14*, 550–553. [CrossRef]
13. Re, F.; Scavia, C.; Zaninetti, A. Variation in contact areas of rock joint surfaces as a function of scale. *Int. J. Rock Mech. Min. Sci.* **1997**, *34*, 254.e212–254.e251. [CrossRef]
14. Nemoto, K.; Watanabe, N.; Hirano, N.; Tsuchiya, N. Direct measurement of contact area and stress dependence of anisotropic flow through rock fracture with heterogeneous aperture distribution. *Earth Planet. Sci. Lett.* **2009**, *281*, 81–87. [CrossRef]
15. Wang, F.; Kaunda, R.B. Artificial neural network modeling of contact electrical resistance profiles for detection of rock wall joint behavior. In Proceedings of the 51st U.S. Rock Mechanics/Geomechanics Symposium, San Francisco, CA, USA, 25–28 June 2017.
16. Ge, Y.; Xie, Z.; Tang, H.; Chen, H.; Lin, Z.; Du, B. Determination of shear failure regions of rock joints based on point clouds and image segmentation. *Eng. Geol.* **2019**, *260*, 105250. [CrossRef]
17. Pirzada, M.A.; Roshan, H.; Sun, H.; Oh, J.; Andersen, M.S.; Hedayat, A.; Bahaaddini, M. Effect of contact surface area on frictional behaviour of dry and saturated rock joints. *J. Struct. Geol.* **2020**, *135*, 104044. [CrossRef]
18. Bandis, S.C.; Lumsden, A.C.; Barton, N.R. Fundamentals of rock joint deformation. *Int. J. Rock Mech. Min. Sci. Geomech. Abstr.* **1983**, *20*, 249–268. [CrossRef]
19. Besl, P.; McKay, H. A method for registration of 3-D shapes. *IEEE Trans. Pattern Anal. Mach. Intell.* **1992**, *14*, 239–256. [CrossRef]
20. Tse, R.; Cruden, D.M. Estimating joint roughness coefficients. *Int. J. Rock Mech. Min. Sci.* **1979**, *16*, 303–307. [CrossRef]
21. Ge, Y.; Lin, Z.; Tang, H.; Zhao, B. Estimation of the appropriate sampling interval for rock joints roughness using laser scanning. *Bull. Eng. Geol. Environ.* **2021**, *80*, 3569–3588. [CrossRef]
22. Barton, N. Review of a new shear-strength criterion for rock joints. *Eng. Geol.* **1973**, *7*, 287–332. [CrossRef]
23. Choi, S.; Jeon, B.; Lee, S.; Jeon, S. Experimental study on hydromechanical behavior of an artificial rock joint with controlled roughness. *Sustainability* **2019**, *11*, 1014. [CrossRef]
24. Nguyen, X.X.; Dong, J.J.; Yu, C.W. Is the widely used relation between mechanical and hydraulic apertures reliable? Viewpoints from laboratory experiments. *Int. J. Rock Mech. Min. Sci.* **2022**, *159*, 105226. [CrossRef]
25. Snow, D.T. A Parallel Plate Model of Fractured Permeable Media. Ph.D. Thesis, University of California, Oakland, CA, USA, 1965.

Disclaimer/Publisher's Note: The statements, opinions and data contained in all publications are solely those of the individual author(s) and contributor(s) and not of MDPI and/or the editor(s). MDPI and/or the editor(s) disclaim responsibility for any injury to people or property resulting from any ideas, methods, instructions or products referred to in the content.

MDPI
St. Alban-Anlage 66
4052 Basel
Switzerland
Tel. +41 61 683 77 34
Fax +41 61 302 89 18
www.mdpi.com

Materials Editorial Office
E-mail: materials@mdpi.com
www.mdpi.com/journal/materials

www.ingramcontent.com/pod-product-compliance
Lightning Source LLC
LaVergne TN
LVHW070421100526
838202LV00014B/1503